Housing Policy, 4th edition

Housing Policy: An Introduction has been completely revised for this, its fourth edition. Its account of housing policy in the UK now gives particular emphasis to the development of policy since Labour's electoral victory in 1997. After examining policy under the Thatcher and Major administrations, this new edition shows that there has been a gradual evolution in housing policy, rather than any radical change to the ways in which the different housing sectors are promoted.

In this edition housing policy issues introduced include:

- the 'brownfield versus greenfield debate' in respect of housing provision;
- the phasing out of renovation grants;
- the continuing shift of emphasis from supply-side to demand-side subsidies;
- the transfer of local authority housing to registered social landlords;
- the partial revival of private rented housing;
- boom, slump and boom in the owner-occupied sector;
- and many other issues ranging from regional policy and housing across the UK to social exclusion, community care and homelessness, as well as the topics covered in previous editions.

Housing Policy: An Introduction provides an accessible introduction by describing policies and explaining their effects, as well as through an analysis of recent changes and their implications, using practical examples. It is ideal for the student tackling housing policy at all levels, but also as a background to current housing policy for researchers and professional practitioners. It provides a strong foundation in housing policy for students taking qualifications in housing, urban studies, social policy and social administration.

Paul Balchin is Reader in Urban Economics and **Maureen Rhoden** is Senior Lecturer in Housing Studies. Both are based at the University of Greenwich.

D0161965

Housing Policy

An introduction

4th edition

Paul Balchin and Maureen Rhoden

London and New York

First published 1984 by Routledge
11 New Fetter Lane, London EC4P 4EE
2nd edition 1989
3rd edition 1995
4th edition 2002

Simultaneously published in the USA and Canada
by Routledge
29 West 35th Street, New York, NY 10001

Routledge is an imprint of the Taylor & Francis Group

Typeset in Times by Keystroke, Jacaranda Lodge, Wolverhampton
Printed and bound in Great Britain by Biddles Ltd, Guildford and
King's Lynn

British Library Cataloguing in Publication Data
A catalogue record for this book is available from the British Library

Library of Congress Cataloging in Publication Data
A catalog record for this book has been requested

ISBN 0–415–25213–X (hbk)
ISBN 0–415–25214–8 (pbk)

Contents

Figures

Tables

Boxes

Preface

The concept of the welfare state was arguably one of the most important influences on public policy in the United Kingdom in the twentieth century. Consolidating and extending many of the ideas of the Liberal governments of 1906–14, the Beveridge Report (1942) highlighted the need to protect 'from the cradle to the grave' all individuals and the family from want, disease, ignorance, squalor and idleness. An improved system of national insurance, the national health service, educational reform, a comprehensive town planning system and Keynesian economic policy were all consequently put in place in the immediate post-war period, and each remained broadly intact throughout the 1950s, 1960s and 1970s irrespective of which party was in power. Housing policy, however, was by comparison erratic and much influenced by the prejudices and nostrums of consecutive governments. However, compared to later divergences in policy, an approach based on consensus often prevailed.

But in the 1980s and early 1990s, the inefficiencies, inequalities and lack of choice characteristic of United Kingdom housing markets became primary causes of concern to most housing specialists. Under the ideology of Thatcherism, housing was singled out to bear the biggest cut in public expenditure at a time when the housing shortage was becoming increasingly severe; housebuilding had plummeted and unemployment had soared; the housing stock remained in a poor condition; council housing had been sold off and council and housing association rents had risen faster than inflation; owner-occupiers (at least until 1991) had received more and more regressive mortgage interest relief and tax exemption; and the unsatisfied needs of low-income women and ethnic minorities, the elderly and the homeless were becoming more and more evident.

This substantially extended edition – completed at the beginning of the second Blair government – provides a comprehensive and up-to-date account of these key economic and financial issues. It should appeal to a broad range of university students on courses in Economics, Housing Studies, Public Administration, Sociology, Surveying and Town Planning at both undergraduate and postgraduate levels.

Although there are several books in print on housing, some are out of date, while others are either highly theoretical or heavily focused on one or two areas of the subject. There is an absence of a wide-ranging book on policy which not only includes a critical analysis of housing issues under the Thatcher, Major and Blair administrations but also attempts to examine housing policy within a macroeconomic context. This book is written to eliminate this deficiency.

In writing this book, however, we were continually aware of being overtaken by economic and political events. Although housing policy since the late 1990s has not been subject to sudden and radical change, it nevertheless does change – if only through stealth. Every effort has therefore been made to ensure that legislative and statistical detail is accurate at the time of going to press.

We must acknowledge very great debts we owe Jill Stewart for making a useful contribution to Chapters 14 and 19 and R. Shean McConnell for his major input in Chapter 16. We are also indebted to other present and past colleagues and to a range of people (especially in the fields of housing management and research) who have stimulated and advised us in the preparation of this book. In addition, we are very grateful to Cherie Apps and Sue Brimacombe, who typed and retyped much of the manuscript, and to Richard Howarth, Sue Lee and Pete Stevens, who produced much of the artwork within the text. Last, but not least, we would like to thank our respective families for their encouragement and patience.

Paul Balchin
Maureen Rhoden
London, Autumn 2001

Abbreviations

ACC	Association of County Councils
ACG	annual capital guideline
ADC	Association of District Councils
ADP	Approved Development Programme
AHP	area housing partnership
AIR	advance/income ratio
ALA	Association of London Authorities
AMA	Association of Metropolitan Authorities
AMC	Association of Municipal Corporations
APPR	advance/purchase price ratio
BEC	Building Employers Confederation
BES	Business Expansion Scheme
CABIN	Campaign Against Building Industry Nationalisation
CCT	compulsory competitive tendering
CEI	comprehensive estates initiative
CES	Centre for Environmental Studies
CIH	Chartered Institute of Housing
CIPFA	Chartered Institute of Public Finance Accountancy
CML	Council of Mortgage Lenders
COI	Central Office of Information
CRE	Commission for Racial Equality
CRI	Capital Receipts Initiative
CSO	Central Statistical Office
CURS	Centre for Urban and Regional Studies
DBRW	Development Board for Rural Wales
DEE	Department of Education and Employment
DETR	Department of the Environment, Transport and Regions
DHSS	Department of Health and Social Security
DIYSO	do-it-yourself shared ownership

DLO	direct labour organisation
DoE	Department of the Environment
DoENI	Department of the Environment Northern Ireland
DSD	Department of Social Development
DSS	Department of Social Security
DTI	Department of Trade and Industry
DTLR	Department of Transport, Local Government and the Regions
EHOA	Environmental Health Officers Association
EU	European Union
FPHM	Faculty of Public Health Medicine
GDP	gross domestic product
GEAR	Glasgow East Area Renewal
GHA	Glasgow Housing Association
GIA	General Improvement Area
GLA	Greater London Authority
GLC	Greater London Council
GoRs	government offices for the regions
HAA	Housing Action Area
HAG	housing association grant
HAMA	Housing Associations as Managing Agents
HAT	housing action trust
HIA	home improvement agency
HIDB	Highlands and Islands Development Board
HIE	Highlands and Islands Enterprises
HIP	Housing Investment Programme
HMO	house in multiple occupation
HMSO	Her Majesty's Stationery Office
HPER	house price/earnings ratio
HPIR	house price/income ratio
HRA	housing revenue account
HSBR	housing sector borrowing requirement
IBG	Institute of British Geographers
IEH	Institute of Environmental Health
ILGS	Institute of Local Government Studies
ILO	International Labour Office
ISMI	income support for mortgage interest
JRF	Joseph Rowntree Foundation
LAW	Land Authority for Wales
LEFTA	Labour Economic, Finance and Taxation Association
LGA	Local Government Association

LHF	London Housing Federation
LPAC	London Planning Advisory Comittee
LRD	Labour Research Department
LSVT	large-scale voluntary transfer
MAFF	Ministry of Agriculture, Fisheries and Food
MHLG	Ministry of Housing and Local Government
MITR	mortgage interest tax relief
MLR	minimum lending rate
MORI	Market and Opinion Research International
MPPI	mortgage payment protection insurance
NAO	National Audit Office
NASS	National Asylum Support Service
NDC	New Deal for the Communities
NFHA	National Federation of Housing Associations
NFHS	National Federation of Housing Societies
NHBC	National House Building Council
NHIC	National Home Improvement Council
NIHE	Northern Ireland Housing Executive
NISRA	Northern Ireland Statistics and Research Agency
NOP	National Opinion Poll
OECD	Organisation for Economic Co-operation and Development
ONS	Office of National Statistics
OPCS	Office of Population Censuses and Surveys
OPT	Organisation of Private Tenants
PAT	policy action team
PEP	Priority Estate Project
PFI	Private Finance Initiative
PMPI	private mortgage protection insurance
PPG	Planning Policy Guidance
PSBR	public sector borrowing requirement
RA	renewal area
RDA	regional development agency
RDG	revenue deficit grant
RIBA	Royal Institute of British Architects
RICS	Royal Institution of Chartered Surveyors
RSI	Rough Sleepers Initiative
RSL	registered social landlord
RTB	Right to Buy
RTM	rent to mortgage
RTPI	Royal Town Planning Institute

SAUS	School of Advanced Urban Studies, University of Bristol
SCA	supplementary credit approval
SDA	Scottish Development Agency
SERPLAN	South East Region Planning Conference
SEU	Social Exclusion Unit
SFHA	Scottish Federation of Housing Associations
SHAC	Shelter Housing Aid Centre
SHG	social housing grant
SRB	Single Regeneration Budget
SSAC	Social Security Advisory Committee
TCPA	Town and Country Planning Association
TMO	tenant management organisation
TSB	Trustee Savings Bank
UDC	urban development corporation
UHRU	Urban Housing Renewal Unit
UP	Urban Programme
UPA	urban priority area
WDA	Welsh Development Agency

1 Introduction

At the General Election of 7 June 2001 the Labour Party secured a second consecutive term in office by winning 413 seats and gaining a majority of 167 over all other parties. However, in housing – as in other areas of policy such as education and health – Labour had gone to the electorate still grappling with a situation not entirely of its own making, but one that it had largely inherited in 1997 from the former Conservative administration. Under the Thatcher and Major governments, Conservative housing policy – between 1979 and 1997 – had dramatically reduced the amount of public expenditure on housing; brought about a marked reduction in capital investment and housebuilding in the public sector; replaced local authorities by housing associations as the major providers of new social housing and – through privatisation and the transfer of housing stock – substantially decreased the supply of local authority dwellings; raised rents ahead of average earnings in both the social and the private rented sectors; imposed an increased burden on mortgagors; left owner-occupiers with a considerably increased risk of repossession and presided over a marked increase in homelessness (Table 1.1). How Labour, under Tony Blair's leadership, subsequently addressed these and other housing issues between 1997 and 2001 undoubtedly had some impact on its electoral performance in 2001, even though greater attention was given at the hustings to other areas of policy.

To achieve an understanding of current housing policy, it is necessary to analyse and comprehend policy historically. This book therefore refers back (where appropriate) to the 1950s and 1960s or earlier, but most of the emphasis is placed on the 1970s to mid-1990s, when the debate on the respective roles of the housing sectors was particularly heated, and on the years since 1997, when, in contrast, a new consensus on housing

Table 1.1 *Key housing indicators, United Kingdom, 1980 and 1996*

	1980	1996	Percentage change
Public expenditure: housing[a] (£bn at 1998/99 prices)	13.1	4.9	–62.6
Local authority capital investment: new build and acquisitions[a, b] (£m at 1997/98 prices)	2,275	67	–97.1
Housing starts: public sector	44,433	1,656	–96.3
housing association	14,911	30,304	+ 103.2
Local authority housing: % total housing stock	30.4	18.0	–40.8
Average rents[b] (£ per week): local authority	7.70	40.00	+419.5
housing association	12.52	48.26	+285.5
private	10.85	50.61	+366.5
Mortgage payments (£ per week)	19.50	54.87	+181.4
Average male earnings (£ per week)	111.4	302.8	+171.8
Mortgage repossessions	3,480	42,560	+1,323
Local authority homeless acceptances[c]	76,342	148,339	+94.3

Source: Wilcox (2000)

Notes:
[a] 1980/81–1996/97
[b] England
[c] Great Britain

emerged. To provide an appropriate background to this analysis, the following introductory review:

- examines housing need, and how it has changed since the early 1980s;
- explores the relationship between welfare regimes and housing policy;
- reviews the development of housing policy since 1945;
- discusses how the housing policies of Labour and the Conservatives have converged – rather than diverged – in recent years; and
- takes note of devolutionary processes in the formulation of housing policy in Scotland, Wales and Northern Ireland.

Housing need

In very crude terms, housing need in Great Britain is largely satisfied. Housing policy since the 1950s has ensured that there has been a substantial growth in owner-occupation and council housing and a decline in private rented accommodation, while the condition of most of the housing stock has greatly improved. By the early 1970s a crude surplus of dwellings over households was achieved (for the first time

Table 1.2 *Number of dwellings and households, Great Britain, 1981–98 (millions)*

	1981	1991	1998	Percentage change, 1981–98
Dwellings	21.08	23.14	24.38	+15.7
Households	20.17	22.39	23.90	+18.5
Surplus	0.91	0.75	0.48	−47.3

Sources: ONS (2000a); Wilcox (2000)

since 1938) and by 1981 reached 910,000, only to decrease subsequently to 480,000 in 1998 (Table 1.2).

The crude surplus in 1998 (albeit lower than in the 1980s) did not, however, indicate the true relationship between supply and need. Of the 24.4 million dwellings in 1998, there were well over a million unfit dwellings or homes lacking basic amenities, dwellings undergoing extensive conversion or improvement, and second homes, while there were about half a million concealed households (such as couples sharing with their parents or in-laws) among the 23.9 million recorded households. There were also great spatial variations in supply and demand, most notably a surplus of cheap housing in much of the North and Midlands, and a shortage of affordable housing in London and elsewhere in the South. Although many dwellings need to be temporarily empty to facilitate household mobility, migration means that dwellings of the right kind or price are not always where they are needed. An ageing population also produces a mismatch between housing occupied and the size and type of dwellings required.

Across the United Kingdom, however, the tenure mix (see Table 1.3) varies from country to country. Table 1.4 shows that – as an outcome of different market conditions and public policy – Northern Ireland and Wales have the highest level of home-ownership in the Union, Scotland has the greatest proportion of dwellings rented from local authorities and housing associations, and England has a disproportionate share of private rented dwellings.

Housing welfare regimes

Although, in the long term, market forces and government intervention are instrumental in determining the specific size of each of the housing

Table 1.3 *Housing tenure, Great Britain, 1950–98*

	Owner-occupied (%)	Local authority (%)	Housing association (%)	Private rented (%)
1950	29.0	18.0	—	53.0
1961	42.3	25.8	—	31.9
1971	50.6	30.6	—	18.9
1981	56.6	30.3	2.2	10.9
1991	66.3	21.1	3.1	9.6
1998	67.6	16.9	5.0	10.6

Source: DoE, *Annual Report*, 1993; DETR, *Housing and Construction Statistics* (various)

Table 1.4 *Housing tenure in the United Kingdom, 1999*

	Owner-occupied (%)	Local authority (%)	Housing association (%)	Private rented (%)
England	68.2	15.6	5.1	11.1
Wales	71.6	15.6	4.1	8.6
Scotland	62.4	25.1	5.7	6.8
Northern Ireland	71.9	21.0	2.6	4.4
United Kingdom	67.6	16.3	5.7	10.8

Source: ONS (2000a)

tenures, the socio-political system that is in operation provides the arena, in the shorter term, in which the relationships between the market and policy develop. Esping-Andersen (1990) suggested that in advanced capitalist countries there were three distinct welfare regimes: social democratic, corporatist and liberal (or neo-liberal). The first group, the social democratic states, are concerned with reforming welfare provision on the basis of universalism and social ownership extended to all classes (Barlow and Duncan, 1994). Within this group (which is currently confined to Scandinavia), countries aim to provide a 'one-nation' system of welfare based on equality of high standards of welfare for all, as opposed to a system based on the satisfaction of minimum basic needs. To an extent, the United Kingdom adopted a social democratic regime in 1945 and maintained it, intermittently, until the mid-1970s. The second group, the corporatist states (such as France, Germany and Austria) attempt to reinforce the rights attached to different classes and professions, and to this end are willing to replace the market as a provider of welfare. However, this differentiated form of welfare provision has not been a feature of welfare provision in the United Kingdom. The third group, the neo-liberal welfare states, provide little more than a means-tested 'safety net' of limited benefits for low-income, working-class state dependants, and include such countries as the United States, Canada, Australia, New Zealand, Ireland and – increasingly since the late 1970s – the United Kingdom, regardless of the political party in power. Although Esping-Andersen (1990) did not include housing tenure in his analysis, it is comparatively straightforward to match the social democratic group with the promotion of various forms of rented housing as alternative sectors available to all, and on a long-term basis. The corporatist welfare group can be identified as those countries in which both the social and private rented sectors are overtly promoted by the state and where either one or the other tenures consequently becomes the dominant sector. The neo-liberal welfare countries tend to be those where owner-occupation is, by far, the dominant sector, and where state intervention in housing is limited to a stigmatised provision of affordable housing for a deprived segment of the population unable to participate in markets (Barlow and Duncan, 1994).

In taking a sympathetic view of Esping-Andersen's thesis, Kemeny (1995) attempts to offer an analysis of the distribution of tenure, particularly *vis-à-vis* the relative importance of social housing. With regard to the maturation of social (non-profit) rented housing (as measured by an inflation-induced decline in the outstanding debt on the existing stock compared to the outstanding debt on newly built, acquired or renovated buildings), Kemeny distinguishes between two rental systems: 'dualist' rental systems, in which the state controls and residualises the social rented sector to protect private (profit) renting from competition – as in the neo-liberal regimes of the United Kingdom, Ireland and the United States; and the 'unitary' system, in which social and private renting system are integrated into a single rental market, with the social democratic and corporatist states of continental Europe being principal examples.

Kemeny (1995) claims that in the dualist system, for reasons of political expediency, the state has introduced social housing or encouraged its development to provide a safety net for the relatively poor. The non-profit rental sector is protected from the profit sector by being segregated from the private market and organised as a stigmatised, residualised and often means-tested sector – a process that is particularly well advanced in the United Kingdom and generally supported across the political spectrum.

There is often no attempt to ameliorate the undesirable effects of the market by creating a balance between the profit motive and social priorities. Within the dualist system, policy intentionally or unintentionally steers all but the lowest-income households towards owner-occupation. For most households there is only a choice between profit renting (at high rent and with little long-term security of tenure) and owner-occupation with its many perceived advantages. Kemeny suggests that in this situation, far from preferences for owner-occupation determining policy, as Saunders (1990) would argue, policies create preferences for home-ownership. In a dualist system, therefore, renting at cost is provided by a state-controlled social housing sector, while private profit renting is left largely to sink or swim.

In unitary rental systems, however, the state encourages social rented housing to compete directly with the private rental sector in order to dampen rents and to provide good-quality housing on secure tenancy terms. Clearly, if the private or profit rental sector is to compete with social or non-profit housing (the maturation process enabling the latter sector to set low levels of rent), the private rental sector will need to be a recipient of subsidies equivalent to those allocated to the social sector in order to ensure an adequate return on investment, while a flexible form of rent control might be necessary across both sectors if the market shows signs of imperfection (a quid pro quo arrangement absent in the United Kingdom). If subsidies are also compatible with those received by owner-occupiers, subsidisation will be tenure neutral, and each of the tenures will be equally attractive financially to a large proportion of households. In a unitary market, therefore, cost-rental social housing competes with the private sector supported by the state on equal or near-equal terms.

Unlike Kemeny, Harloe (1995) is fairly critical of Esping-Andersen's (1990) model of welfare provision. To him, the three regimes represent too rigid a classification, and he suggests that while systems of social housing seem to be converging on a residualist model, the pace of change varies from country to country depending on different economic, social and political circumstances. It could likewise be argued that Esping-Andersen's (1990) approach lacks historicity, exaggerates path dependence, and leaves no choice for reform and path changes. It might also be suggested that in constructing a model of welfare provision it is necessary to examine the development of welfare state regimes, taking into account both path dependence and path change. During the inter-war period, for example, there emerged 'Lib-Lab' regimes in Britain (meaning a combination of the policies of the Liberal and Labour parties, or support given in Parliament to a Labour government by the Liberals) under which the Housing Acts of 1919, 1924 and 1930 ensured that council housing satisfied the housing needs of as many households as possible, regardless of income, while – at the same time – Rent Acts decontrolled much of the private rented sector. After the Second World War a 'Lab-Lib' regime was introduced which continued to facilitate the expansion of the council housing stock, but, in the private sector, retained rent control in its various forms, on and off, until the 1980s. Only from the 1980s did a neo-liberal regime fully emerge that aimed to demunicipalise much, if not all, of the council housing stock, and reintroduced market rents in the private rented sector. It is of note that throughout most of the century, owner-occupation was promoted under each regime, with subsidisation reaching its peak during the neo-liberal 1990s.

Development of housing policy since 1945

Evidence suggests that there have been four major stages in the development of housing policy in Britain since the Second World War although, from time to time, different stages overlapped (Boelhouwer, 1991). The first stage of policy development was dominated by the urgency of reducing the serious housing shortages in the late 1940s and early 1950s. The second stage, beginning in 1955, incorporated large building programmes, including a major concern for improving the quality of the housing stock. Initially marked by a concentration on slum clearance and redevelopment, from 1969 it focused on improvement and repairs. The third stage, commencing in 1971, was largely characterised by shifts of emphasis from supply-side expenditure to demand-side subsidies. These benefited the reasonably well housed at the expense of the inadequately accommodated, in the belief that such a shift would reduce state intervention in the production process and, instead, promote the development of the free market. The fourth stage, commencing in the mid-1980s, saw the reappearance of quantitative and qualitative housing shortages and an increase in state involvement – if not in expenditure.

It is interesting to note that there was little close relationship between the different stages of housing policy and governmental change (Figure 1.1), although it might be suggested that until the mid-1970s policy developed within the context of a social democratic/'Lab-Lib' welfare regime, while subsequently it has been conditioned by a neo-liberal approach to welfare provision.

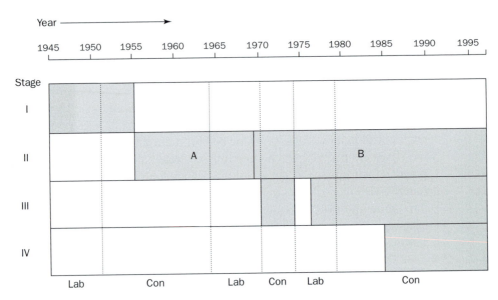

Figure 1.1 *Development of housing policy, United Kingdom, 1945–97*

I High degree of government involvement, particularly in order to alleviate housing shortages.
II Greater emphasis on housing quality:
  A Improvement in the quality of the stock by slum clearance programmes and substitute new
    construction.
  B Emphasis on maintenance and improvement instead of slum clearance.
III Greater emphasis on problems of housing distribution and targeting specific groups, and the
    withdrawal of the state in favour of the private sector.
IV Reappearance of quantitative and qualitative housing shortages; state involvement increases.

Lab: Labour; Con: Conservative.

The economics of Thatcherism

From the 1950s to the 1970s, Conservative governments believed to a
greater or lesser extent in the practicality of Keynesian demand
management policy, in the importance of a mixed economy and in the
idea of 'one nation'. After 1979, however, the new Thatcher brand of
Conservative policy attempted to substitute monetarism as a means of
controlling the macroeconomy, aimed to extend the free market economy
by 'rolling back the frontiers of state intervention' and sought to bring
about an irreversible shift in the balance of power from the poor to the
better-off.

Since the Conservative government's overriding macroeconomic priority
was to 'squeeze inflation out of the system', and since in its view
inflation was solely the result of excessive increases in the money supply,

so the reduction in the money supply seemed a logical route to follow. However, over the period 1979–83 it failed to achieve this objective. Although sterling M3 (then the principal money supply indicator) decreased by 5 per cent in 1979/80, in each successive year to 1982/83 it increased by 4.4, 2.3 and 7 per cent. By the end of this period the supply of money had increased by 80 per cent more than the average level in the period 1974–79. During the first year of the Thatcher government, retail prices rose by 21 per cent and the rate of inflation remained in double figures until 1982. The government attributed this to its failure to control money supply, and adopted an alternative form of monetarism: inverted Keynesianism. At the time of stagflation, the government thus applied tighter fiscal measures and a rigid pay policy in the public sector, and relied upon unemployment to keep wages low in the private sector. The standard of living consequently fell by 2.7 per cent over the period 1979–82. Throughout most of the 1980s, however, the economy was deflated by high interest rates – the demand for money, rather than its supply, being constrained by Treasury policy. Bank base rates fluctuated between an 8.5 per cent low and a 16.5 per cent high and had a deterrent effect on borrowing. High interest rates also maintained a high exchange value of sterling – keeping the price of imports down and subsequently lowering the rate of inflation.

As a further means of reducing the rate of inflation and of applying policies of privatisation, cutbacks in public expenditure were planned throughout each of the Conservative periods of government well into the 1990s. Largely owing to the need to pay larger welfare payments at times of high unemployment and economic recession, public expenditure in fact increased from £90 billion to £281 billion between 1979/80 and 1993/94. But in real terms this was an increase of only 27 per cent (or less than 2 per cent per annum) and involved cuts in the planned spending programmes of most government departments. Although, for example, in cash terms expenditure on social security, health and personal social services, education and defence was permitted to rise by 333, 292, 220 and 149 per cent respectively, real increases amounted in turn to only 76, 60, 34 and 1 per cent. In the case of housing, substantial cuts in public expenditure (in both cash and real terms) were introduced soon after the first Thatcher government came to power. In its public expenditure White Paper of 1980, the government announced what in effect became a 48 per cent cut in planned public spending on housing in real terms, 1979/80 to 1983/84 – housing bearing 75 per cent of the projected 4 per cent cut in public expenditure in this period, according to

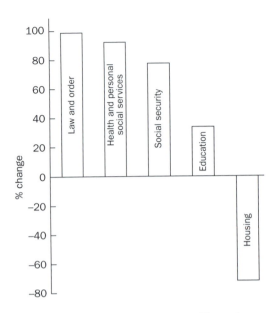

Figure 1.2 *Public expenditure on specific services, 1980/81–1999/2000 (at 1998/99 prices)*
Source: Wilcox (2000)

the House of Commons Environment Committee (Forrest and Murie, 1988; Hamnett, 1988a). Even more swingeing cuts were subsequently announced in the 1990s. Public expenditure on housing thus decreased in real terms by about 73 per cent over the period 1980/81–1998/99 (Figure 1.2). Moreover, whereas social security and health and personal social services were absorbing an increasingly large share of public expenditure, respectively 34 and 21 per cent in 1999/00 (compared to 24.2 and 14 per cent in 1980/81), housing's share plummeted to 1.2 per cent in the same year (compared to 5.6 per cent in 1980/81).

The cutback in public expenditure on housing had a devastating effect on council housebuilding (which all but ceased by 1993/94) and on homelessness, while severely reduced central government subsidies resulted in local authority rents soaring at a rate well above that of inflation. In view of these effects, why was the housing programme singled out for such draconian cuts? Hamnett (1988a) examined two explanations. The first postulated that after the 1976 sterling crisis and under International Monetary Fund pressure, there were strong economic reasons for cutting anything that could be cut in order to curb inflation and to maintain the international value of the pound; the second suggested that 'the Conservatives cut expenditure on council housing with enthusiasm as part of their wider goal of rolling back council housing in favour of owner-occupation'.

Regarding the first explanation, while public expenditure on the housing programme was undoubtedly reduced, public spending on housing benefits (rent allowances and rebates) to private- and social-sector tenants had risen dramatically from £280 million in 1979/80 to £3,540 million in 1987/88, while mortgage interest relief to owner-occupiers had soared from £1,639 to £4,850 million over the same period. But neither of these transfers, however, was included in the housing programme. Housing benefit was (and is) an item in the expanding programme of the Department of Social Security (DSS), while mortgage interest relief is not included as an item of public expenditure at all but is equivalent to tax forgone by the Inland Revenue. Clearly, therefore, cuts in the housing programme *per se* cannot be justified on the grounds of public sector austerity. The country can evidently afford welfare support for tenants and tax relief for home-owners, but not for housing investment or other forms of housing subsidy (Forrest and Murie, 1988).

The second explanation, favoured by Hamnett, is ideological rather than economic. From 1979 into the 1990s, Conservative governments presided over the greatest onslaught on the direct provision of local authority housing since its inception, and simultaneously promoted the growth and dominance of owner-occupation. Apart from making considerable reductions in capital expenditure on council housebuilding, the size of the social sector was systematically reduced through the application of a mandatory 'Right to Buy' policy and other processes of privatisation. Thus, by the end of the 1980s 'the cumulative impact of Conservative policies since 1979 . . . [was] such that the state housing sector in Britain . . . [was] stopped dead in its tracks and thrown into sharp reverse for the first time in 60 years' (Hamnett, 1988a). Council housing, as a proportion of the total housing stock, thus declined from 31.5 to 24.4 per cent in the period 1979–88 (and to 16.9 per cent by 1998). Meanwhile, aided by a range of tax benefits and a favourable economic climate for investment, the owner-occupation sector expanded its share of the total housing stock from 55.3 per cent in 1979 to 61.7 per cent in 1988 (and to 67.6 per cent in 1998).

Housing policy: from divergence to convergence

At both the 1983 and 1987 elections the Conservatives, far from making any pledge to increase housing investment in the public sector, committed themselves to an extension of their 'Right to Buy' scheme and the development of shared ownership. The party also pledged that it would raise the ceiling for mortgage interest relief. In favouring owner-occupation, the Conservatives inevitably reflected the mood of much of the electorate. For this and many other reasons, the 1983 election reinforced the *status quo*. Thatcherism remained intact. At the 1987 election the Conservatives (recognising that two out of three homes were by then owner-occupied) pledged to keep the existing system of mortgage interest relief, and stated that they would offer council tenants a choice of landlord, renovate (and demunicipalise) run-down council estates through the medium of housing action trusts (HATs), and deregulate new private lettings. While these latter proposals may not have gained many extra votes, the party's high-profile promotion of owner-occupation unquestionably secured the support of a very significant proportion of the electorate. In 1992 the Conservatives proposed a new rent to mortgage scheme and pledged that they would continue 'Right to Buy' discounts and the transfer of local authority estates to other landlords, and that they would maintain mortgage interest relief. However, as in 1987, once back in office the Conservatives continued to relegate housing to a diminishing role as far as claims on public expenditure were concerned, and housing policy remained a victim of Treasury constraint.

But even more seriously, Conservative housing policy in the long term, with its biased, profligate and irresponsible support for the owner-occupier and neglect of the tenant or homeless, had demonstrated that the Thatcher and Major governments adhered to the view that

> the role of the state is closely related to the power of the dominant class . . . the state is not a neutral organisation standing above society adjudicating between rival interests. Ultimately it is an instrument of class rule, serving to maintain the power of the dominant class and the subordination of other classes.
>
> (Bassett and Short, 1979)

Housing policy from 1979 to 1997, perhaps more than any other manifestation of Conservative thinking, was the means by which an increasingly divided society was being created.

Reacting to the Conservatives' antipathy to council housing, the Labour Party at its 1982 conference passed a resolution that if returned to office it would restore levels of housing investment to those pertaining in 1975/76 in real terms, 'so as to increase dramatically the supply of public sector dwellings for rent'. But Mr Allan Roberts MP (1982) claimed that despite homelessness increasing and council waiting lists doubling, housing did not seem to be a political issue or a vote-winner for Labour. He argued that the party was

> losing the debate in the country on public expenditure *versus* private consumption and by default the critics of council housing [were] being let off scot free. But there [had] never been a time when policies to revitalise the public rented sector were so relevant to the real housing needs of the nation.

With regard to the impending election he warned that housing could be a deciding factor as to whether Labour won or lost, and pleaded that if the party were to win votes it had to make housing an issue immediately, putting forward a socialist alternative to the free-market housing policies of the Conservatives. However, at the 1983 General Election, although the party's manifesto pledge that Labour, if returned to office, would encourage and assist local authorities to begin a massive programme of housebuilding and improvement, its manifesto also commited the party to end enforced council house sales and to phase out mortgage interest relief above basic rate. These latter pledges – interpreted as being antagonistic to home-ownership – may have been a major reason why the party lost the 1983 election.

At the 1987 election, Labour ended its opposition to council house sales and pledged that it would maintain the 'Right to Buy'. It also showed its general support for owner-occupation by declaring that it would maintain mortgage interest relief, at least at the basic rate of income tax. However, the party's main priorities would be to launch a housebuilding and private and public sector renovation drive as part of its jobs programme and to combat the problems of bad housing, overcrowding and homelessness. At the election of 1992, Labour reaffirmed its support for the 'Right to Buy', proposed targeting mortgage interest relief at first-time buyers, and outlined new ways in which councils and housing associations could help the homeless, but, unlike in earlier elections, made no commitment to increase the low level of public expenditure on housebuilding and renovation.

Returning to office in 1997, Labour demonstrated both its reluctance to invest in social housing and its support for owner-occupation. Although it had critically drawn attention in its manifesto to the Conservatives' virtual abandonment of social housing, and was willing for capital receipts from the sale of council houses to be reinvested in building new housing and rehabilitating old stock, it supported 'effective schemes to deploy private finance to improve the public housing stock and to introduce greater diversity and choice' (Labour Party, 1997: 26), thereby favouring further privatisation. In respect of owner-occupation, Labour was concerned that the 'two thirds of families who own their own homes [had] suffered a massive increase in insecurity over the [previous] decade, with record mortgage arrears, record negative equity and record repossessions' (Labour Party, 1997: 25). It thus pledged that it 'would reject the boom and bust policies [that had] caused the collapse of the housing market . . . [and would] work with mortgage providers to encourage greater provision of more flexible mortgages to protect families in a world of increased job insecurity' (Labour Party, 1997: 26). Labour also saw the need to tackle the problems of gazumping, which had reappeared with the then current upturn in house prices. It favoured boosting home-ownership still further by introducing 'commonhold', a new form of tenure that would enable people living in flats to own their own homes individually but own the whole property collectively; and also intended simplifying the rules restricting the purchase of freeholds by leaseholders.

Clearly, there were some vestiges of Labour housing policy not driven by market forces. Although the party, in its manifesto, valued the revival of the private rented sector, it aimed to 'provide protection where it was most needed; for tenants in houses in multiple occupation' (Labour Party, 1997: 26). Labour was also very concerned that homelessness had more than doubled under the Conservatives, that there were, in 1997, 40,000 families in England in expensive temporary accommodation, and that local authorities no longer had a duty to find permanent housing for homeless families. It therefore pledged that it would 'place a new duty on local authorities to protect those who are homeless through no fault of their own and are in priority need' (Labour Party, 1997: 26). Young people without homes living rough on the street were also a major concern to the party. Labour therefore set out to attack the problem in two ways: by using the phased relief of capital receipts from council house sales to increase the stock of housing for rent, and by deploying its welfare to work programme to lead the young unemployed into work and financial independence.

Labour – mainly through the medium of the Department of the Environment, Transport and the Regions (DETR) – thus set out to put its housing policy into effect. However, since its manifesto commitments, although generally applicable to the United Kingdom, were specifically targeted at England, distinctive interpretations of manifesto pledges emerged in Scotland and Wales as a consequence of devolution, and in Northern Ireland as a result of the 'peace process'. No longer was housing policy determined solely by London-appointed ministers at the Scottish Office, Welsh Office and Northern Ireland Office.

Devolution of housing policy

With the establishment of the Scottish Parliament in 1999, Scotland, which already had a pattern of tenure and house purchase system markedly different from that of the rest of the United Kingdom, gained the opportunity, through its powers of primary legislation, to formulate even more distinctive housing policies than hitherto. However, housing soon became 'a highly politicised arena and [became] one of the more contentious areas of policy for the Scottish Executive' (Earley, 2000: 28). There were two key areas of major concern: first, the need to establish a single social housing tenancy north of the Border 'with common rights and in particular a universal Right to Buy' (Earley, 2000: 28); and second, the need to introduce a New Housing Partnership programme, under which local authorities were to transfer their housing stocks to housing associations, which in the case of Glasgow City Council would involve the transfer of 100,000 dwellings.

In Wales, following the devolution of government in 1999, housing policy continues to evolve and develop. Although the newly established National Assembly only has powers to amend secondary legislation (rather than create primary legislation), it nevertheless published a framework for a national housing strategy in 1999, and in May 2000 received policy recommendations from four expert Task Groups (National Assembly, 2000). Taking account of these recommendations, and policy development in England, the National Assembly will undoubtedly find time to produce Welsh secondary legislation whenever necessary (Williams, 2000).

As a result of the peace process, new administrative arrangements came into being in Northern Ireland in 1999. Housing policy henceforth was to be formulated by the province's Department of Social Development (DSD) in consultation with strategic bodies such as the Northern Ireland Housing Executive (NIHE). This was intended to provide a framework for 'developing housing policy and funding to meet needs in the social and private sectors' (Stevens, 2000: 15), but – as in Scotland and Wales – initiatives continued to be influenced by policy developments emanating from Whitehall. Two housing issues were of particular concern to the DSD: first, the transfer of the Department of the Environment Northern Ireland's (DoENI) monitoring and regulatory responsibilities to the NIHE; and second, the transfer of responsibility for new building in the social sector from the NIHE to housing associations.

Structure and contents of the book

Derived from earlier texts – *Housing Policy and Housing Needs* (Balchin, 1981) and the first three editions of the present work – this book is divided into three parts and twenty inclusive chapters. Part I focuses on the supply of new and renovated housing, and contains chapters on housing investment, housebuilding and rehabilitation policy. Part II then examines housing markets and housing tenure, with chapters on housing finance, private rented housing, local authority housing, privatisation and stock transfer, housing associations and owner-occupation. Part III subsequently considers a number of single issues in housing policy: specifically, affordability, regional disparities, regeneration, social exclusion, housing and community support, housing and the elderly, gender and housing, black and Asian households and housing, and homelessness. A concluding chapter reviews the housing policy aims of the Blair administration, following Labour's second successive electoral win in June 2001, and a postscript outlines a number of housing initiatives announced later in the year.

Further reading

Birchall, J. (ed.) (1992) *Housing Policy in the 1990s*, London: Routledge. A series of analytical reviews of Conservative housing policy in the late 1980s and an examination of the possible future effects of this policy.

Black, J. and Stafford, D. (1988) *Housing Policy and Finance*, London: Routledge. A comprehensive introduction to the field, providing a free-market rationale for the development of housing policy in the 1980s.

Cowan, D. and Marsh, A. (eds) (2001) *Two Steps Forward: Housing Policy into the New Millennium*, Bristol: Policy Press. An up-to-date review of analytical essays examining housing policy largely within a managerial and social context.

Currie, H. and Murie, A. (1998) *Housing in Scotland*, Coventry: Chartered Institute of Housing. An analysis of contemporary policy concerns in Scotland.

Darke, J. (ed.) (1992) *The Roof over Your Head*, London: Spokesman. A blueprint for Labour's future housing policy, as proposed at the beginning of the Major years of government.

Department of the Environment, Transport and the Regions/Department of Social Security (2000) The Housing Green Paper, *Quality and Choice:*

A Decent Home for All, London: HMSO. Essential reading on the evolution of housing policy in the early years of the twenty-first century.

Glennister, H. and Hills, J. (ed.) (1998) *Paying for Welfare: Towards 2000*, Oxford: Prentice Hall. An excellent analytical review of developments of public spending across a range of public sectors.

Harriott, S. and Matthews, L. (1998) *Social Housing: An Introduction*, Harlow: Longman. Provides a fairly up-to-date introduction to the field.

Kemeny, J. (1992) *Housing and Social Theory*, London: Routledge. An explicitly theoretical explanation of the relationship between rental tenures.

Labour Housing Group (1984) *Right to a Home*, Nottingham: Spokesman. A radical blueprint for Labour's future housing policy – as suggested during the middle Thatcher years of government.

Malpass, P. (ed.) (1986) *The Housing Crisis*, London: Croom Helm. Containing much original research, this book examines what seemed to be the growing housing crisis of the mid-1980s and the reasons for it.

Malpass, P. and Means, R. (eds) (1992) *Implementing Housing Policy*, Buckingham: Open University Press. Very accessible analytical reviews of housing issues current in the early 1990s, during a period of Conservative government.

Malpass, P. and Murie, A. (1999) *Housing Policy and Practice*, 5th edn, Basingstoke: Macmillan. A useful complement to the present book, providing as it does a good account both of current structures and of the historical development of housing policy.

Marsh, A. and Mullins, D. (1998) *Housing and Public Policy*, Buckingham: Open University Press. Provides a useful introduction to policy, although it pre-dates recent developments.

Smith, M. (1998) *Guide to Housing*, 3rd edn, London: Housing Centre Trust. An uncritical though wide-ranging survey of the housing system in Britain, including useful chapters on Scotland and Northern Ireland.

Williams, P. (ed.) (1996) *Directions in Housing Policy*, London: Paul Chapman. An excellent collection of essays on the key housing issues of the mid-1990s.

Part I   **Supply of new and renovated housing**

2 Housing investment

The supply of new and renovated housing is inextricably linked to the level of housing investment. While in recent years there has been an increase in investment in the private housing sector (interrupted only by the house-price slump in the owner-occupied market in the early 1990s), investment in the social rented sectors tended to fall throughout most of the 1980s and 1990s, a result of cuts in public expenditure and the promulgation of the free market within an increasingly neo-liberal welfare state. In analysing that scenario, this chapter:

- focuses on quantitative and qualitative deficiencies in the housing stock;
- indicates the broad relationship between investment and housebuilding;
- explains how resource expenditure is dependent upon borrowing;
- examines the connection between investment cuts and the repair, improvement and replacement of the housing stock;
- discusses the impact that monetarism had on housing investment, and suggests how the presentation of Treasury figures probably misinformed constructive debate on the level of housing investment; and
- concludes by reviewing policy developments in the 1980s and 1990s, and under Labour, post-1997.

Quantitative and qualitative deficiencies in the housing stock

In the mid-1980s Britain was faced with four major and interrelated housing problems (NFHA, 1985). First, there was a serious housing shortage (in net terms), as evidenced most clearly by the number of homeless households. Second, the condition of housing was deteriorating, with the private rented sector containing the highest proportion of substandard dwellings. Third, although comprising a minority of the local authority stock, unsatisfactory council estates accommodated a concentration of the poorest tenants and ethnic minority groups in dwellings that were inappropriately designed, badly constructed and difficult to manage and maintain. Fourth, because housing is compartmentalised into sectors and tenures, there was a denial of choice, mobility and diversity. Communities were therefore increasingly polarised, most notably between owner-occupiers and tenants, generating tension and divisiveness within society. Five years later, these problems remained and were of even greater cause for concern. The *Inquiry into British Housing, Second Report* (JRF, 1991) attributed the problems to two underlying factors: inadequate investment in rented housing and a heavy financial bias towards owner-occupation – problems that remained endemic throughout the 1990s and beyond.

Investment and housebuilding

Since the late 1930s there has been a virtual absence of investment into the building of private rented housing, and very little private investment in the modernisation of existing property, while more recently, investment in new social housing plummeted. In the local authority sector, investment in new building (and acquisitions) decreased in real terms from £2.2 billion in 1980/81 to only £47 million 1998/99 (a reduction of nearly 98 per cent), while investment by housing associations fell by nearly 9 per cent over the same period (Table 2.1). Housebuilding in the social sector consequently diminished – from around 56,400 starts in 1980 to 21,900 in 1999, while the number of homeless accepted by local authorities reached a peak of 179,410 in 1992 and remained well over 100,000 throughout the rest of the decade. Although capital expenditure on the renovation of council housing (housing revenue account stock) was maintained, there were substantial cuts in public investment in home-ownership, private renovation and the housing element of the Urban Programme (UP). Taking into account capital receipts, net public investment in housing fell by over three-quarters during the last twenty years of the twentieth century. In contrast, investment in the owner-occupied sector was affected by cyclical changes in market conditions, with booms in 1988 and 1997 (when respectively 221,400 and 162,400 houses were built) and slumps in 1980 and 1992 (when output fell to only 98,900 and 120,200 starts).

Table 2.1 *Local authority housing gross investment, England, 1980/81–1998/99 (£ million at 1997/98 prices)*

	1980/81	1990/91	1998/99	Percentage change 1980/81–1998/99
New build and acquisitions	2,275	636	47	−97.9
HRA stock renovation	1,512	2,013	1,519	+0.5
Housing association	384	226	331	−13.8
Private renovation	594	571	483	−18.7
Home ownership	1,379	206	108	−92.2
Urban programme	20	49	0	—
Total	6,164	3,701	2,488	−59.6
Capital receipts	2,341	2,801	1,570	−32.9
Net investment	3,823	900	918	−76.0

Source: DETR (2000c)

However, with the number of new households increasing, the decrease in the total number of housing starts from well over 300,000 per annum in the 1970s to around 173,000 in 1999 inevitably created acute shortages. Although governments relied increasingly on the private sector to be the dominant supplier of new housing, and despite the proportion of owner-occupied houses in Britain increasing from 56.0 per cent in 1980 to 67.6 per cent in 1998, the bias towards owner-occupation (and the low priority given to rented housing throughout much of this period) either created shortages for those who could not afford to access the owner-occupied sector or provided the only option for those who could afford to buy.

Borrowing and resource expenditure

For most private house buyers, mortgage finance is a necessity. During the boom and slump years of the late 1980s, and throughout the 1990s, net advances secured on dwellings increased from £19.7 billion in 1985/86 to £39.9 billion in 1988/89, before decreasing to £15.2 billion in 1994/95 and rising to £38.4 billion in 1998/99 (Table 2.2). But less than 10 per cent of advances each year were used to purchase newly built houses; the rest was spent on the purchase of 'second-hand' dwellings. Thus only a small proportion of mortgage finance facilitated *resource expenditure* (on housebuilding); the larger proportion funding *transfer expenditure* – the owner-occupied market acting very largely as a 'capital guzzler', with very little increase in its size at the end of the day except through the privatisation of council stock.

Table 2.2 *Housing investment United Kingdom, 1985/86–1998/99 (£ million cash)*

	Private sector net advances[a]	Gross capital expenditure	
		Local authorities	Housing Corporation[b]
1985/86	19,658	3,742	957
1986/87	27,183	3,808	978
1987/88	29,554	4,253	1,033
1988/89	39,967	4,611	1,116
1989/90	33,930	6,194	1,393
1990/91	25,973	4,161	1,705
1991/92	18,415	3,786	2,408
1992/93	16,647	3,689	3,212
1993/94	19,213	4,125	2,432
1994/95	15,165	3,862	2,095
1995/96	19,106	3,732	1,728
1996/97	23,834	3,546	1,579
1997/98	25,229	3,229	1,127
1998/99	38,382	3,319	1,116

Source: Wilcox (2000)

Notes: [a] Data for calendar years 1985, 1986, etc.
[b] Includes data also relating to Housing for Wales, Scottish Homes and the Northern Ireland Housing Executive.

Capital expenditure by local authorities and the Housing Corporation (via housing associations) is, however, very largely resource expenditure – spending on housebuilding and renovation. Whereas in cash terms, investment by local authorities remained fairly constant over the period 1985/86–1998/98 (although falling in real terms), investment by the Housing Corporation in England (and by equivalent bodies elsewhere in the United Kingdom) increased dramatically to a peak in 1992/93 before falling back to a relatively modest level in 1998/99 (Table 2.2).

Because at least 90 per cent of private sector advances involved transfer expenditure, total housing investment in recent years (and consequently the housing sector borrowing requirement) was largely concerned with exchanging existing and often quite old property, rather than with facilitating new housebuilding or renovation. These advances often exceeded the total public sector borrowing requirement.

Therefore, since only a minority of total borrowing is undertaken to increase the size and improve the condition of the housing stock, it is important that housebuilding and rehabilitation are as productive as possible. In the public sector it was only in the 1970s that the government began to consider how this could be achieved.

After considering the question 'how much housing do we need and how can resources be used to supply it?', the Green Paper *Housing Policy: A Consultative Document* (DoE, 1977a) innovated housing investment programmes (HIPs) – a significant change in the system of allocating Exchequer funds to local authorities. HIPs are determined by the government calculating national demand (by adding together local demand), and the total is then used as a basis for local authority bids for loan allocations – the total amount available being restricted by the Treasury's cash limits within the public sector's total borrowing requirement. Allocations are in 'blocks' of two: block 1 consists of new housebuilding, slum clearance, the improvement of council housing and the acquisition of dwellings for continued housing use; block 2 comprises private sector improvement grants and gross lending to private persons for house purchase and improvement. Within both blocks, local authorities can switch expenditure between programmes and can also reallocate spending between blocks of up to 10 per cent of the value of the smaller block. Over- or underspending of up to 10 per cent can be carried forward from one year to the next, decreasing or increasing the following year's expenditure.

This new system involved funds being allocated generally according to a formula combining a national index of housing stress with the spending records of authorities – subsidies ceasing to be fixed amounts per dwelling. Local authorities had to justify their programmes according to shortages and needs not catered for in the private sector. The Green Paper thus made public housing mainly a 'residual' rather than a 'general needs' tenure, thereby undermining all the principles on which social ownership of housing was based.

Fluctuations in investment

Under Labour, although total government expenditure remained broadly unchanged from 1974/75 to 1979/80, public spending on housing fell from a peak of £7,154 million to £4,520 million (at 1980 prices) – a drop of 37 per cent – and housing expenditure as a proportion of total government spending fell from 10 to 5.9 per cent. Investment in new and rehabilitated housing was severely reduced, but housing subsidies actually increased – families in need of housing, in effect, being penalised to the benefit of those reasonably accommodated.

Although the Conservatives' White Paper *The Government's Expenditure Plans, 1980/81* (Treasury, 1980) decreased total planned public spending by 5 per cent in real terms over the four years from 1980/81 (inclusive), planned housing expenditure was reduced by 38 per cent to £2,792 million (cash) by 1983/84 – housing bearing three-quarters of all government cuts (House of Commons, 1980). As it turned out, at the end of these four years housing expenditure had decreased by (only) 28 per cent to £3,262 million, but this nevertheless represented a reduction in its share of total public expenditure from 4.8 per cent in 1980/81 to a mere 2.6 per cent in 1983/84, an appallingly low level in view of the housing problems inflicting the United Kingdom's towns and cities. But even before these expenditure cuts were announced, the Conservatives, on regaining power, reduced former Labour HIP allocations to English local authorities in real terms by 11 per cent (1979/80) and a further 33 per cent (1980/81). These cuts were more drastic than local authorities had anticipated and resulted in overspending on local housing programmes to the extent of £180 million by October 1980. The Secretary of State, Mr Michael Heseltine, therefore imposed a moratorium on council housebuilding for a period of six months.

From October 1980 to the spring of 1982, HIP allocations were periodically reduced (for example by 15 per cent in December 1980), forcing local authorities both to cut back on housebuilding (from 107,400 starts in 1978 to 37,200 in 1981) and to sell off their housing stock to sitting tenants at prices up to 50 per cent below market values in order to supplement their HIP allocations. Table 2.3, however, shows that the reduction in spending on new dwellings was more than made up by the increase in expenditure on improvement throughout most of the 1980s, reflecting a shift of emphasis in policy – notwithstanding the reduction in net capital expenditure.

As well as monetarism having an impact on investment in the 1980s, cuts in housing expenditure pushed up council rents from an average of £6.48 per week in 1979/80, £13.97 in 1983/84 to £18.50 in 1988/89, with many Conservative-controlled councils charging rents in excess of £20 per week – increases out of all proportion to the increase in average wages or the retail price index. Rents began to subsidise rates in many local authority areas, because in abolishing the 'no profit' rule for council housing, local authorities (especially the shire districts) were able to make financial surpluses.

Table 2.3 *Capital expenditure on housing by programme, 1979/80–1987/88 (£ million cash)*

	1979/80	1983/84	1987/88	Change 1979/80– 1987/88 (%)
Local authority gross capital expenditure:				
New dwellings	1,078	743	590	−45
Improvement investment	723	1,148	1,590	+120
Improvement grants	134	1,064	535	+299
New Towns gross capital expenditure:				
New dwellings	149	10	5	−97
Improvements	4	19	9	+125
Other	668	149	106	−84
Gross capital expenditure	2,756	3,133	2,835	+3
Receipts from the sale of land and housing	−76	−1,081	−1,674	
Net capital expenditure	2,680	2,052	1,161	−57

Source: Treasury (1988)

The abysmal level of housebuilding had clearly become unacceptable politically (if only temporarily) in 1982. There was a limit to the degree to which the government was prepared to squeeze housing investment. Local authorities, however, were underspending their HIP allocations – with a moratorium on housebuilding, new cost yardsticks with inevitable delays in application, and capital receipts from the sale of housing and other assets all being contributory factors. Also, local authorities were reluctant to use their allocations to the full if it meant that they would overshoot their current expenditure (on wages, services and maintenance) and risk reduced block grants the following financial year as a penalty – as threatened by the Secretary of State. However, by October 1982 it was predicted that investment underspending would amount to over £1 billion by the end of 1982/83 (with the possibility of the government undershooting its public sector borrowing requirement of £9.5 billion). The Secretary of State therefore encouraged local authorities to commit themselves to extra capital expenditure of £1 billion on housebuilding, improvement and the clearance of derelict land by March 1983. An extra £109 million was even allocated by the government to the 164 English local authorities to help them spend more than their existing entitlement before March 1983, but these measures did little to prevent the public sector borrowing requirement for 1982/83 slumping to £7.5 million, and underspending for the year amounting to £800 million.

Partly because of pressure from the construction industry, but also taking into account the previous year's underspend and capital receipts from the sale of council and housing association dwellings, the government increased gross capital expenditure to £3,133 million in 1983/84 (in contrast to £2,406 million in the previous year). To assist the construction industry further, the Minister of Housing and Construction, Mr John Stanley, in December 1982 announced minimum housing investment levels for 1984/85. On the assumption that average rents rose in line with prices during that year, and that councils received £1,300 million from the sale of assets, he assured the industry that local authorities could expect to receive at least 80 per cent of their 1983/84 provision in the following year – a concessionary two-year commitment by the Treasury which it had been unwilling to make before.

But in December 1984 the Treasury reversed its support for housing investment. It announced that from April 1985 local authorities could spend only 20 per cent of their receipts from council house sales in any one year (compared to 40 per cent before). This effectively froze a further £1.2 billion of the £6 billion of accumulated local authority receipts. In response, the Building Employers Confederation (BEC) claimed that this reduced resource would result in 75,000 council houses not being built and in 48,000 fewer improvement grants being awarded – the Trades Union Congress describing the cuts as 'the economics of the madhouse' (McLaughlan, 1984).

In the late 1980s and early 1990s, the Treasury permitted local authority and Housing Corporation capital expenditure to increase gradually from £2,452 million in 1985/86 to £3,926 million in 1992/93 at current prices – an increase of 60 per cent – but over the same period the total housing programme increased by 169 per cent – again illustrating the redistribution of expenditure away from housebuilding and renovation to demand subsidies. In real terms, however, total government expenditure on housing remained broadly constant during the period in question, increasing only marginally (at 1992–93 prices) from £6 billion in 1985/86 to £6.2 billion in 1992/93, although it must be remembered that this level of spending was considerably below that of 1979/80, when housing programme expenditure reached £13.3 billion (at 1992–93 prices).

Within the context of economic criteria or ideological commitment, the question must therefore be asked, why did the government make such a draconian long-term cut in housing expenditure? Was it due to the amelioration of housing problems from the mid-1970s onwards? Was it due to profligacy in the mid-1970s and the subsequent adoption of monetarism/Thatcherism? Was it due to the way in which the Treasury presented figures or was it due to the belief that, since owner-occupation had become the dominant tenure, housing investment could be left mainly to the private sector? These suggested answers will be examined in turn.

Repair, improvement and replacement of the housing stock

As much as a quarter of Britain's housing dates from the period 1875–1914, a time during which there was a shift of emphasis from investment in manufacturing industry to residential development at home or investment overseas, a time when Britain failed to adapt to the electrical–mechanical technology and necessary institutional changes of the second Industrial Revolution. As Desai (1983) points out:

> From the 1870s onwards the British economy began to lose its competitive edge in third markets to the USA and Germany. Instead of responding to this new challenge by changing the structure of industry, the economy increasingly took on the character of a *rentier* economy. The domination of finance capital over industrial capital dates from this period.

A similar surge of investment in housing occurred during the 1920s and 1930s when the inter-war depression deterred not only manufacturing investment at home but also investment overseas. But this time, finance capital facilitated the growth of the owner-occupied sector, in addition to the expansion of the private rented market.

Housing is to a great extent a commodity like any other good or service and, like any other commodity, investment is necessary for its production. Bassett and Short (1979) explain that since housing development is a lengthy process (in contrast to the manufacture and sale of most manufactured commodities) and the value of each housing unit is normally a multiple of average earnings, numerous 'chains of capital' are required to facilitate its availability. Thus in addition to 'building industrial capital' (part of the primary circuit of capital) being required to secure its development, 'secondary' (or finance) capital is necessary to enable landlords or owner-occupiers to acquire the completed commodity. Secondary capital can be either 'commercial capital' to facilitate the purchase of housing to rent (loans being repaid out of rent income) or 'interest-bearing capital' (mortgage loans, with repayments out of income and interest relief). In recent years the secondary circuit (especially interest-bearing capital) has expanded out of all proportion to the growth of the primary circuit. It is this, together with the mismatch between the desire of housing developers to maximise profit and prevailing economic trends and policies that frustrate this desire, which explains why at least half of Britain's housing stock is over fifty years old, and deteriorating faster than it is being repaired, improved or replaced. By 1981 there were nearly 4 million dwellings in England (alone) each on average requiring repairs costing over £2,500 (including a million over £7,000), 1.2 million unfit and 994,000 lacking one or more basic amenities. The situation was exacerbated by the very low level of housebuilding: 153,700 starts in 1981 – one-third less than in 1979 and two-thirds less than in 1968 (the post-war peak). In no sense could it have been said that housing problems were diminishing.

In the late 1970s, however, far more resources were being attracted into increasing the size of the total housing stock than into renewal. In 1977 £2,500 million was spent on renewal, but £3,000 million on net increases to stock, private sector housebuilding masking the shift of emphasis in public policy away from new housebuilding to rehabilitation (then the main form of renewal). On conservative estimates at least £4,200 million should have been spent on renewal (Stone, 1970), which together with the £3,000 million being spent on stock addition would have resulted in an output of £7,200 million – considerably in excess of the capacity of the construction industry, estimated to be about £6,600 million given full employment.

The *English House Condition Survey 1991* (DoE, 1983), showed that although local authority housing (which accounted for about 30 per cent of the total housing stock) contained only 6 per cent of the total number of unfit dwellings and 5 per cent of those in serious disrepair (private rented housing being in a considerably worse condition), these low proportions masked very substantial renewal problems. The Association of Metropolitan Authorities (AMA, 1983, 1984, 1985) estimated that the repair and renovation of council houses built mainly in the 1920s and 1930s (using traditional methods of construction) would cost £9,000 million to implement, that the repair costs to council houses built in the 1950s and 1960s (using non-traditional techniques such as steel frames and reinforced concrete) would amount to £5,000 million, and the repair costs to council houses constructed in the 1960s and 1970s (using systems building techniques) would cost a further £5,000 million to put

Box 2.1

A low level of housing investment

By European standards, the proportion of total investment and gross domestic product (GDP) spent on housing investment in the United Kingdom in the 1980s and 1990s was deplorably low (Table 2.4). To be on a par with those countries where housing investment was a top priority it would have been necessary for the United Kingdom to have doubled its capital expenditure on housing. Although it might be argued that this is unnecessary since other countries urbanised later than Britain, started from a poorer housing base, and therefore need to devote more resources to housing, it must be remembered that it is simply because Britain's housing stock is old and, in part, poorly maintained that enormous resources are needed for renewal.

Table 2.4 *Housing investment and GDP: international comparisons*

	Gross fixed investment in housing as a percentage of GDP, average 1985/96	GDP per capita ($) 1996
Germany	6.3	28,860
Canada	5.9	19,200
Japan	5.6	41,080
Netherlands	5.2	25,850
Italy	5.1	19,930
Australia	5.0	20,370
France	5.0	26,290
Ireland	4.8	17,450
New Zealand	4.8	15,850
United States	4.2	27,590
Sweden	4.2	25,770
Belgium	4.1	26,440
United Kingdom	3.5	19,810

Sources: OECD (1998); World Bank (1998)

right. In total, the £19,000 million needed to be spent on repairs can be contrasted with the government's expenditure plans for public sector renovation in 1985/86 – £1,312 million, a severely inadequate figure. In addition, in the private sector the cost of repairs amounted to £25,000 million in 1985 (Carvel, 1985) – a sum which dwarfed the amount of

government support for private sector renovation and clearance in that
year: £608 million.

The above renovation costs may have been underestimated. Mr Peter
McGurk (Director of the Institute of Housing) (1987) claimed that the
total sum that needed to be spent on repairs and improvement was £50
billion; Shelter (1987) suggested that the cost could be £54 billion, while
the AMA (1986) (using different criteria than hitherto) put the figure at
£75 billion. Clearly, investment in housing was at an all-time low and the
implications of an increasingly unrepaired and unimproved housing stock
should have been matters of considerable concern. The alternative (which
was becoming increasingly evident) was greater scarcity, increasing
homelessness, soaring house prices, widening social divisions and a
decaying environment.

The failure to renew the housing stock adequately was 'analogous with a
household living beyond its means, but not spending enough on repairs'
(Kilroy, 1981a). For years, Britain was 'over-housing' itself by not
spending enough on repairing, improving and replacing its old dwellings,
but by increasing the size of stock (particularly in the owner-occupied
sector). With a total shift of emphasis to renewal, housing aspirations
would need to be satisfied mainly within the built-up areas, and not least
within the inner cities rather than in greenfield locations. On the
assumption that some renewal would have taken the form of clearance
and redevelopment (replacement) rather than rehabilitation (repairs and
improvement), excess demand would have necessitated an adequate
supply of land to be forthcoming through changes in compulsory
purchase procedure and the valuation of urban sites. But households,
especially those on average or above average incomes with good job
prospects, would have needed to have settled for slightly lower-quality
housing and possibly less attractive residential environments if 'we were
to get the housing we required at a price we could afford' (Kilroy, 1981a).
Instead, there has been greater scarcity, increasing homelessness,
widening social divisions and a decaying urban environment.

Disinflation and housing investment

There can be little doubt that monetarism had a severe effect on the level
of housing investment. Government expenditure on housing fell, in real
terms, by 19 per cent from 1974/75 to 1979/80 and then by a further
53 per cent by 1992/93, with both the Labour and the Conservative

governments using housing and the construction industry to disinflate the economy. Apart from the effects of the reverse multiplier, there were two reasons why housing and construction were selected for this task.

First, by the late 1970s the number of dwellings was beginning to exceed the number of households. A deficit of 600,000 dwellings in England and Wales in 1966 turned into a crude surplus of 500,000 by 1976, and the government may have thought that this would reduce waiting and transfer lists and the number of homeless households. Ministers may also have been badly advised about the extent of unfit housing and the numbers of dwellings needing repairs, and the locational and price-range mismatch of demand and supply in the owner-occupied sector. In the 1980s, when the crude surplus was turning rapidly into a substantial net deficit, it was tragic that the government – in pursuing a disinflationary housing investment policy – probably still believed that there was a real surplus. This was in spite of many reports to the contrary; for example, the AMA (1982) predicted that there would be a net deficit of 500,000 dwellings in England and Wales by 1986 if current policies continued, and subsequently Niner (1989) predicted a shortage of 3 to 4 million dwellings by the year 2001.

Second, since monetarist disinflationary policy was intended to be partly implemented by a reduced level of public expenditure, allegedly to release resources for private sector industry (made 'leaner and fitter' by high interest rates and the shake-out of 'surplus' labour), it was hardly surprising that Thatcherism regarded council housing as a residual tenure, suitable only for welfare purposes. In consequence, public sector housing bore 75 per cent of all government planned expenditure cuts, 1979/80 to 1983/84, and continued to decrease throughout most of the 1980s. In order to boost owner-occupation, up to 75 per cent discounts were given to council tenants to buy their own homes, and owner-occupiers benefited from mortgage interest relief and capital taxation exemption – none of these forms of assistance were deemed 'expenditure' and they were therefore unaffected by public spending cuts. Indeed, mortgage interest relief increased from £865 million in 1975/76 to £7,700 million in 1990/91, and capital gains tax exemption in the latter year was probably worth £3,000 million.

Presentation of Treasury figures

The impression has often been given that public spending on housing not only has been absorbing a growing amount of funds, but has also crowded out investment in manufacturing. To be sure, expenditure did increase – from £3 billion in 1963 to £7 billion in 1974 – and therefore, at least until the early 1980s

> the government . . . concentrated its public expenditure cuts on the housing programme. [But] it has been provided with a rationale for doing so by the old methods of accounting for public expenditure which makes it appear that there has been a sustained period of expansion in housing expenditure when actual investment has been falling.
>
> (Shelter, 1982a)

Kilroy (1981b) offered the following explanation. If an owner buys a house with, say, a £20,000 mortgage at 15 per cent costing £3,000 per annum, expenditure amounts to only £3,000 per annum (the £20,000 being a 'liability' and not spending as such). But if a local authority undertook the same transaction, the Treasury would view the total outlay of £23,000 as 'cash' spending, not having drawn the normal accountancy distinction between current account and capital account, and regarding the total outlay as a potential increase in the government's borrowing requirement. This partly explains the apparent escalation of government housing expenditure until the mid-1970s, and the subsequent backlash.

Kilroy argued, however, that even ignoring this accountancy oddity, increased public spending on housing did not mean that there was more investment. Investment by housing authorities on new homes and local authority improvement actually fell – from about £3,800 million to £1,950 million, 1968–79 (at 1980 prices) – although this was partly offset by more spending on the acquisition of land and dwellings, housing association activity, and improvement grants to private owners (the latter amount rising from £50 million to £850 million over the same period). Another item, capital transfers (loans and disposals), increased temporarily to over £2,000 million in 1974 because of help given to the building societies faced with a liquidity crisis, but from 1976 into the 1980s spending was largely negative owing to repayments and the sale of council houses. It was the increase in transfer payments which was mainly responsible for the big increase in public spending on housing. General subsidies rose throughout the 1970s, with a jump from £1,100 million to £1,700 million in 1972/73–1973/74 as a result of the

Conservatives' Housing Finance Act of 1972 (which pushed up rents and necessitated a big increase in rent rebates and allowances), and a further jump in 1974/75 to £2,300 million when the new Labour government stabilised rents and increased general subsidies to local authorities. The option mortgage subsidy remained fairly constant in the late 1970s at about £200 million per annum, but mortgage tax relief increased from £865 million in 1975/76 to £1,639 million in 1979/80. Although this form of assistance was not deemed to be 'public expenditure', in principle it was, and had the same effects as if it were a cash subsidy.

Public sector investment was of course boosted by the receipts of council house sales, but since building societies had to fund a significant proportion of these sales, council tenants wishing to buy faced competition from would-be mortgagors in the private sector. But although net loans for house purchase increased from 1 to 4 per cent of the GDP, 1960s to early 1980s, and increased further to 6 or 7 per cent by 1990, most of this was exchange expenditure involving the transfer of houses from one owner to another without any production taking place. By contrast, in public sector housing, the borrowing requirement increases only when investment, involving mainly resource expenditure, is undertaken. But whereas in the 1970s the debt in the owner-occupied sector trebled, in the public housing sector it only doubled – exchange being facilitated more than production. Yet while the former debt is geared to interest charges on the sum outstanding, the latter debt reflects the front-loading of interest charges in the mid-1970s, now past. It was ironic that public sector investment programmes in the 1980s were not growing in proportion to the erosion of the debt outstanding, while at the same time mortgage interest relief (and the private debt) continued to rise.

In the 1980s and 1990s the public housing debt further decreased in real terms, while the mortgage debt continued to soar. But although the government controls borrowing by local authorities, it does not (directly) control building society and bank borrowing, yet both the local authorities and the financial institutions channel personal savings into housing. There is no reason to suppose that the housing component of the public sector borrowing requirement is economically more 'damaging' than the much higher housing sector borrowing requirement (HSBR) (Griffiths and Holmes, 1984).

A massive amount of housing investment will clearly be required in the early years of the twenty-first century – including a significant amount of investment by the public sector (alone or in partnership with private financial institutions), particularly in the context of housing renewal. But, as Kilroy (1981b) pointed out, since large-scale housing investment will have to be financed by a substantial amount of borrowing, capital expenditure should be accounted separately and insulated from current spending (although the revenue account should not be balanced by an open-ended call on taxation). Kilroy thought that this would not be impossible since loan charges were declining nationally (after a peak in the mid-1970s), 'inflation accounting' could bring substantial savings by enabling the 'principal' element in loans to be paid off more slowly (the interest charges on projects under construction being capitalised), and cross-subsidisation could be adopted to balance the needs and resources of different areas. With these accountancy changes, local authorities could become 'masters in their own house' even if there were large fluctuations in needs and resources, and once again local authorities could become leading providers of affordable rented housing.

Private sector investment

Cuts in public housing expenditure in the late 1970s and in the 1980s may have been based on various myths about the comparative advantage of private sector housing finance over public sector investment – myths which reinforced the ideological policy aim of the Conservatives to support owner-occupation at the expense of council housing. First, there was the myth that owner-occupation is self-financing. But this is inconsistent with the very large increase in the GDP spent on the exchange of homes (albeit an increase in paper values rather than productive worth). Second, it was thought that the increase in housing wealth is 'ploughed back' when owners sell and buy. But a substantial part of realised value leaks into general consumption, and has to be attracted back as savings into the building societies by competitively high interest rates. Other institutional rates, therefore, have to be similarly high, possibly to the detriment of industrial investment. Since exchange rates are consequently higher than they would be otherwise, domestic suppliers to the home market and exporters are therefore doubly disadvantaged. Whereas in most other industrial countries, the savings market helps productive industry, in the United Kingdom it finances consumption and the transfer of housing. Third, there is a myth that the public housing debt has grown almost out of control. In money terms it did increase, 1965–80, but in real terms fell continually from 1968, and inflation eroded the existing debt. In contrast, the mortgage debt increased eightfold in money terms, and doubled in real terms over the same period.

Kilroy (1981a) argued, however, that there should be less concern about 'who gets more' than about what effects the different forms of assistance have in resource terms. Although interest relief and tax exemptions are intended to help both the buyer and the housebuilding industry, there is little evidence that they do so. By 1980, tax concessions had pulled house prices up sixteenfold since the Second World War, whereas retail prices went up only ninefold and stocks and shares, sixfold. Although in 1960 dwellings and land accounted for only 23 per cent of personal wealth, they accounted for about 50 per cent by 1980, whereas company shares and securities as a proportion of personal wealth slumped from 23 to 11 per cent. It must be obvious that this had an adverse effect on both the creation of wealth in the productive economy and the elimination of housing shortages and improvement of stock.

Policy developments, 1980s and 1990s

The Labour Party in the early 1980s firmly committed itself at the minimum (if returned to office) to arrest the deterioration in both the availability of housing and the condition of the stock. *Labour's Programme, 1982* (Labour Party, 1982a) stated that a massive public housebuilding and renovation programme would have a priority claim on resources. The party's 1983 election manifesto stated that it would achieve this through an immediate 50 per cent increase in housing investment programmes, with the emphasis being on the urgent repair and replacement of run-down estates. At the 1987 General Election, Labour pledged that if it won it would spend £2 billion (released from the 'frozen' receipts of council house sales) on a substantial housebuilding and repair programme – a programme which would also help to reduce unemployment by 1 million in two years; and at the 1992 election proposed establishing a Housing Bank to facilitate the balanced use of local authority capital receipts and investment at attractive rates of interest.

The Liberal–Social Democratic Alliance at the 1983 election also proposed an expansion of housebuilding in the public sector. It promised to spend an extra £500 million in this area during the first year of an Alliance government to ensure a steady expansion of building and more jobs. In 1987, like Labour, the Alliance pledged that it would restore to local authorities the right to use the proceeds from council house sales on replacing and repairing the housing stock; and at the 1992 election the Liberal Democrats (more cautiously) proposed relaxing controls on local authority capital receipts, especially for new council house and housing association building, and creating a new Partnership Housing sector to mobilise both private and public investment.

The Conservatives in their 1983 and 1987 manifestos, in contrast, implied that investment would only be increased if an emphasis were placed on the extension of owner-occupation. Therefore, although improvement grants would continue to occupy an important role in their housing policy, the party's aim was to increase property ownership by extending 'Right to Buy' provision (see Chapter 8). With regard to public spending it was clear that exchange expenditure was to be more important than resource expenditure, notwithstanding a significant increase in improvement. Doubtless the party assumed that in the owner-occupied sector the market could be relied upon to ensure an increasing quantity of homes – a progression towards the realisation of a 'property-owning democracy'. Bias against public expenditure on housing was further manifested by continuing cuts throughout the 1980s. Total public spending on housing was decreased from £3,262 million to £2,450 million, 1984/85–1987/88, and net capital spending was reduced from £2,052 million to £1,161 million over the same years. The government clearly believed that people preferred to own their own homes rather than rent, and that local authorities no longer needed to supply housing for ordinary families. At the 1987 election, the Conservatives, in stating that home-ownership had 'been the great success story of housing policy' since 1979 and that they would retain the current system of mortgage interest relief, implied that they would rely mainly upon housebuilding and renovation in the private sector to satisfy people's expanding housing aspirations. In 1992, however, some direct references to investment were made in the party's manifesto. The Conservatives pledged that they would continue with the Estate Action and HAT programmes with the aim of concentrating resources on the worst council estates.

The Conservatives, however, were in general failing to respond to a number of recommendations proffered by housing specialists and independent bodies in the middle and late 1980s. For example, Hilditch (1985) proposed that the HIP system should be 'turned on its head'. Instead of restricting those local authorities that wished to spend, a national needs index should be used, he suggested, to determine the 'minimum level of investment for each district – the core allocation which every authority would be obliged to spend'. The implications of low investment, he argued, were very severe and the need for increased investment would expand substantially throughout the 1990s. Wilson (1986) pointed out that the number of houses over 100 years old would have increased from 1.75 million in 1971 to 3.25 million by 1990, and he stressed the urgency of increasing the supply of housing in the face of a predicted 20 per cent growth in the number of households (1986–2000), to house the homeless and to replace housing beyond repair; while Mr Peter McGurk (Director of the Institute of Housing) (1987) claimed that 100,000 fewer public housing units were being constructed each year than the number required, and that £50 billion needed to be spent on repairs and maintenance to stop the housing in question being turned into slums.

In 1986 a consensus solution to the prevailing housing crisis was emphasised by a diverse group of organisations, namely the Association of County Councils (ACC), the Association of District Councils (ADC), the Association of Metropolitan Authorities (AMA), the Institute of Environmental Health Officers, the Royal Institute of British Architects (RIBA), the Royal Institution of Chartered Surveyors (RICS) and the housing pressure group Shelter. The group proposed not only that both public and private spending should be undertaken on a consistent and continuous basis to allow for the effective planning of a major housing programme, but that local authorities should be permitted to spend their capital receipts on housebuilding and renovation (a view adopted by both the Labour and the Alliance parties at the 1987 election).

Shelter (1986) proposed a substantial increase in public expenditure on housing to restore the allocation of improvement and repair grants to 1983/84 levels (grant expenditure had fallen from £1,064 million in 1983/84 to £528 million in 1986/87), to accelerate the rate of public sector renovation to clear a backlog of £19 billion of repairs, and to increase the rate of local authority and housing association housebuilding by respectively 60,000 and 40,000 units per annum. Shelter claimed that although the cost of its programme would amount to £7.75 billion (compared to public housing expenditure of £2.64 billion in 1986/87), it would not be inflationary or excessively expensive since the Treasury would save on benefit payments and gain extra tax revenue.

By 1991, in England alone, however, some 1,498,000 dwellings, or one in thirteen homes, were found to be unfit, and one in six homes needed urgent repairs costing more than £1,000 (DoE, 1993). According to Leather *et al*. (1994), at 1994 prices the cost of tackling the problems of unfitness would amount to £9 billion, the cost of undertaking urgent repairs to private sector homes amounted to £22 billion, while for private and housing association homes the outstanding cost of the repairs necessary for mortgage purposes would be at least £70 billion.

Despite this dire scenario, the last Conservative government of the twentieth century announced major cuts in public expenditure on housing in November 1996, only a few months before a General Election the following year. In its spending plans for 1997/98, the capital budget of the Housing Corporation was reduced by more than a third, while local authority credit approvals were axed – measures that were designed principally to help counter the inflationary impact of reduced income tax and consumer-led growth. The implications for housing investment and housing need were clear. Local authorities would be able to continue with their renovation programmes only if they were able to raise the necessary funds from accelerated programmes of stock transfer, while cuts in DETR funding meant that registered social landlords (RSLs) would find it increasingly difficult to comply with the government's medium-term policy of providing 60,000 new social lettings a year (Wilcox, 1997). The private rented sector, moreover, could not be expected to provide an alternative form of low-income housing since the government was intent on restricting the levels of rent that would be eligible for housing benefits (Wilcox, 1997).

Conclusion

In opposition, the Labour Party was concerned with the shortage of low-cost rented housing and its consequences such as overcrowding, the uneconomic use of bed and breakfast accommodation and homelessness. The party thus had a commitment to provide decent housing at affordable rents – through both new building and renovation, to satisfy social need and to generate employment.

Soon after returning to power in May 1997, Labour therefore introduced the first stage in a programme that enabled capital receipts (totalling £5 billion) to be – in effect – released to boost housing investment. Under the Local Authority Capital Receipts Act of 1997 (which amended the Local Government and Housing Act of 1989 to allow the issue of credit approvals that took account of receipts set aside for the repayment of debt), local authorities were permitted to borrow £200 million in 1997/98 and £700 million in 1998/99 for repairing the existing stock and building new houses. While these sums, together, were sufficient for 60,000 houses to be repaired and 28,000 homes to be built, they need to be seen within the context of a 33 per cent reduction in gross capital spending over the previous seven years and a 60 per cent cut since 1980/81.

In 1998, housing investment received a boost when – under the *Comprehensive Spending Review* (Treasury, 1998) – the Chancellor of the Exchequer authorised local authorities to spend more than £3 billion of council house receipts on a rolling programme of maintenance and new housebuilding to 2001/02, in addition to the £800 million earmarked from housing receipts. However, although capital spending was set to rise year on year (particularly on repairs and renovation), it was clear that

> total housing investment over Labour's first term . . . [would be only] £5–£6 billion, 32 per cent lower in real terms than it was during the [previous] government's five-year term. Most critically, spending on new houses . . . [would] remain at record low levels, with the Housing Corporation's development budget frozen at rock bottom.
>
> (*Roof*, 1998: 5)

Clearly, the number of new houses being built over the three years to 2001/02 would be well 'below all accepted estimates of need' (*Roof*, 1998).

It was not until the beginning of the new millennium that Labour demonstrated that it was serious in its intention to increase capital expenditure. The *Spending Review* (Treasury, 2000) set out the government's plans to provide an extra £1.6 billion for housing investment over the period 2000/01–2003/04. The government also intended to boost the housing private finance initiative (PFI), facilitating an injection of an extra £600 million into council housing over the two years 2002/03 and 2003/04. Clearly, the government hoped that it would meet its 'Green Paper commitment to bring all social housing up to a decent standard within a decade, with one-third of homes that currently fail to reach those standards improved by 31 March 2004' (Raynsford, 2000).

As Table 2.5 indicates, much of the extra investment will – through the medium of increased housing-based credit approvals and major repairs allowance – contribute to an increased supply of new and renovated local authority housing, but, in consequence of a substantial increase in

Box 2.2

Investment and improved efficiency in construction: the social housing sector

To ensure that there is sufficient investment to meet future housing need, it is imperative that efficiency in the housebuilding industry is improved, possibly along the lines proposed by the Construction Industries Task Force (DETR, 1998a). Chaired by Sir John Egan, the Task Force recommended that the industry, each year, should aim to:

- reduce capital costs and construction time by 10 ten per cent;
- reduce defects and accidents by 20 per cent;
- increase productivity, turnover and profits by 10 per cent; and
- increase predictability by 20 per cent.

To ensure that social housing organisations comply with Egan, a Housing Forum was set up to:

- run demonstration projects;
- bring together working groups on partnering and benchmarking; and
- disseminate information to members and others.

It was intended that – under the Housing Corporation as a regulator – all housing associations would be '100 per cent Egan compliant' by 2004, at least as far as getting changes in thinking and practice under way is concerned, although the associations would not necessarily be expected to have hit all their numerical targets. However, by 2001, housing associations should have been able to have shown that they were benchmarking to enable their performance to be measured against standard measures in the construction sector as a whole (Bayley, 2000). The associations would have been expected to have fulfilled any two of the following:

- been members of a benchmarking club;
- trained key staff in performance indicators;
- gained chartered status from the Construction Clients Forum; and
- met indicators on health and safety, and diversity.

Subsequently, housing association projects would have had to comply with any two of the following efficiency proxies:

- involved the use of partnering;
- shown a commitment to the principles of standardisation, pre-assembly and modularisation;
- acted as a Housing Forum demonstration project;
- facilitated feedback from residents, suppliers and contractors.

At the time of writing, there was no clear definition of '100 per cent compliance'. The term could have meant a project fulfilling the current proxies more extensively, or alternatively satisfying other requirements.

Aside from Egan, the DETR was keen to improve efficiency in the housebuilding industry through the reintroduction of prefabricated methods of construction. Undoubtedly 'bad experience with systems building in the 1960s . . . equated "non-traditional" with "problem" in the minds of the public, and probably helped to institutionalise a deep conservatism among house builders and buyers' (Bayley, 2000: 28–29). However, the government enabled the Housing Corporation to use £40 million from its approved development programme for schemes using newly approved systems of prefabrication. It was recognised that it would be possible to bring about savings from the use of prefabrication only if factories were producing in quantity, and this would be possible only if there was – again – adequate demand for new systems of prefabrication within the social housing sector.

Table 2.5 Spending Review *2000: housing expenditure, 2001/02–2003/04 (£ billion)*

	Plans agreed prior to Review, 2001/02	Total spending plans			
		2001/02	2002/03	2003/04	Total
Local authority capital of which	2,305	2,305	2,465	2,545	7,315
Housing basic credit approvals	2,305	705	793	842	2,340
Major repairs allowance	0	1,600	1,512	1,403	4,515
Housing Corporation of which	890	995	1,158	1,460	3,613
Approved Development Programme	691	789	940	1,236	2,965

Source: Treasury (2000)

funding through the Housing Corporation's approved development programme (ADP), there will also be a very large increase in the supply of new affordable housing provided by RSLs. In its attempt to increase investment, the government also aimed to get the best value for money and to raise standards by promoting the increased use of modern construction techniques and greater productivity in line with the Egan Report (DETR, 1998a).

Housing investment could be enhanced still further if local authorities – like RSLs and private landlords – were permitted by the Treasury to borrow in the market, instead of being dependent upon basic credit approvals. A possible mechanism to faciliate this could involve local authorities borrowing against the value of their rent stream or major repairs allowance, thereby putting them on the same footing as housing associations (Hebden, 2000). However, although this would confer greater financial freedom on local authorities, such a reform might also require the government to inject additional revenue resources into housing if local authorities are to start rebuilding their housing stocks.

Further reading

Aughton, H. and Malpass, P. (1994) *Housing Finance: An Introduction*, 4th edn, London: Shelter Publications. Contains a useful introductory summary of housing investment in England.

Ball, M. (1996) *Investing in New Housing: Lessons for the Future*, Bristol: Policy Press. In the context of the housebuilding slump of the mid-1990s, this publication presents a case for counter-cyclical housing investment policies.

Gibb, K., Munro, M. and Satsangi, M. (1999) *Housing Finance in the UK*, 2nd edn, Basingstoke: Macmillan. Contains comprehensive, though concise, coverage of housing investment within the social housing sectors.

Hills, J. (1991) *Unravelling Housing Finance*, Oxford: Clarendon. Probably the best book on housing investment and finance, although beginning to show its age.

Terry, R. and Joseph, D. (1998) *Effective and Protected Housing Investment*, York: Joseph Rowntree Foundation/York Publishing Services. Contains the findings of research that suggest that the current system of housing investment fails to satisfy housing needs, and that it is necessary to make further changes in housing policy to stimulate investment, making it more effective and protecting it in the long term.

Wilcox, S. (2000) *Housing Finance Review 2000/2001* (and subsequent editions), York: Joseph Rowntree Foundation. An invaluable – and annually updated – source of historic and current data on housing investment across Britain, supported by useful analyses.

3 Housebuilding

Although the level of investment is the fundamental determinant of the volume of housebuilding, housing supply is also substantially influenced by the cyclical nature of housebuilding and by government policy. Within this context, this chapter:

- considers the relationship between housebuilding and housing needs;
- examines the housebuilding cycle in both the social and the private sectors;
- discusses the government's response to the cycle during the 1980s and 1990s;
- analyses the relationship between new housing need, regional imbalances in need, and housing need in the South East;
- explores the arguments for and against greenfield versus brownfield housing development; and
- concludes by reviewing housebuilding under Labour, post-1997.

Housebuilding and housing needs

Since the late 1970s it has been increasingly recognised that the rate of housebuilding has been failing to satisfy needs. The Green Paper *Housing Policy: A Consultative Document* (DoE, 1977a) recommended that in order to keep pace with the 'baby boom' of the 1960s, replace unfit housing and facilitate household mobility, there was a need for 310,000 housing starts per annum until the end of the century. In response to the *English House Condition Survey 1981* (DoE, 1983), which showed that there were more than 2 million houses unfit for human habitation, numerous further estimates were made of the volume of housebuilding that was required to meet housing needs over the following two decades. The Housing Research Foundation, for example, suggested that an extra 3 million houses would be required by the year 2000 (Carvel, 1984b); the Building Employers Confederation estimated that there was an annual shortfall in output of about 100,000 units per annum (Simpson, 1985); and the School of Advanced Urban Studies of the University of Bristol claimed that a public sector housebuilding programme of 162,000 dwellings per annum was necessary to reduce homelessness alone (Ezard, 1987), apart from housebuilding needed to replace the accommodation of those inadequately housed.

The continuation of the housing shortage into the 1990s prompted the *Inquiry into British Housing: Second Report* (JRF, 1991) to emphasise the need to build 228,000–290,000 houses per annum up to the year 2001 and, to stress that, within this total, 100,000 should be built for rent. The RICS (1992) similarly called for the construction of 100,000 social sector homes per annum (over 60,000 more than current output) to reduce homelessness and to meet the demand for affordable housing. Although this would have cost about £2 billion per annum (gross) (in addition to the £4 billion currently being spent in England on social housing), finance could have been found if local authorities had been permitted to spend their £8 billion of capital receipts from the sale of council houses and from the phasing out of mortgage interest relief. Since 100,000 new jobs would have been created in the construction industry as a result of housebuilding on the scale prescribed, the net cost of provision would have been substantially reduced because of savings in benefit payments to unemployed labour and the extra tax paid by both building employees and companies. The RICS also proposed that a National Housing Investment Bank be established to attract investment funds and to provide loans to builders for the construction of low-cost housing. But the structure of the housebuilding industry, the housebuilding cycle and government policy all militated against housing supply being brought into line with housing needs, particularly in the 1980s and 1990s.

The housebuilding industry had an output exceeding £5 billion per annum at the peak of the building boom in the late 1980s, but it was exceptionally fragmented. Although there were about 120,000 firms in the construction industry in the late 1980s, there were fewer than 40 housebuilding firms employing 1,200 or more workers, and although the largest undertook the development of over 11,000 dwellings per annum, the typical medium firm built little more than 1,000 houses each year. But 95 per cent of Britain's builders were small, family-run concerns employing fewer than 50 workers. They operated locally and worked on minor contracts. They were often solely involved in small-scale infilling.

Many did little more than collect together skilled craftsmen at the right time and employ labour-only subcontractors.

The housebuilding cycle

The cyclical nature of housebuilding has been evident over the past century but has been particularly pronounced since the 1960s (Figure 3.1). Following a period of continual growth in the 1950s, housebuilding in the private sector faced a period of instability. There were peaks in the number of housing starts in 1967, 1972, 1978 and 1988, interspersed in quick succession by troughs in 1966, 1970, 1974, 1980 and 1992. The social sector declined severely from the 1950s to the mid-1960s (owing to inadequate support given by Conservative governments), but this was followed by peaks in 1967 and 1975 with troughs in 1973, 1981 and 1991 (Table 3.1). However, the overall trend has been downward since 1967, reaching a nadir of 152,900 starts in 1981 (less than in any year since the late 1940s). Even though the number of starts increased to 252,300 in 1988, output was still very low compared to the number of starts in the 1960s and 1970s.

Output in the social sector decreased to a mere 26,500 in 1991 – lower than in any peacetime year since the First World War. The social sector, however, is subdivided mainly into local authority and housing

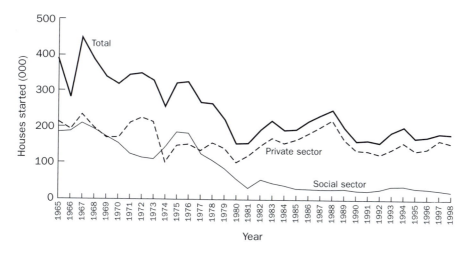

Figure 3.1   *The housebuilding cycle, Great Britain, 1965–98*
Sources: MHLG; Scottish Development; DETR, *Housing and Construction Statistics*

Table 3.1 Houses started, Great Britain, 1964–98 (000)

Year	Social sector	Private sector	Total
1964	178.6	247.5	426.1
1965	181.4	211.1	392.4
1966	185.9	193.4	379.3
1967	213.9	233.6	447.5
1968	194.3	200.1	394.4
1969	176.6	166.8	343.4
1970	154.1	165.1	319.2
1971	136.6	207.3	343.9
1972	123.0	227.4	350.4
1973	112.8	214.9	327.7
1974	146.7	105.3	252.0
1975	173.8	149.1	322.9
1976	170.8	154.7	325.5
1977	132.1	134.8	266.9
1978	107.4	157.3	264.7
1979	81.2	144.0	225.2
1980	56.4	98.9	155.3
1981	37.2	116.7	153.9
1982	53.0	140.5	193.5
1983	48.0	169.8	217.8
1984	40.2	153.7	193.9
1985	34.1	163.1	197.2
1986	32.9	180.1	213.0
1987	32.8	196.8	229.6
1988	30.9	221.4	252.3
1989	31.1	169.9	201.0
1990	27.2	135.4	162.6
1991	26.5	135.0	161.5
1992	36.4	120.1	156.5
1993	44.0	140.4	184.4
1994	42.8	158.4	201.2
1995	34.2	134.2	168.4
1996	30.1	144.8	174.9
1997	26.7	162.1	188.8
1998	21.5	154.9	177.4

Source: MHLG; Scottish Development; DETR, *Housing and Construction Statistics*

association sub-sectors. Table 3.2 not only shows that housebuilding in the local authority sub-sector has diminished to almost zero in recent years, but that more council houses have been sold off to their tenants than the number built.

Table 3.2 *Houses started in the social sector and the sale of council houses to their tenants, Great Britain, 1980–98 (000)*

	Local authorities and New Towns	*Housing associations*	*Completed sales*
1980	41.5	14.8	89.6
1981	26.0	11.6	114.9
1982	34.8	18.3	200.4
1983	34.6	14.3	142.1
1984	27.4	12.7	102.8
1985	22.0	12.5	94.3
1986	20.4	13.1	90.9
1987	19.9	12.9	105.6
1988	16.4	14.5	163.8
1989	15.2	15.9	186.0
1990	8.6	18.6	128.7
1991	4.1	22.4	74.9
1992	2.6	33.8	65.1
1993	2.2	41.8	61.4
1994	1.4	41.4	66.6
1995	1.1	33.0	50.6
1996	0.6	29.4	45.2
1997	0.4	26.4	58.0
1998	0.2	22.3	55.9

Source: DoE, *Housing and Construction Statistics*

The private sector cycle

In the private sector, the housebuilding industry is mainly speculative, and in the 1970s and 1980s about half of all speculative housebuilding was undertaken by firms employing fewer than 100 workers. Houses were mainly built in expectation of being sold during or shortly after construction. There are essentially two sorts of housebuilding firms: those which rely largely on profits realised on land held for up to ten years and those which depend on profits made from a quick turnover, resulting almost entirely from the construction of dwellings.

The private housebuilding industry – particularly because of the preponderance of small firms – is very sensitive to fluctuations in cost. It is often argued that at times of high interest rates and a tight monetary policy, the number of housing starts falls to a relatively low level (for example in 1974, 1980–81 and 1990–92), and at times of low interest rates and relaxed monetary policy, the number of housing starts rises to a high level (for example in 1972, 1978 and 1986–88). In these boom years prices initially agreed between the builder and the prospective buyer are often revised upwards – the buyer being 'gazumped'. During recession some builders have to cut their profits and accept a lower price than they anticipated, while others become bankrupt since their minimum selling prices remain above the maximum bid prices of potential buyers. There is thus a causal, but usually lagged, relationship between changes in interest rates and fluctuations in the rate of housebuilding. While the 'best' three years for housing starts (1972, 1978 and 1988) followed years in which interest rates were comparatively low, the 'worst' three years for starts (1974, 1980 and 1992) came after years in which interest rates were relatively high (Table 3.3).

Although housebuilding is cyclical in a macroeconomic sense, with booms and slumps normally following each other in response to changing rates of interest, it became apparent in the 1980s that major fluctuations in housebuilding also coincided with the 'political cycle'. In the election year of 1983, the base rate had been brought down to a comparatively low 9.83 per cent and a mini-housebuilding boom occurred, and in 1987 (another election year) a base rate of 9.74 per cent stimulated a boom the following year. In the early 1990s, however, the link between political and housebuilding cycles appeared to have been severed. A base rate of 9.42 per cent in the election year of 1992 failed to produce a notable boom in 1993.

Effects of the cycle, 1967–93

After the post-war peak year of 1967 the total number of starts declined by 57 per cent during 1967–84 to a level lower than in any year in the period 1950–79. Notwithstanding a number of peaks and troughs, the general trend was very distinctly downward. Apart from a massive fall in output, the slump had the following characteristics: a reduction in employment; a consequential increase in unemployment pay and other welfare payments and forgone tax revenue and national insurance contributions amounting in total to over £20 billion by the mid-1980s; under-capacity (with a loss of skilled workers, a decline in the number of apprenticeships, a low level of investment in new plant, extensive destocking and closures in the supply industries); a substantial increase in bankruptcy within the housebuilding industry; reduced productive

Table 3.3 *Relationship between housing starts and rates of interest, 1970–99*

Year	Bank rate/ minimum lending rate/base rate (%)	Average mortgage rate (%)	Housing starts (000s)	
1970	7.00	8.50	165.1	
1971	5.00	8.00	207.3	
1972	9.00	8.50	227.4	Boom
1973	9.33	9.59	214.9	
1974	12.33	11.00	105.3	Slump
1975	10.47	11.00	149.1	
1976	11.11	11.06	154.7	
1977	8.94	11.05	134.8	
1978	9.04	9.55	157.3	Boom
1979	13.68	11.94	144.0	
1980	16.32	14.92	98.8	Slump
1981	13.27	14.01	116.7	
1982	11.93	13.30	140.5	
1983	9.83	11.03	169.8	
1984	9.68	11.64	153.7	
1985	12.25	13.47	163.1	
1986	10.90	12.07	180.1	
1987	9.74	11.43	196.8	
1988	10.09	11.19	221.7	Boom
1989	13.85	13.66	169.9	
1990	14.77	15.10	137.1	
1991	11.70	12.47	136.7	
1992	9.42	10.60	120.1	Slump
1993	5.50	7.94	140.4	
1994	6.25	7.84	158.2	
1995	6.50	7.48	134.1	
1996	6.00	6.51	143.6	
1997	7.25	7.58	162.4	Boom
1998	6.25	7.29	157.4	
1999	5.50	6.49	157.6	

Source: DoE, *Housing and Construction Statistics*

efficiency; and recurring constraints on building society lending (at least until the mid-1980s). In the early 1980s both the Social and Economic Committee of the European Commission and the Building and Civil Engineering Economic Development Committee recommended policies to effect a stabilising or expansionary impact on construction so that the industry could plan ahead and increase the level of investment and training. But policies were not forthcoming; quite the reverse. The government was set on combating inflation, almost regardless (or so it seemed) of the effects of its deflationary measures on the housebuilding industry (and industry in general).

By 1987/88, however, there was a marked upturn in housebuilding in the private sector. Apart from lower interest rates (coinciding with the election of 1987), the surge in housebuilding was also attributed to rising real wages (associated with cuts in direct taxation) and the more liberal lending policies of the mortgage institutions, which in turn stimulated speculative housing development. Overall, the housebuilding industry in 1988 had its best year since the late 1970s, and according to the Building Employers' Confederation, a survey of 600 building firms found that 67 per cent were working at full or almost full capacity and that 76 per cent (rightly) anticipated an increase in their workload in 1988 (Warman, 1987).

By 1992, however, housebuilding had slumped to its lowest level since the recession of 1981 and the industry again suffered in much the same way as it had in the early 1980s, with similar knock-on effects on the economy as a whole.

The government's response to the housebuilding cycle

Deflationary monetary policies substantially decreased social sector demand in the early 1980s. Because local authorities spent £180 million above official guidelines during the first quarter of 1981, the Conservative government imposed a six-month moratorium in November 1981 on all council housebuilding, improvements and repairs. But housing was already penalised by having to bear 75 per cent of all planned public expenditure cuts 1979/80–1983/84. Not only would this make it impossible for the industry to supply the 310,000 houses deemed necessary by the 1977 Green Paper, but an under-provision of at least half a million homes by the late 1980s seemed certain. It was very clear that in the early 1980s

> The Tories no longer . . . gave . . . housing the priority they did in
> 1950 when the party conference demanded 300,000 houses and
> Harold Macmillan, as housing minister, provided them after the party
> was returned to power.
>
> (Bowen, 1980)

Was this apparent lack of concern due to a Thatcherite desire to revert to a Victorian *laissez-faire* and self-help environment characteristic of much of the nineteenth century, or was it because the government was miscalculating the effects of its policy and was being poorly advised? Unwillingness to face facts suggests the former explanation. The House of Commons Environment Committee (House of Commons, 1981a) reported that it was unable to dissuade the government from its refusal to analyse the impact of its housing cuts and claimed that 'the denial of this background information to the government's housing policy precludes properly informed public debate and inhibits the progress of work with which parliament has charged the committee' – specifically, an investigation of the effects of government housing policy. Even if it had been the latter explanation, the result would have been the same. Housing investment programmes were being used as a residual item to reconcile budgetary arithmetic.

Table 3.4 shows that within a reduced level of housebuilding in the public sector in England in the period 1980–98, there was a shift of emphasis away from new building to repairs and maintenance. In the 1990s, with housing associations becoming the main providers of new social sector housing, local authorities virtually gave up building houses for general needs – confining new supply to special needs only. 'I can see no arguments for generalised new build by councils, now or in the future. . . . The next great push . . . should be to get rid of the state as a big landlord', proclaimed Mr William Waldegrave, Housing Minister (1987) – a prelude to the introduction of large-scale voluntary transfers, housing action trusts and other means of disposing of the local authority stock under the Housing Act of 1988 (see Chapter 8). But although the value of new housing output decreased by 98 per cent over the period 1980–98, the value of renovation work increased by 0.5 per cent, but in recent years even this was cut back markedly from a peak in 1990. Similar trends occurred in the rest of Great Britain, and it was clearly evident that although renovation work remained fairly buoyant over most of the 1980s and 1990s, it could not compensate the housebuilding industry for the massive cutback in public expenditure on new housebuilding. Consecutive governments clearly believed that the best way of helping the industry was to get rates of inflation and interest down to a low level in the hope that this would provide a sound base for expansion, particularly in the private sector. Since annual expenditure was still trifling when compared to the estimated £20 billion that needed to be spent on repairs to the council stock, it was unrealistic for the

Table 3.4 *Public sector housebuilding, England, 1980–98*

	Year			Percentage change,
	1980	1990	1998	1980–98
Output (£ million at 1997/98 prices)				
New housebuilding	2,227	636	47	–97.9
Renovation to stock	1,512	2,013	1,519	+0.5
Total output	3,739	2,649	1,566	–58.1
New housebuilding completions (000)				
Local authorities	27.9	6.5	0.2	–99.3
Housing associations	13.1	14.1	16.3	+24.4
Total completions	41.0	20.6	16.5	– 59.8

Source: Wilcox (2000)

government to rely on long-term low inflationary economic growth to solve the public sector's housing problems, particularly since this would inflict a very considerable and escalating cost burden on future generations of investors and tenants.

It was reported that the Major government was even giving consideration to ending housebuilding in the social sector completely (Shelter, 1994). Had this materialised, the housing associations would have ceased to be 'providers' of new social housing, and housebuilding would have been subsequently determined entirely by the vagaries of the free market and solely within the private sector.

New housing need in England

Notwithstanding regional variations in housing need, there is currently a substantial national shortage of housing in both the social rented sector and segments of the open market. Holmans (1995) suggested that in order to meet housing need over the period 1991–2011, about 240,000 new homes a year would be required, with about 40 per cent being in the social sector, an estimate compatible with the projected 4.4 million growth in the number of households, 1996–2016 (DoE, 1995a). However, based on the government's revised 1998-based projections, Holmans and Brownie (2001) subsequently estimated that the need for new affordable housing (including dwellings available in the private rented sector to those who can claim housing benefit) would amount to only 82,500 units

Table 3.5 *Estimated new building and conversions required each year to accommodate newly arising housing need, 1996–2016*

	Affordable and private rented (with housing benefit)	Owner-occupiers and private rented (non-housing benefit)
North East	5,000	2,500
North West	7,500	14,000
Yorkshire & Humberside	7,000	11,000
East Midlands	6,000	12,000
West Midlands	7,500	10,000
Eastern	10,500	18,500
London	18,000	25,000
South East	12,500	32,000
South West	8,500	18,500
Total	82,500	143,500

Source: Holmans and Brownie (2001)

per annum from 1996 to 2016, while the figure for owner-occupier and other rented homes would be 143,500 a year (Table 3.5).

Predictions of housing need, however, have to be treated with some caution. At the end of the 1990s, both vacancies and re-lets in the social sector increased significantly; the number of homeless households had decreased markedly since reaching a peak in 1992; local authority waiting lists consisted increasingly of younger single people, not a high-priority group; and there was a growing number of difficult-to-let dwellings across the social housing sector (see Bramley, 1998). The need for social housing was tempered by the resurgence of the private rented sector during the house price slump of the early 1990s; increased sharing; and young people deferring household formation by residing longer in their parental home. Perhaps more than any other factor, stark regional variations in housing need render it very difficult to predict national requirements, not least because households in need of accommodation in areas of housing shortage are unlikely to migrate to areas of housing surplus if employment opportunities in those areas are limited.

Regional imbalances in housing need

Whereas in the northern and Midland regions in 1998 there was a substantial surplus of social dwellings, there were marked deficits in East Anglia, Greater London and the rest of the South East (Figure 3.2). Thus,

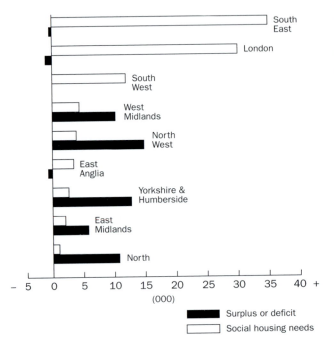

Figure 3.2 *Housing surpluses and deficits, England, 1998*
Source: Bramley (1998)

although nationally there was a surplus of 58,344 units, only 8 per cent of needs were concentrated in the three northern regions, but as much as 82 per cent of the total national requirement was concentrated in just three southern regions (Bramley, 1998).

Clearly, in areas of surplus there would be justification in injecting resources into the further provision of social housing. Indeed, 'there may even be more merit in strategies which seek to take the worst housing stock out of the system completely through demolition' (Bramley, 1998: 25). However, in areas of housing deficit there is a continuing need for further investment in social rented housing, both through the medium of conventional grant support and through the greater use of planning powers, 'particularly . . . where need coincides with high land values' (Bramley, 1998: 25).

New housing need in the South East

In the late 1990s the government issued a guideline building target for the South East which reflected the projected need for 1.4 million new dwellings between 1991 and 2016 (DoE, 1996a). Fearful that this number of dwellings would necessitate substantial greenfield development, SERPLAN, the South East Region Planning Conference, proposed that the government should scale down the number of houses permitted to be built to only 914,000 over the same period (Hetherington, 1999). This, in the view of SERPLAN, would require an annual output of only 33,000 dwellings, ideally with development concentrated in a string of new townships and in an expanded Milton Keynes.

Government inspectors, however, rejected this proposal in October 1999 and recommended that 1.1 million homes would be required over the 25-year period (or 55,000 per annum) and that growth should be concentrated in new developments of 80,000–100,000 homes around Ashford, Crawley, Milton Keynes and Stansted (Hetherington, 2000a).

This recommendation, in turn, was rejected by the Secretary of State, Mr John Prescott, in February 2000, and the former policy of 'predict and provide' was replaced by a more flexible system looking no more than five years ahead, with quinquennial reviews thereafter. Thus, for the period 2001 to 2006, the Secretary of State approved plans for the development of 215,000 new houses in the South East (outside of Greater London) and 115,000 dwellings in the capital, equivalent to an annual provision of 43,000 and 23,000 houses respectively. Within this new form of control the government favoured a major expansion of the Thames Gateway (which contained a considerable amount of brownfield land) and Ashford as a growth point close to the Channel Tunnel, although it signalled that it would be prepared to approve incursions into green belt land in order to curb piecemeal development elsewhere.

SERPLAN, however, opposed the government's plans for 43,000 new houses per annum to 2006, preferring instead a lower annual target of 33,400. In response, the Secretary of State in December 2000 announced that – under the governments's planning guidance for the region – 39,000 new houses would be built each year, 6 per cent more than recommended by SERPLAN, and that this new number would not use any more land than SERPLAN's 33,400 (McCarthy, 2000).

Greenfield versus brownfield debate

Both the Department of the Environment (DoE; the forerunner of the DETR) (1995a) and the Town and Country Planning Association (1996) recommended that in relation to the projected 4.4 million increase in the number of households (1991–2016), 50 per cent of all new housing should be developed on brownfield sites and 50 per cent on greenfield land, a view initially shared with Labour on its return to government in 1997.

The arguments in favour of brownfield versus greenfield development are, of course, fairly evenly matched, in weight if not necessarily in number. Advocates of brownfield development cite the abundance of derelict land throughout England, and argue that housing development on such land would:

- enhance the environment since derelict areas in British towns and cities would be cleared, and the countryside would be better preserved;
- reduce the volume of commuting and lower exhaust emissions;
- protect the green belt since urban encroachment into rural areas would be reduced; and
- result in more vibrant cities since – with densities being raised, for example, to Parisian levels of around 20,000/km^2, compared to only 8,000/km^2 in London – more capital would be invested in both private and public services.

Supporters of greenfield development, however, point out that:

- Brownfield sites are not distributed evenly across England. Most derelict urban land is in the North and Midlands, whereas most new housing that will be required in the South and East. Counties with the highest projected demand – Somerset, Berkshire and Suffolk – have little or no developable brownfield land.
- Most new housing units developed in the inner city would be unafford-able to new households, and even if they became affordable (possibly as an outcome of subsidised regeneration), many new households would prefer a detached house in the country to a flat in the city for the same price.
- There are over 700,000 empty houses in England, mainly in cities, which if occupied would reduce the need to develop new housing in the inner cities.
- The inner city environment needs protecting just as much as the rural environment (perhaps to an even greater extent, since rural areas are much more extensive), and many inner city residents might place a greater priority on space, peace and quiet, and traffic-free areas than on its antithesis: high-density 'town cramming'.
- The urban poor might be the principal victims of brownfield develop-ment since they could be squeezed into the worst sites and into the worst housing.
- The costs of site preparation in greenfield areas are comparatively low (there is little contaminated land).
- Housebuilding in the countryside is often highly profitable – because of land banking and land speculation – and thus there is little need for supply-side subsidies (as there is for much brownfield development), although there are calls on public expenditure for infrastructure

development, notwithstanding the tendency for many developers to offer financially constrained local authorities facilities such as new schools, leisure centres and swimming pools in return for planning permission ('planning gain').

After reviewing the respective advantages and disadvantages of brownfield and greenfield development, and taking into account the revised predictions of housing need, the DETR in 1998 declared that brownfield development should account for 60 per cent of new housing development, 1996–2016. In respect of London, however, it was recognised that virtually all the 630,000 new households required in the capital during this period could be accommodated by building on brownfield sites, provided there were drastic reductions on parking and car use. In these circumstances, densities could be as high as 500 habitable rooms/ha compared to the present densities of 125–250/ha (DETR/LPAC, 1998).

An Urban Task Force chaired by Lord Rogers subsequently indicated that even under current policy, as many as 55 per cent of new dwellings could be accommodated on recycled brownfield land by 2021, compared to only 49 per cent in 1994 (DETR, 1999). There would, however, be wide regional disparities in the use of brownfield land for new housing development. Table 3.6 reveals that in the West and East Midlands, Yorkshire and Humberside, the North West and London, well over 60 per cent of new housing development would almost certainly take place on brownfield sites, whereas in the rest of the South East and in the South West most development would continue to make use of greenfield land.

It should be recognised, however, that a range of problems need to be resolved before brownfield development can take place on the scale required even in the more suitable regions. Research undertaken by Sims (2001) has shown that owners often have unrealistic aspirations regarding the value of their brownfield sites and that this can slow down the development process and increase its cost; contamination is only one of the many physical characteristics that may prove an obstacle to development; developers are concerned that high-density development – as proposed by the Urban Task Force – could lead to 'town cramming' ; and that brownfield problems are, in general, becoming more complex and time-consuming to resolve. Clearly, these problems will need to be addressed and solutions found before brownfield land can accommodate 60 per cent of new housing development by 2016.

Table 3.6 *Projected additional households likely to be accommodated in new housing on brownfield land, 1996–2021*

Region	No. of households projected to form, 1996–2021	Additional dwellings, 1996–2021[a]	Additional households, 1996–2021 (%)[b]	New dwellings on brownfield land in 1994 (%)
North East	100,000	59,000	59	52
North West	300,000	189,000	63	57
Yorkshire & Humberside	300,000	193,000	64	50
East Midlands	300,000	195,000	65	37
West Midlands	300,000	242,000	81	47
Eastern	500,000	245,000	49	53
London	600,000	367,000	61	81
South East	900,000	352,000	39	47
South West	500,000	245,000	49	34
Total	3,800,000	2,087,000	55	49

Source: DETR (1999)

Notes: [a] Estimate of the number of additional dwellings likely to be accommodated on brownfield land under current policies, 1996–2021.
[b] Percentage of additional households likely to be accommodated in new dwellings on brownfield land under current policies, 1996–2021.

The principal change in policy would, of course, be to build at much higher densities. The average density of all residential developments across England is as low as 25 units/ha, in contrast to 40 units/ha in the United States. The main outcome of this extravagant use of land is that over the past 20 years, the built-up area of England has doubled, owing to the lack of control on greenfield development, and that –as the inner cities deteriorate, people migrate to the outer suburbs or the countryside, causing a polarisation between rich and poor ghettos (Rogers, 2001a). Higher densities in the past were clearly compatible with a high quality of urban design and affluent lifestyle, as for example in Georgian Bath, Edinburgh and parts of west London, and there is no reason a priori why this should not be the case in the twenty-first century. With higher densities, London has room for an extra 570,000 homes without using a single greenfield site (Rogers, 2001a). Clearly, to build at a higher density,

> we must build on the vast swathes of redundant industrial land – brownfield sites that scar English towns. We must re-use, recycle and regenerate our city areas before touching the countryside. 'Brownfield first' should be the commitment of all local authorities, whether urban or rural.
>
> (Rogers, 2001b)

Box 3.1

Land supply and town planning

Town planning legislation in the United Kingdom has since its inception been closely related to housebuilding. Until the Town and Country Planning Act of 1968, *blueprint planning* ensured that the relationship between planning and housing development was viewed largely in its physical context, and solutions to related problems were similarly physical – involving zoning, density controls, building regulations, planning standards and design. But after the 1968 Act it was increasingly recognised that the complex problems relating to planning and housing development could not be addressed – and solutions could not be found – in purely physical terms. Continual reference to economic and social considerations was deemed necessary. Whereas blueprint planning provided the framework in which the housing market operated, the market very largely provides the framework in which the successor to blueprint planning, *process planning*, is undertaken. A substantial increase in the demand for housing thus calls for a response from process planning, often necessitating a revision of plans.

In the 1980s and 1990s, however, central government reverted to a predominantly land-use approach to planning. Under a comparatively relaxed regime, there was a tendency for local authorities to 'police' new housing demand, for example 'allowing new development on infill sites or on the edge of larger towns and villages, but preventing development in the open countryside and in designated areas such as green belts' (O'Leary, 1997: 133). Housebuilders consequently negotiated with local authorities for the release of land to permit the development of expansion schemes on the edge of towns or villages, or searched for vacant under-used sites in urban areas. Older houses in large grounds were also targeted because they could be redeveloped to provide low-rise flats (O'Leary, 1997).

With the winding-down of the New Towns programme in the 1970s and 1980s, and with an emphasis on market-led regeneration in the inner cities (see Chapter 13), it was inevitable that the demand for housebuilding land in the latter years of the twentieth century was highly fragmented. There were simultaneously changes in planning practice which – to an extent – recognised the increasing problems of housing supply. Circulars and policy guidance notes were increasingly used 'to provide direction to local authorities and the development industry on how central government expects the planning legislation to work in practice' (O'Leary, 1997: 139). Circular 7/91 (DoE, 1991), for example, encouraged local authorities to take the need for affordable housing into account when formulating their development plans and making planning decisions, and to negotiate with developers to ensure that some affordable housing is included in their schemes. PPG3, *Housing* (DoE, 1992a), recommended the involvement of housing associations in the provision of housing for rent or shared ownership in an attempt to ensure that the benefits of affordability could be enjoyed by both initial and subsequent occupiers. Further guidance on the relationship between planning and affordable housing was provided by the revised circular on this issue (DoE, 1996b).

continued

It is sometimes argued that the planning system, far from ensuring there is an adequate supply of land for affordable housing, actually creates a shortage of development land and inflates prices, a view expressed by the House Builders Federation and supported by research by Evans (1988). However, Bramley et al. (1995) suggest that if land allocations for housing in development plans were significantly increased across the country, the effect on prices would be only slight since prices are affected primarily by market changes (and particularly by fluctuations in demand) rather than by planning controls.

However, if development in the foreseeable future occurs at higher density, it is far from certain whether the housing needs of the poor, the homeless and even those on average wages will be satisfied. With local authorities selling off their housing stocks and land to housing associations and local housing companies, households will be more and more dependent upon the private sector. But in the absence of local authority housing in city centres, and with land values being increasingly inflated by commercial development, there will be an escalation of rents and house prices that will squeeze out existing residents – a concern expressed by the Urban Bishops Panel (a subcommittee of the House of Bishops of the Church of England) in a memorandum submitted to the Environment, Transport and Regional Affairs Select Committee of the House of Commons in February 2000.

Conclusion

Despite its very large majority at the 1997 General Election and a massive budget surplus, the Labour government in the late 1990s – in its quest for financial prudence – was reluctant to boost public investment in social housing. Instead, there were major cuts in local authority capital expenditure, and the approved development programme of the Housing Corporation was frozen at the level inherited from the Conservatives. As a result, the number of new houses built by local authorities and housing associations fell from a 1990s peak of 41,800 starts under the Conservatives in 1993 to a 1990s trough of 21,700 under Labour in 1998, or from an average of 35,800 per annum during the 1992–97 Conservative administration to an annual average of only 23,300 under Labour in the period 1997–99.

There were not only substantial Exchequer funds that could have been injected into social housing, but a significant proportion of these funds derived from housing itself. Over the tax years 1997/98–1999/2000, the government saved around £10 billion from the reduction in mortgage interest relief and increase in stamp duty on housing sales. 'Added to existing spending and private finance, that could almost have paid for 100,000 new houses a year for a full two terms' (Dwelly, 2000a: 8). However, even if public investment in social housing had doubled to about £2 billion per annum, this would still have been less in real terms than the previous Conservative government had spent in 1992/93 and 1993/94 (Dwelly, 2000a).

It can only be concluded that the Labour government – at least as much as its Conservative predecessor – was no longer committed to the large-scale provision of social housing, was intent on pulling out of housing investment as far as possible, and had become wedded to the notion that in an increasingly free-market economy it was better to subsidise people (where necessary) rather than bricks and mortar. Labour also recognised that in many of its heartlands – for example in the inner cities of the North and the Midlands – there was now a low level of demand for social housing, and that in those areas, demolition rather than new housebuilding might take priority.

It is thus very questionable whether the predicted need for at least 70,000 affordable rented homes each year will be met over the period 1996–2016. It is perhaps worth recalling that in Great Britain as a whole, housing starts in the social sector in the 1990s under both the Conservatives and Labour averaged less than half of this number. Taking into account regional surpluses and deficits in housing provision, there will have to be at least a doubling in the capital and revenue resources in real terms if the need for affordable housing across the country is to be satisfied over the period to 2016. However, rather than announcing in its *Spending Review* (Treasury, 2000) that this level of investment will be forthcoming, the government instead planned to undertake a more modest level of public spending, complemented by recently revised planning guidance for housing (PPG3; DoE, 1992a) that reinforced the powers of local authorities merely to secure the provision of affordable housing as a declared proportion of new private development (see Raynsford, 2000).

Further reading

Ball, M. (1983) *Housing Policy and Economic Power*, London: Methuen. Contains a useful analysis of the economics of housebuilding in the owner-occupied sector – essential for an understanding of the housebuilding cycle.

Ball, M. (1988) *Rebuilding Construction: Economic Change in the Construction Industry*, London: Routledge. Provides an essential overview of the attributes and performance of the construction industry, including a useful examination of the role of housebuilding.

Cheshire, P. and Sheppard, S. (1998) 'Estimating the demand for housing, land and neighbourhood characteristics', *Oxford Bulletin of Economics and Statistics*, Vol. 60, p. 3. Sets out clearly the factors involved in estimating the future demand for housing land in relation to land requirements and neighbourhood development.

Department of the Environment, Transport and the Regions (1995) *Projection of Households in England to 2016*, London: HMSO. An essential reference source for examining both national and regional housing needs in the early part of this century.

Hillebrandt, P. (2000) *Economic Theory and the Construction Industry*, 3rd edn, Basingstoke: Macmillan. This well-established textbook incorporates some useful analysis of housing economics within the context of the construction industry.

Holmans, A.E. (1995) *Housing Demand and Need in England, 1991–2011*, York: Joseph Rowntree Foundation. A major contribution to the debate on future housing requirements. Should be read alongside the DETR (1995) publication.

Holmans, A.E. and Simpson, M. (1999) *Low Demand: Separating Fact from Fiction*, Coventry: Chartered Institute of Housing/Joseph Rowntree Foundation. Provides a useful insight into the causes, attributes and effects of the low demand for housing in some of the English regions.

Holmans, A.E., Morrison, N. and Whitehead, C. (1998) *How Many Homes Will We Need?*, London: Shelter. A continuation of the debate on future housing needs – focusing particularly on the regional dimension.

4 Housing rehabilitation policy

Housing renewal can involve *either* the demolition and redevelopment of dwellings, often as part of a slum clearance scheme, *or* the rehabilitation of housing by means of repair, improvement, conversion or modernisation. Although both forms of renewal take place in each of the housing sectors to a varying degree, demolition and redevelopment was undertaken overwhelmingly by local authorities particularly from the 1950s to the early 1970s, while rehabilitation took place mainly in the private sector in the 1970s and early 1980s. 'Partitioned renewal', as Merrett (1979) described this dichotomy, was superseded by the privatisation of funding on an increasing scale across all housing sectors from the mid-1980s into the last decade of the twentieth century.

It is within this context that this chapter:

- examines the scale of the renewal problem;
- discusses the history of housing rehabilitation policy to 1969;
- considers the Housing Act of 1969;
- reflects on proposals for reform 1972–74;
- reviews rehabilitation policy under Labour 1974–79, and under the Conservatives in the 1980s;
- highlights some of the inadequacies of private sector rehabilitation in the 1990s; and
- concludes by considering rehabilitation under Labour, post-1997.

Scale of the renewal problem

The number of houses demolished or closed in Britain declined from an annual rate of 61,785 to a mere 6,622 over the period 1975/76 to 1997/98 (Table 4.1). This was not only a reflection of cuts in central government

Table 4.1 *Houses demolished or closed, Great Britain, 1975–98*

	1975–76	1979–80	1985–86	1991–92	1997–98
England and Wales					
Demolished:					
In clearance areas	41,772	23,747	5,279	1,221	847
Elsewhere	3,939	2,472	2,354	447	332
Closed	5,416	4,207	1,397	703	358
Total	51,127	30,426	9,030	2,371	1,537
Scotland					
Unfit	9,964	4,001	939	—	—
Others	694	1,572	679	1,816	5,085
Total	10,658	5,573	1,618	1,816	5,085
Great Britain	61,785	35,999	10,648	4,187	6,622

Source: DoE, *Housing and Construction Statistics*

expenditure but a major shift of emphasis away from demolition and redevelopment to improvement and repairs. However, as was pointed out in Chapter 2, if comprehensive rehabilitation had been carried out in the mid-1990s the cost would have been exceedingly high, an estimated £70 billion at current prices (Leather *et al.*, 1994), a sum at least ten times that being spent on rehabilitation annually during the last decade of the twentieth century.

In the 1980s it was claimed that the housing crisis was caused by the collision of three trends: property developed during the Victorian building boom was reaching the end of its natural life (by the mid-1980s there were over 3 million houses in Britain over 100 years old); jerry-built housing put up in the 1930s was due for demolition; and post-war council estates were in need of extensive renovation (Best, 1985). Taking into account the number of dwellings requiring major repairs and improvement, there was a substantial net deficit of housing in Britain of well over 2 million dwellings by the mid-1990s. Britain therefore faced the prospect of a new generation of slums in the near future. Even in the penultimate decade of the twentieth century, Lord Scarman, speaking at the launch of the United Nations International Year of Shelter, warned that

Our children and grandchildren would find themselves condemned to live in a slum unless something was done to meet the huge and accumulating bill for repairs to our housing stock.

(Scarman, 1987)

This appalling situation was in part the result of Britain's rapid urbanisation during the nineteenth century, but it was very largely the product of decades of neglect and ineffective housing policy.

History of housing rehabilitation policy to 1969

The substantial increase in population in the first half of the nineteenth century, the growth in the proportion of the population living in urban areas and the almost complete absence of controls over building and public health resulted in unprecedented congestion and squalor. Within the new industrial towns of the coalfields and in much of London, the rationale of housing development was to get as many dwellings as possible on to a site at the least cost and as close as possible to places of work. From the 1840s there was a partial movement away from *laissez-faire* attitudes concerning the urban environment but, except for some public sector infilling and development by charitable bodies, housing within the inner areas remained largely neglected until after the Second World War.

In the immediate post-war period, housing rehabilitation, or 'patching up', as it was disparagingly called, was discouraged. The White Paper *Capital Investment in 1948* (Treasury, 1947) and Circular 40/48 (MoH, 1949) restricted patching up with the aim of steering resources to council housebuilding. But thereafter until the late 1960s, rehabilitation was recognised as a useful means of attempting to reduce the housing shortage – albeit as a stopgap process, with the emphasis being placed on new housebuilding. The Housing Act of 1949 introduced improvement grants for private owners and improvement subsidies for local authorities, and following the White Paper *Housing – The Next Step* (MHLG, 1953), such assistance was extended by the Housing Repairs and Rent Act of 1954, the Housing Act of 1957, the House Purchase and Housing Act of 1959, and the Housing Acts of 1961 and 1964 – the last-mentioned enabling local authorities to declare 'improvement areas'. Most of this legislation linked improvement to rising rents since normally grants were limited to 50 per cent of the cost of rehabilitation. Progress was at first slow. In the years 1949–58 only 159,869 dwellings in England and Wales were improved with grants, but in the following 10-year period 1,147,875 grants were awarded. However, the annual number of grant-aided improvements fell from 130,832 in 1960 to 108,938 in 1969.

In the aftermath of Labour's victory at the 1964 General Election, the 1965 White Paper *Housing Programme 1965–69* (MHLG, 1965) confirmed the party's half-million annual housebuilding programme. It seemed as though rehabilitation was again to be relegated to a minor role. But devaluation of sterling in 1967 and accompanying disinflationary fiscal and monetary policies resulted in housebuilding in the United Kingdom falling from a record 404,356 starts in 1967 to 366,793 in 1969, and the *House Condition Survey, England and Wales 1967* (MHLG, 1969a) showed that there were 1.8 million unfit dwellings in England and Wales (not 0.8 million, as previously thought); 2.3 million lacked one or more basic amenities; and 3.7 million, although not unfit, required repairs costing £125 million. Out of a total housing stock of 15.7 million, the proportion of dwellings requiring improvement was not inconsiderable.

The decline in housebuilding and the recognition of the extent of the problem of poor housing inevitably resulted in a shift of emphasis to rehabilitation. Local authority housing programmes needed to be changed. The White Paper *Old Houses into New Homes* (MHLG, 1968) therefore proposed that 'Within a total of public investment in housing at about the level it [had] reached a greater share should go to the improvement of older houses.' There were obvious relative cost advantages of rehabilitation as opposed to redevelopment. Needleman (1965) proposed that usually the best way to improve the minimum standard of housing with the available resources was to concentrate upon rehabilitation up until at least the 1980s, a conclusion also reached by Stone (1970). Hillman (1969) indicated the comparative cost advantage of rehabilitation. It would cost, for example, £600 million per annum to construct 200,000 local authority houses, but only £115 million per annum (half paid through rates and taxation) to improve 230,000 homes (realistic annual costings in 1969 for the early 1970s).

The Housing Act of 1969

Stemming from the 1968 White Paper, the Housing Act of 1969 was intended substantially to hasten the pace of rehabilitation. The Minister of Housing and Local Government, Mr Anthony Greenwood (1969), clearly stated that a primary goal of the provisions of the Act was to raise the standard of living of those residing in areas of bad housing which had the potential for improvement:

> The idea of continuing to live year after year without basic amenities
> . . . is totally unacceptable at this period in our history. Older people
> may have got used to a settled way of life, and in many cases it would
> be wrong to disturb them, but we should do everything we can to see
> that children are brought up in better than substandard conditions.

Improvement was to be encouraged by increased grants and a relaxation of the conditions attached to them. For private housing, standard grants were to be raised from £155 to £200 to assist in the provision of the standard amenities. Discretionary grants were also increased – from £400 to £1,000 – to enable such work as essential repairs, damp-proofing and rewiring to be done in addition to the installation of the standard amenities. Discretionary grants could also be used for conversions – up to £1,000 per dwelling being available for a conversion into two flats or up to £1,200 per dwelling for conversion into three or more flats. In all cases the grants would have to be matched pound for pound by the applicant's authorised expenditure and the discretionary grant would be conditional on improvement up to a 12-point standard.

The government contribution towards the improvement of local authority and housing association properties was £1,000 per dwelling if improvements alone were to be done and £1,250 per dwelling if acquisition was necessary in addition. The grants were calculated at 50 per cent of the cost of improvement and acquisition. The government also provided local authorities with the equivalent amount of assistance for the purchase of houses for improvement as they received as a subsidy for building new houses under the Housing Subsidies Act of 1967. The Housing Act of 1971 adjusted the amount of the improvement subsidy available to private owners, local authorities and housing associations. Discretionary grants paid by local authorities to private owner-occupiers and landlords in Development and Intermediate Areas were increased to 75 per cent of the approved expense of improvement works, but the government's share of the cost of improving local authority housing fell to three-eighths of the cost, the local authority having to meet the remaining five-eighths (75 per cent and 25 per cent in Development and Intermediate Areas). Where local authorities acquired properties for conversion or improvement, their acquisition costs were included in the approved expense for improvement and contributions, together with the costs of works, and were subject to higher cost limits. Housing associations could receive either cash grants in the same way as private owners or contributions in the same way as local authorities, but, outside of Development and Intermediate Areas, the proportion of government contribution was higher than that to local authorities, one-half instead of three-eighths of the total approved cost.

It is important to note that as a result of inflation the real increase in the grant over the period 1949–69 was about 67 per cent rather than approximately 233 per cent – the absolute increase. But the principal incentive for improvement may have been not the size of the grant but that it did not have to be repaid.

Further encouragement was given to improvement by the rent provisions of the Housing Act of 1969. If grant-aided improvements were carried out in the case of regulated tenancies, then the rent could rise to a new fair rent level as certified by the rent officer (under the provisions of the Rent Act of 1968), with the increase phased over three equal annual stages. Where improvements to controlled tenancies reached the qualifying standard, the local authority would issue a qualification certificate. Upon receipt of the certificate a landlord could apply to the rent officer, who would determine a fair rent. Increases in rent would have to be phased over five years.

General Improvement Areas

The 1969 Act placed great emphasis on the declaration of General Improvement Areas (GIAs) – areas which in scale could vary widely from small and compact areas of, say, 300 houses up to larger areas of between 500 and 800 houses. Under the Act, local authorities received reserve powers enabling them to acquire property compulsorily, especially if owners threatened the success of a whole scheme by failing to take up improvement grants. But the improvement of housing was not sufficient. The improvement of the residential environment was also important. Government grants of one-half of the cost of environmental improvement were to be available up to a limit of £50 per house in the area. The Act enabled local authorities to improve amenities and to acquire land for this purpose. The importance of area improvement was previously recognised by the Denington Committee (1966). Its report, *Our Older Homes: A Call for Action*, argued that a dwelling is not satisfactory unless it stands in a satisfactory environment, and that when improving a dwelling, the local authority should also attempt to improve the environment.

It was thus acknowledged by the 1969 Act that the resources and effort devoted to rehabilitation would produce a better return if they were directed to whole areas rather than to individual and dispersed dwellings.

Proposals for reform, 1972–74

Since there was much concern about some of the consequences of the 1969 Act, a House of Commons Expenditure Committee (Environmental and Home Office Subcommittee) on House Improvement Grants was appointed in 1972 to consider:

1 who was receiving grants, how and for what;
2 the effects of improvement grants on the housing stock, its condition, its ownership and its price;
3 whether the legislation was achieving its purpose.

A principal supplier of evidence to the Committee, the Royal Town Planning Institute (RTPI, 1973), emphasised that:

> the problem of 'gentrification' . . . poses important questions about what and who the present legislation is for? Is it intended to upgrade existing areas of sub-standard housing or is it to improve the conditions of the existing residents of these areas? This becomes an urgent problem when those tenants displaced as a result of improvement, by the need for vacant possession by landlords or higher rents, suffer a decline in their housing standards.

The Association of Municipal Corporations (AMC, 1973), suggesting recommendations, regarded it reasonable and helpful if local authorities were empowered to

> impose conditions when making improvement grants as to repayment on sale within a defined term of years . . . to nominate a tenant to occupy a dwelling improved with a grant (and) to prescribe the type of applicant for tenancy, or the nature of the letting.

The White Paper *Better Homes: The Next Priorities* (DoE, 1973b) subsequently proposed that where improvement grants were awarded to landlords, local authorities should have the right to insist that the dwelling was let for at least seven years, and if a property was sold after improvement (assuming the owner had vacant possession), the grant should be repaid with compound interest to the local authority. It was also proposed that local authorities should have discretion to compel owners to improve their properties (and if they refused, the authorities should undertake the work themselves and impose a charge). Local authorities should also be given powers to buy up empty properties using compulsory purchase orders. Perhaps the most important aspect of the White Paper was the proposal to declare Housing Action Areas (HAAs). These would be inner stress areas of about 400–500 houses where developers and large landlords had displaced lower-income tenants, where house prices were rapidly rising and where communities were being broken up and homelessness was increasing.

Rehabilitation policy under Labour, 1974–79

A Labour government was returned to office in March 1974. Introducing its new Housing Bill, the Secretary of State for the Environment, Mr Anthony Crosland (1974), stated:

> I have for long been a passionate opponent of indiscriminate
> clearance, which I believe has gone too far . . . in many areas.
> I believe that indiscriminate clearance can be appallingly destructive
> of existing communities and frequently a very expensive solution.

To emphasise the continuing shift of emphasis of public policy, the Housing Act of 1974 therefore subsumed the Housing and Planning Bill which the former government had introduced just prior to the election. The Act prevented recipients of improvement grants from selling their properties (or leaving them empty) within five years unless the grant was repaid to the local authority at a compound rate of interest (this contrasted with the three-year limit prescribed by the previous Bill). A seven-year restriction within the HAAs was also imposed (as the previous Bill had required). Local authorities were empowered to demand the improvement of individual rented properties – a nine-month period being imposed on landlords for this purpose. If landlords failed to improve their tenancies, the local authority was able to purchase the properties with a compulsory purchase order if necessary and then to hand the municipalised accommodation back to the original tenants. To alleviate the 'disincentive' effect on owners of the new conditions to grant approval, grants and limits of eligible expenses were increased. Improvement grants were raised from £1,000 to £1,600, or, where a building of three or more storeys was being converted, from £1,200 to £1,850. These amounts represented 50 per cent of the increased level of eligible expenses – £3,200 and £3,700 in respect of the above cases. Within the GIAs, improvement grants were increased to £1,920 and £2,220, and in the HAAs they were raised to £2,400 and £2,775. Intermediate grants were introduced replacing standard grants. These were at a higher rate and equal to 60 per cent of eligible expenses up to £700. Repairs or replacement grants were to be made available at 60 per cent of eligible expenses up to a maximum of £800. The principle that improvement grants should give a house a life of at least 30 years was retained. After improvement a dwelling should, if practicable, have all standard amenities and meet a 10-point standard.

By 1979, however, one could only have been pessimistic about the possibility of improvement grants having a significant impact on the condition of housing. Although the number of GIAs and HAAs had increased from respectively 964 and 78 in 1975 to 1,217 and 384 in 1978, it was doubtful whether they were having more than a patchy effect on housing in the inner urban areas. It might have seemed that over the long term considerable improvements had been made in the quality of the housing stock. The *House Condition Surveys 1967, 1971, 1976* (MHLG, 1969a; DoE, 1973c, 1978a) showed that the number of unfit houses had decreased from 1,836,000 in 1967, to 1,244,000 in 1971, and 894,000 in 1976; and the number of dwellings lacking one or more basic amenities had fallen from 3,943,000 to 2,866,000 and 1,633,000 in the same years – falling again to 1,400,000 in 1977 according to the *National Dwelling and Housing Survey Report* (DoE, 1978b). Yet in total, Shelter (1979a) estimated that there were still 3.1 million substandard houses in England and Wales in 1976, which consisted of not only those which were unfit, or lacked basic amenities, but also those which required repairs costing over £1,000. Together with an additional 110,000 houses which had become substandard as a result of obsolescence, 420,000 dwellings would have been needed each year to have dealt with this backlog – the 213,000 improvements in 1979 being a lamentable contribution. It was particularly regrettable that according to a Department of Environment report, improvement in the HAAs was progressing at a disturbingly low rate (Shelter, 1979b). It was likewise regrettable that proportionately, more money was being spent on improvement in small provincial towns, where improvement was least necessary, than in the inner urban areas of greatest need (NHIC, 1979).

Apart from the possible inadequacies of grant provision and area improvement policy there were underlying economic factors which constrained the rate of rehabilitation. The first of these was the concept of the 'prisoner's dilemma'. Davis and Whinston (1961) argued that it would be against the interest of individual owners to invest in improving their own properties if their neighbours did not similarly and simultaneously improve theirs. Unless a whole row of properties were improved, the individual improved property would not produce such an attractive return or appreciation as other forms of investment, and meanwhile the owners of unimproved properties could be benefiting from these more rewarding investments. The second factor was the 'valuation gap', the extent to which the total cost of rehabilitation incurred by the owner exceeded the resultant increase in value (if any). The National Home Improvement Council (1980), referring to mortgaged properties, reported that without improvement grants, rehabilitation would have produced valuation gaps in all regions except the South East (excluding Greater London) and South West, with an average gap nationally of £1,200. But with grants, only the West Midlands continued to show a gap. The council's report, however, considered only properties which had been improved. It did not quantify the extent to which rehabilitation may have been deterred by likely valuation gaps. In the late 1980s the valuation gap was undoubtedly a deterrent to improvement in many parts of Britain. Whereas in London £40,000 spent on improving a £100,000 property could increase its value to £200,000, in Sheffield, for example, £30,000 spent improving a £10,000 house would raise its value only to £20,000.

Rehabilitation under the Conservatives after 1979

In an attempt to hasten the pace of rehabilitation, the Conservatives, in the Housing Act of 1980, enabled both private and public sector tenants (as well as landlords and owner-occupiers) to apply for improvement grants; abolished the five-year repayment rule in relation to the only or main residence of an owner-occupier; extended the availability of repair grants (with regard to substantial repairs to pre-1919 dwellings); made it easier for low-income households (owners and tenants) to obtain grants towards the cost of basic amenities – grant aid being paid for less comprehensive improvement than previously allowed; made the grant system more flexible by giving the Secretary of State powers to fix (for different cases) eligible expense limits, percentage grant rates and rateable values, etc.; and enabled the Secretary of State to underwrite part of any losses incurred by local authorities acquiring and improving dwellings for resale.

In December 1980 new grant levels were announced. Improvement grants were raised to £6,375 (in respect of HAA dwellings and those which were unfit, lacking amenities or in need of substantial repairs) or £2,750 in the case of other dwellings (and up to £8,625 and £3,725 respectively in Greater London) – the two levels of grant amounting to 75 and 50 per cent of eligible expenditure. Intermediate grants were raised to £950 for amenities and £1,250 for 'minor' repairs (£2,500 and £3,500 respectively in Greater London). Like intermediate grants, they amounted to 50 per cent of eligible expenditure, but in April 1982 repair grants and intermediate grants were increased to 90 per cent of eligible expenditure, at first for the financial year 1982/83 but later extended to 1983/84. Meanwhile the government made an extra £75 million available for improvement grants and increased its contribution to local authority repairs from 90 to 95 per cent.

The government clearly committed itself to rehabilitation as one of the principal planks in its housing policy – the number of grants awarded in the private sector rising from 95,000 in 1980 to 320,000 in 1984, and in the local authority and New Town sector from 79,000 in 1981 to 434,000 in 1994. From 1983 to 1985, more private sector dwellings were rehabilitated than the number built, but throughout the whole period 1980–96, rehabilitation was on a larger scale than new housebuilding in the local authority and new town sector (Table 4.2).

Housing Action Areas

Even by 1979 it was recognised that area improvement policy had been unsuccessful. In the 471 HAAs declared up to the end of that year, only one-third of households (and one-fifth in London) took up improvement grants, since even with 75–90 per cent grants, households still had to find large cash sums because eligible expenditure limits were too low. Compulsory improvement powers (under the Housing Act of 1974) had rarely been used because they were cumbersome and time-consuming, and building societies rarely lent in HAAs. The HAA initiative had been 'too little and too late'. By 1980 the 1974 Act (according to DoE calculations) should have included up to 1 million houses in HAAs, but there were only about 150,000. Declaration was proceeding very slowly. In the three years 1979–81, only respectively 87, 75 and 50 HAAs were declared, making totals of 471, 546 and 596. GIA declarations were also taking place slowly – 64, 80 and 57 in the same three years, totalling 1,281, 1,361 and 1,418. Qualitatively, HAAs were also disappointing. Niner and Forrest (1982), looking at six HAAs (three in London and three elsewhere), concluded that although there were high rates of improvement in these specific areas, population turnover was high and improvement policy might have been no less disruptive than clearance and redevelopment. Owner-occupiers in the HAAs suffered particularly

Table 4.2 Renovation with the aid of grant or subsidy, Great Britain, 1980–96

	Private sector housing		Local authority and new town dwellings		Housing association dwellings		Total renovations	Total housing starts
	Number of grants[a]	Housing starts	Number of grants/ dwellings	Housing starts	Number of grants/ dwellings	Housing starts		
	(000)	(000)	(000)	(000)	(000)	(000)	(000)	(000)
1980	95	99	100	42	18	15	213	156
1981	94	117	79	26	14	12	187	155
1982	139	141	109	35	22	18	270	194
1983	293	172	127	35	18	14	438	221
1984	320	158	123	27	21	13	464	198
1985	200	166	157	27	13	12	370	205
1986	163	181	208	23	15	13	386	217
1987	159	197	241	20	13	13	413	230
1988	157	222	250	20	13	14	420	256
1989	145	170	258	18	15	16	418	204
1990	138	137	325	17	12	18	475	172
1991	60	137	270	10	8	22	338	169
1992	30	120	267	5	10	34	307	159
1993	22	140	364	3	8	42	394	185
1994	20	158	434	2	8	41	462	201
1995	19	134	406	2	12	33	437	169
1996	16	145	421	1	—	29	437	175

Source: DETR, *Housing and Construction Statistics* (various)

Note: [a] Grants awarded under Housing Act 1985 and earlier Acts.

at the hands of private sector builders, although most were well satisfied with their local authority improvement grant service. Monck and Lomas (1981), in their study of HAAs, thought that inflation would continue to be an obstacle to private initiative and, apart from a selective increase in improvement grants, municipalisation followed by local authority rehabilitation might offer the best solution to the improvement problem, although there would need to be greater central government control regarding the choice of areas. Monck and Lomas recognised, however (unlike Niner and Forrest), that improvement could ossify a community, and that households (especially with young children) might prefer to relocate (possibly because of street traffic, lack of gardens, delinquency and the reputation of local schools). The desire to preserve (possibly non-existent) communities might conflict with the aim of improving the 'well-being' of residents.

Enveloping – a new initiative

Local authorities faced with the need to undertake renewal have traditionally had a choice of either clearance and redevelopment, or rehabilitation. With the shift of emphasis to rehabilitation in the 1970s, a further choice often had to be made between municipalisation (taking into account the purchase price of the properties plus improvement costs, and repairs and maintenance costs for 30 years) or providing up to 90 per cent improvement grants and leaving the properties in private ownership. Since it was less expensive, authorities normally chose the latter, but the result was often 'pepper-potting' (a diffused distribution of improvement), and administrative expenditure reduced its cost-effectiveness. The 'prisoner's dilemma' and the valuation gap inhibited the pace of privately initiated improvement. To combat these weaknesses, Birmingham City Council, for example, adopted 'enveloping' in 1977 whereby with the aid of £2 million from the government's construction programme it renewed or repaired roofs, chimneys, guttering, pointing, windows and paintwork in an attempt to encourage owners to take up grants to rehabilitate interiors. Initially concentrating on two HAAs, by 1983 over 2,000 properties had been enveloped and work had started on a further 2,000.

From 1 April 1983, the government began paying 75 per cent grants (and local authorities the remaining 25 per cent) towards the cost of enveloping, up to £10,000 per house – a sum of £750 million being allocated for this purpose over a 10-year period. Over 155,000 HAA homes could benefit and more HAAs could be declared to take advantage of enveloping, although the process could be extended to other areas. Builders might gain from economies of scale and local authorities from lower administrative costs than those incurred in pepper-potting.

Enveloping, however, posed numerous problems of equitability. Clearly, enveloping benefited owner-occupiers and private landlords, whose property might rise in value, for example by £10,000 per house (even before they invest 90 per cent improvement grants and their own capital), while owners blighted by clearance schemes see the value of their properties fall to virtually zero. Also, it is questionable whether on top of mortgage interest relief and tax exemptions, owner-occupiers should

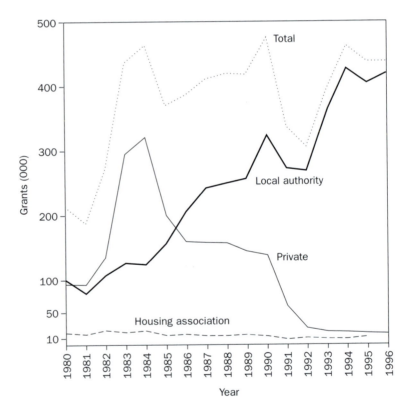

Figure 4.1 *Improvement grants to private owners and tenants, and social sector renovations, Great Britain, 1980–96*

Source: DoE, *Housing and Construction Statistics*

Note: Housing association renovations for Scotland, 1991–96, omitted.

further benefit from state 'hand-outs' while council rents were rising because of the reduction of subsidies. There was also a danger that enveloping might occur in areas which would subsequently deteriorate quickly if there were few applications for improvement grants (possibly because elderly households are unwilling to risk disruption and, like some people belonging to ethnic minorities, do not wish to borrow to supplement a grant), and authorities in their desire to abandon clearance policy might attempt to preserve the unpreservable.

The rehabilitation boom

By 1983 the Conservative government seemed eager to give rehabilitation a boost. Support for private sector renovation therefore increased from £573 million in 1982/83 to £1,064 million in 1983/84 (Table 4.3) – the number of grants to private owners and tenants rising from 135,000 in 1982 to 320,000 in 1984, a record, and considerably higher than the previous peak in 1974 (Figure 4.1 and Table 4.2). In both the private and the public sectors, renovation greatly exceeded the volume of new housebuilding. Renewal policy in 1979–83, however (despite fluctuations in public expenditure), had not been based just on grant-aided rehabilitation and area improvement. Homesteading had been promoted (households being offered empty run-down council houses at very reduced prices on condition that they rehabilitate), and improvement for sale by local authorities and housing associations had also been encouraged.

But many changes in the rehabilitation process were needed. Much more municipalisation and housing association acquisition was required.

Table 4.3 *Public expenditure on housing renovation grants and subsidies 1982/83–1987/88 (£ million cash)*

	1982/83	1983/84	1984/85	1985/86	1986/87	1987/88
Support for private sector renovation	573	1,064	891	608	528	535
Local authority renovation	962	1,148	1,280	1,312	1,363	1,590
New Town renovation	15	19	23	9	8	9
Total renovation	1,550	2,231	2,194	1,929	1,899	2,134

Source: Treasury (1988)

Owners should have been issued with compulsory improvement notices, and if they were unable or unwilling to comply, should have relinquished their property into social ownership by means of reinvigorated compulsory acquisition powers. But such course of action was frustrated, first by the cost and time it took to attempt to persuade landlords to undertake improvements (compulsory improvement procedure taking a minimum of 18 months), and second by public expenditure cuts; for example, the first Thatcherite Budget of June 1979 severely cut expenditure on public acquisitions. Although municipalisation accounted for less than 10 per cent of the £5,372 million allotted to housing, it bore 28 per cent of the immediate £300 million cut and further disproportionate cuts subsequently.

Since the GIAs and HAAs made only a minimal contribution to the satisfaction of housing needs – in part because area renewal was rarely 'pursued with the commitment and resources which it demanded' (Leather and Mackintosh, 1993) – it seemed likely that the Environmental Health Officers Association's proposal to set up 'Housing Renovation Areas' would be widely adopted to deal with improvement and related problems in a flexible and integrated way, possibly by combining GIA, HAA and clearance area powers. The London Borough of Hammersmith and Fulham was the first authority to apply a version of this proposal – a 'Housing Improvement Zone' set up in an extensive area which was not suitable for the declaration of GIAs or HAAs but which contained a high proportion of substandard housing, an intention being to harness private sector capital in addition to public resources. There were, however, problems relating to equity of benefit, a factor which encouraged many other authorities to proceed with caution.

The rate of improvement may have been slow because many households found the grant system complex and the number of different grants confusing, and the procedure was often chaotic since local government officers had to process up to 40 bits of paper before awarding one improvement grant. The Shelter Housing Aid Centre (SHAC) therefore recommended that only one 'unified grant' should be available, and this should be for the purpose of improving dwellings up to a new 10-point standard, but local authorities should have the discretion to determine the conditions attached to it and award it on a means-tested basis from 50 to 100 per cent of cost.

Even in 1984 (long after gentrification first became a cause of concern), harassment by neglect and the abuse of the grant system were rife throughout much of west London. Properties were left to deteriorate to a condition of serious disrepair, leaving tenants with little choice but to vacate. Since properties with vacant possession were worth over two and a half times their value with sitting tenants (say £200,000 instead of £80,000), landlords were eager to dispossess their tenants and undertake rehabilitation to increase values still further. The *Sunday Times* (1983) reported that in the City of Westminster, property companies and non-resident landlords used about £2,500,000 worth of grants to rehabilitate properties prior to selling them on long leases of 99 or 125 years (a practice also apparent in the Royal Borough of Kensington and Chelsea). Although Section 23 of the Housing Act of 1974 compelled landlords to repay their grants with interest if they sold their property within five years from the receipt of a grant, both Westminster and Kensington and Chelsea councils did not seek repayment since they equated leasing with letting, rather than selling, a view the Conservative-controlled council was forced to modify in the wake of Labour opposition. But within the same boroughs, landlords took advantage of another loophole in the 1974 Act. Section 74 stated that on completion of improvement work, the property had to be let at a regulated (fair) rent, but landlords, by delaying work on the completion of improvement, let out most of the property (already improved) as holiday or business lets at considerably above regulated rent. Westminster City Council subsequently limited improvement grants to owner-occupiers and developers in HAAs where the properties were going to be relet at fair rents. The circumventions of the 1974 Act clearly called for new legislation since local authorities alone could not be expected to adopt remedies to the problems associated with gentrification. Gentrification in London was particularly a cause for concern since it meant that essential service workers such as nurses, teachers, postmen, bus drivers, etc. were finding it increasingly difficult or impossible to live near their work and vacancies in these occupations were becoming increasingly unfilled.

Legislation thus seemed essential to ensure that improvement grants were repaid when a leasehold interest was established within five years of the receipt of a grant; that rent registration was an automatic condition of a grant award on rented property; and that landlords were no longer able to qualify for multiple grants without public scrutiny of their private resources and requirements.

But the main problem was not abuse or inequity. It was that housing improvement expenditure in real terms was still seriously lagging behind the formation of new slums, and this was at a time when housebuilding was at a considerably lower level than in the late 1960s when the shift of emphasis to rehabilitation occurred.

The rehabilitation slump, privatisation and targeting

Despite worries about the scale of rehabilitation, the Treasury became sceptical in the 1980s about the cost-effectiveness of improvement grant expenditure. Local authorities had awarded almost £6 billion of improvement grants over the period 1969 to 1990 (Leather and Mackintosh, 1993), while annual expenditure on grants had increased from £200 million in 1979/80 to £650 million in 1983/84. Undoubtedly grants were sometimes awarded to people who did not need them (notably developers and large landlords), and much improvement might have been undertaken without grants. But although a decrease in the total sum available (rather than adopting a more selective approach) would certainly have worsened the condition of housing, especially in the inner cities, the government was intent on reducing public expenditure on private sector rehabilitation. In November 1983 it announced that from April 1984 repair grants would be cut from 90 to 75 per cent (except in the case of 'hardship', where 90 per cent grants would still be available). The cut was condemned in the House of Commons by a Labour motion which argued that it would hit householders, increase unemployment in the construction industry and impede economic recovery. Local authorities were faced with over 500,000 applications in the grant system by the end of 1983. Being unaware of how much they would have to spend on grants in 1983/84, many authorities (for example Barnsley, Birmingham, Camden, Doncaster, Hackney, Manchester, Sheffield and Wigan) stopped accepting applications for discretionary improvement and repair grants, and many applications already received could not be dealt with for several years at the current rate of processing.

On the demand side, as well as on the supply side, government policy adversely affected the rate of improvement. As if to deter rehabilitation, and with the aim of collecting £450 million in a full year, the budget of March 1984 extended value added tax (at 15 per cent) to home improvement and alteration works. The AMA forecast that this would cause a loss of 10,000 jobs in the construction industry (where unemployment was already at record levels), and the Building Employers Confederation (BEC) feared that it would fuel the 'black economy' since tax-dodging operators would be at an advantage over bona fide firms – with adverse consequences upon the quality of improvement work.

In respect of its improvement policies, the DoE was not only under pressure from the Treasury but was also being encouraged by the National Federation of Housing Associations (NFHA) to rationalise grant provision. In its *Inquiry into British Housing,* the NFHA (1985) stressed that although there were three vital areas for government intervention – encouraging voluntary action, dealing with the very worst stock and helping the worst-off households – its basic philosophy was that the main responsibility for the repair and improvement of private sector housing should be with individual owners.

It was therefore considered in government that much greater selectivity in the award of grants was essential. The Green Paper *Housing Improvement: A New Approach* (DoE, 1985b) thus proposed the following changes. First, improvement and repair grants should be means-tested, mandatory and related to renovation up to a new standard of fitness (on the assumption that a third of recipients in 1985 could have afforded to have paid for the work themselves, while other less well-off households were ineligible since they lived in properties with rateable values above the qualifying limit – £400 in London and £225 elsewhere in 1985). Second, as an alternative, interest-free discretionary loans for work beyond the new fitness standard should be available, with the local authority taking a financial stake in the improved property – for example, a loan of £5,000 on a house valued at £30,000 would allow the local authority to claim 10 per cent of the eventual sale price of the property. Discretionary loans would also be means-tested, but for both forms of assistance the eligibility ceiling on the applicant's income and capital would be low – although in respect of income, not as low as the supplementary cut-off level (£3,000 in 1985). Third, grants and loans should be available for the renovation of any pre-war housing (and not just for pre-1919 stock in the case of repair grants).

While it was important that grants should not be awarded unduly often to high-income claimants or distributed very thinly over extensive areas of housing of variable condition, critics of the Green Paper rightly argued that the aggregate effect of means-testing would be a reduction in the total volume of rehabilitation. The BEC had called on the government in 1984 to commit itself to a £6 billion improvement over 10 years (involving grant expenditure of £600 million per annum in real terms) (Carvel, 1984c), and the NHIC (1985) argued that unless public investment increased, a new house condition survey would indicate that a substantial number of dwellings in serious disrepair would degenerate into slums. Contrary to advice of this sort, support for private sector renovation was reduced from £608 million in 1985/86 to £535 million in 1987/88 (Table 4.3) and the number of grants paid to private owners and tenants fell from 200,000 to 159,000, 1985–87 (Table 4.2).

In response to the 1985 Green Paper, the White Paper *Housing: The Government's Proposals* (DoE, 1987) declared that improvement grants should be targeted towards the worst housing and to households in greatest need. Houses below a new standard of fitness would qualify for a single mandatory grant, while above this standard, grant assistance would be discretionary – related to the cost of necessary repairs and improvement and the ability of households to finance the work from income and savings. To avoid the possibility of windfall gains being realised from grant-aided improvement, discretionary grants would be repayable on a sliding scale when properties were sold within three years if the selling price exceeded a specific level. Help should be directed to areas of greatest need. Within newly designated Renewal Areas (which would replace GIAs and HAAs), local authorities would be encouraged to undertake a mix of rehabilitation and redevelopment.

Even before these proposals were (in an amended form) implemented, the government began to cut back its expenditure on grants to private landlords. In January 1989 mandatory repair and intermediate grants were reduced from 75 to 20 per cent of the cost of work on the grounds that under the Housing Act of 1988, landlords could charge higher rents for new assured tenancies, thereby reducing their need for grant aid.

The extent to which housing remained in a poor condition was indicated in the *English House Condition Survey 1986* (DoE, 1988) (Table 4.4). Although there had been a significant reduction in the number of dwellings lacking basic amenities since the survey of 1981, the number

Table 4.4 *The condition of housing in England, 1981–91*

| | Number of dwellings (000) | | | |
| | 1981–86 fitness standard | | 1991 fitness standard | |
Condition	1981	1986	1986	1991
Dwellings lacking amenities	862	543	463	205
Unfit dwellings	1,116	1,053	1,660	1,500
Dwellings in serious disrepair	1,178	1,113	n.a.	1,420

Source: DoE (1983, 1988, 1993)

of unfit dwellings and those in serious disrepair remained much the same. The survey also showed that conditions were far worse within the private rented sector than in owner-occupation or social housing, and that low-income households and the elderly suffered the worst housing. Based on this and on the White Paper *Housing: The Government's Proposals* (DoE, 1987), the Local Government and Housing Act 1989 targeted grants towards the worst housing, and to households in greatest need.

The 1989 Act replaced the grant system of earlier legislation with a new and largely mandatory regime of grants, and reformed the system of area improvement. Renovation grants, house in multiple occupation (HMO) grants, common parts grants, minor works grants and disabled facilities grants were introduced and all were means-tested. Owner-occupiers and tenants would be unlikely to qualify for grant assistance unless their incomes and savings were no higher than the level which would render them eligible for income support or housing benefit; and landlords would not qualify unless (without a grant) their outlay on improvement and repairs failed to exceed their rental income (but that with a grant their return would be sufficient only to service a loan at 3 per cent over base rate for 10 years). With regard to area improvement, GIAs and HAAs were to be superseded by Renewal Areas (RAs). Each would have a 10-year life after declaration and contain between 500 and 3,000 dwellings, of which 75 per cent would be privately owned and 75 per cent would be in poor condition, and over 30 per cent of households in each RA would need to be in receipt of housing benefit.

Funding constraints

Despite the availability of more valuable improvement grants, owners (and tenants) still needed to find quite substantial sums of money to supplement their grants – for example, normally £5,500 in the case of improvement grants – and tenants were concerned about their security of tenure and the extent to which any improvement they might undertake would result in large increases in rent.

Although in total, improvement grant expenditure increased from £134 million in 1979/80 to £1,064 million in 1983/84, HIPs were slashed in 1981 and many local authorities temporarily stopped giving improvement grants (unlike the mandatory intermediate grants, these were discretionary).

The inadequacies of private sector rehabilitation: the 1990s and into the new millennium

The impact of the 1989 Act on the volume of rehabilitation in the private sector was undoubtedly very mixed. While the number of grants targeted at low-income households and the elderly increased notably in the 1990s to a peak in 1996 (Table 4.5), the total number of grants awarded in that year (112,566) was considerably less than the number awarded at the time of the 1980s peak (320,000 in 1984). Most private owners, moreover, were ineligible for grant assistance regardless of the condition of their housing, and the overall pace of grant-aided rehabilitation in the private sector remained depressed. However, because of the slump in the housing market in the early 1990s, the extent to which means-testing contributed to the low level of rehabilitation was uncertain, but in the late 1990s – when there was a housing boom – there was an increase in grant aid and minor works assistance under the Housing Grants, Construction and Regeneration Act of 1996, despite the replacement of most mandatory grants by discretionary assistance (Table 4.6).

In England, by 1996 about 1.5 million dwellings were unfit (equivalent to 7.5 per cent of the housing stock), with the average cost of remedial work amounting to around £3,300 per dwelling (Table 4.7). In 1996 the private rented sector remained in a particularly poor condition (17.9 per cent of its stock was unfit) and – at almost £6,000 a dwelling – it was the most

Table 4.5 *Grants paid under the Local Government and Housing Act 1989, Great Britain, 1990–99*

	Mandatory grants	Discretionary grants	Minor works assistance	Total of all grants and assistance
1990	691	330	8,118	9,139
1991	30,127	3,879	36,076	70,082
1992	55,308	6,173	33,110	94,591
1993	59,275	6,166	30,614	96,055
1994	64,447	6,044	35,100	105,591
1995	68,283	5,429	36,676	110,388
1996	68,862	5,527	38,177	112,566
1997	40,247	5,091	13,132	58,470
1998	7,783	1,742	295	9,820
1999	1,932	443	168	2,543

Source: DETR, *Housing and Construction Statistics*

Table 4.6 Grants paid under the Housing Grants, Construction and Regeneration Act 1996, Great Britain, 1997–99

	Mandatory grants	Housing repairs assistance	Total of all grants and assistance
1997	19,016	23,955	40,971
1998	44,830	57,785	102,615
1999	49,822	70,627	120,449

Source: DETR, *Housing and Construction Statistics*

costly in which to undertake remedial work, although in percentage terms the degree of unfitness had decreased in all sectors over the period 1986–96. In absolute terms, the owner-occupied sector contained the worst housing, while the social housing sectors contained far fewer but broadly similar proportions of dwellings in poor condition.

As in England, the private rented sector in Scotland contained proportionately the greatest amount of poor housing (4.1 per cent was unfit in 1996), while in absolute terms there was – as in England and Wales – more unfit housing in the owner-occupied sector than in any other sector, but proportionately the degree of unfitness was not very different from that in the social housing sector (Table 4.8). It can be

Table 4.7 Housing unfitness in England, 1986–96

	Owner-occupied	Private rented	Local authority	Housing association	Vacant	Total unfit stock
Percentage						
1986	6.6	25.4	6.8	4.9	28.1	8.8
1991	5.5	20.5	6.9	6.7	22.7	7.6
1996	5.3	17.9	6.8	3.9	27.7	7.5
Number (000)						
1986	769	361	281	23	228	1,662
1991	715	333	265	41	145	1,498
1996	721	318	227	35	221	1,522
Percentage change, 1986–96						
	–6.2	–11.9	–19.2	52.2	–3.1	–8.4
Average cost of remedying unfitness in 1996 (£)						
	5,498	5,972	3,346	4,506	—	3,301

Sources: DoE (1993); DETR (1998b)

Table 4.8 *Housing below tolerable standard, Scotland, 1996*

	Owner-occupied	Private rented	Public rented	Housing association and co-operative
Number (000)	11	7	3	0
% tenure	0.9	4.1	0.5	0.2

Source: Scottish House Condition Survey 1996

assumed that the unit cost of remedial work in the private rented sector was higher than in each of the other sectors.

In Wales the private rented sector contained the worst housing (18.4 per cent was unfit in 1997/98), and the mean cost of remedial work was greater in this sector (at £1,883) than in other parts of the housing stock. In absolute terms, the largest amount of unfit housing was in the owner-occupied sector, but proportionately it was broadly comparable to that in the social sector (Table 4.9).

Table 4.9 *Housing unfitness in Wales, 1997/98*

	Owner-occupied	Private rented	Social housing	All occupied dwellings
Number of unfit dwellings	828,400	80,900	248,000	1,157,300
% tenure	7.6	18.4	8.2	8.5
Mean repair costs per dwelling (£)	952	1,883	654	953

Source: Welsh House Condition Survey 1997/98

Housing in Northern Ireland, according to 1996 House Condition Surveys, was in a similar state of repair to housing in England. In the province, 7.3 per cent of properties were considered to be unfit, and more than half of these were either owner-occupied or private rented (Stevens, 2000a).

With regard to Great Britain as a whole, the condition of housing was such at the end of the twentieth century that there was a £37 billion backlog in repairs across all tenures (Leather, 2000). Taking into account Northern Ireland, the backlog in the United Kingdom undoubtedly exceeded £40 billion.

In England local authority investment in private sector renovation was cut dramatically in the late 1990s as a consequence of the 1996 Act. Whereas in the early 1990s, local authority investment in private renovation peaked at £613 million in 1993, by 1998 it had been cut back to £326 million (at 1997/98 prices) – a decrease of nearly 50 per cent (Wilcox, 2000). However, on a very small scale there was an increase in the funding of home improvement agencies (HIAs) – from £4.4 million in 1996/97 to £6.1 million in 2000/01 – to help the elderly, disabled and other vulnerable people to remain in their homes by organising necessary repair and improvement work, and help was also provided through the Home Energy Efficiency Scheme.

Although such sums of money were dwarfed by the £16 billion spent annually by home-owners on repairing and renovating their houses (DETR, 2000a), a sizeable minority of owners were either unwilling or unable to keep their homes in a reasonable state of repair. This was not surprising since – as the *English House Condition Survey, 1996* had revealed – households in the lowest income quartile comprised 47 per cent of those in poor housing, 46 per cent of all households in homes with amenities in need of modernisation, and 44 per cent of homes without central or programmable heating (DETR, 1998b).

Since the fear of 'cowboy builders' or poor worksmanship was a further reason why people were deterred from taking adequate care of their homes, the government encouraged the National House Building Council (NHBC) to improve its new-home warranty scheme to provide a safeguard against the failure of builders in the scheme to comply with building regulations and the NHBC's own additional standards. In addition, the government set up the Cowboy Builders Working Group and commissioned it to recommend ways to drive out bad practice. The group subsequently proposed that a *quality mark* scheme should be introduced for work concerning domestic repair, maintenance and improvement, and that builders who registered with the scheme would be assessed with regard to their technical competence, financial soundness, and training in both management and craft skills.

Conclusion

In the new economic climate, the Labour government at the beginning of the new millennium confirmed its view that it was 'neither possible nor desirable to provide public money to tackle all the problems of poor condition housing in the private sector' but that its aim was 'to provide better opportunities for people to maintain and repair their own homes from their own resources where they are able to, and to help those that cannot do so' (DETR, 2000a: 39). It also believed that whereas in the past, local authority help to home-owners had been driven by a desire to preserve properties, henceforth the focus should be on protecting people, and that priority should be given to those cases:

- where the owner of a dwelling in poor condition cannot afford repairs;
- where housing in poor condition is having a negative impact on the wider area; and
- where the authority has a scheme for area improvement.

The government was concerned that under the Housing Grants, Reconstruction and Regeneration Act of 1996, large sums of money in the form of renovation grants went to people with considerable assets, but where no help was available to others who were equally or more deserving. Many recipients – although income-poor and thus eligible for grants – had considerable capital in the form of equity locked up in their homes. The Housing Green Paper (DETR/DSS, 2000) thus proposed that:

- local authorities should be given greater discretion over how they give grants;
- the loan-giving powers of local authorities should be broadened to enable authorities to give preferential or interest-free loans for improvement; and
- local authorities should be given new powers to make payments to third parties such as home improvement agencies to help lever private finance for home improvement.

Since it is recognised that in many parts of the country there are concentrations of owner-occupied housing in poor condition, local authority strategic investment is often undertaken within the spatial context of renewal areas or by means of group repairs to whole blocks of buildings. However, in some areas the problem is less about improving unfit housing and more about tackling low demand. Yet even if the problems contributing to low demand (social, economic and environmental) fall outside the ambit of housing policy,

> carefully targeted investment in the housing stock, as part of a package of other measures, can help prevent low demand from spiraling into wholesale abandonment. Many local authorities have

found that declaring a renewal area as part of an overall regeneration strategy acts as a catalyst for action.

(DETR/DSS, 2000: 35)

To help tackle the problems associated with low demand, the Green Paper proposed that local authorities should have more freedom over where to declare RAs, and how to carry out group repair. With regard to renewal area powers, the DETR proposed removing the conditions relating to the size of the area, the proportion of properties in private ownership, and the proportion of unfit properties in the area. In respect of group repair powers, it was proposed that the conditions relating to the proportion of unfit dwellings be removed and that flats would no longer be excluded from group repair schemes. It was also proposed that greater discretion be given to local authorities to determine how much participants in group repair should contribute, in order for local authorities to renovate whole blocks.

It was recognised, however, that in some areas of chronic low demand, crumbling houses and substantial valuation gaps, renewal may not be a viable long-term solution, and that clearance might be the only option. Local authorities would thus need to resort to powers contained within existing legislation to purchase and remove surplus housing, while relocation grants – available in clearance areas – would help home-owners purchase properties nearby. The Green Paper suggested that not only should local authority powers to award relocation grants be linked to their general powers within renewal areas, and be in line with the government's proposed reforms to housing renewal grants legislation, but that local authorities – in respect of relocation grants – could also be given greater discretion to determine eligibility criteria, amounts and grant conditions.

Local authorities were concerned, however, that private sector renovation grant subsidies in England and Wales were to be merged into a 'single housing pot' in 2000/01, alongside credit approvals for public sector housing improvements, prior to the creation of a single capital pot in 2001/02. Under these new arrangements – and in the face of competition from social housing programmes – it could become more difficult for local authorities to maintain (or increase) the proportion of their capital expenditure on private sector renewal (see Dwelly, 1999). Earlier, in Scotland, private renovation grants had been merged with the general local authority capital pot in 1995/96, resulting in a 63 per cent cut in grant expenditure, from £120 million to £45 million, by 1998/99, and a diversion of around £160 million into spending on education, social work and transport (Dwelly, 1999).

Arguably, government attempts to bring about a once-and-for-all improvement in housing standards in the private sector – through a succession of housing Acts since 1949 – had been far 'too expensive in an era of mass low-income home-ownership' (Leather, 2000b: 164). The 1989 Act – based on this belief – means-tested grant applicants, and the 1996 Act removed the mandatory element in grant provision. Although local authorities still had an obligation to provide grant aid to low-income home-owners, they lacked the resources to exercise this function.

It was not surprising, therefore, that little additional money was allocated to private sector renovation in the *Spending Review* (Treasury, 2000), except in the case of increased funding for disabled facilities grants and HIAs.

Clearly, not only does Britain have a 'combination of a substantial number of low-income home-owners in an ageing housing stock' (Leather, 2000b: 165), but an increasing proportion of home-owners are 'asset rich but income poor'. Leather (2000b) therefore suggests that if the government were willing to meet interest payments, some low-income home-owners might be able to borrow from a financial institution to meet the cost of renovation and/or repairs, using their property as security. In some circumstances this already applies to home-owners on income support in respect of essential repairs and improvement work, but the provision could be extended to more comprehensive forms of renovation. Leather also argues that the middle-aged and elderly on low incomes could be persuaded to draw on some of the equity in their homes to fund renovation (see p. 378), while younger owner-occupiers could be encouraged to accumulate savings specifically to meet repair and maintenance costs.

Clearly, within the context of urban regeneration, both public and private resources are required to provide employment, improve the physical environment, enhance the standard of a wide range of local services, stimulate community involvement and reduce crime. Thus in the foreseeable future it is unlikely that public resources will be directed at the renovation of privately owned housing stock on the scale required, hence necessitating a much larger contribution from private sources than hitherto. This will be more – rather than less – crucial in the private rented sector, if the government makes the payment of housing benefit to landlords of substandard dwellings conditional upon the improvement of the property (see p. 455).

Unlike the rate of rehabilitation in the private sector, renovations in the local authority sector remained at a high level in the late 1980s and throughout the 1990s (Table 4.2). But faced with the need to spend over £13 billion to rectify poor housing conditions in the local authority sector, the Conservative government in the mid-1990s had every

---

Box 4.1

Stock transfers and the renovation of local authority housing

The cost of renovating the local authority housing stock is substantial – £19 billion over 10 years according to the Housing Green Paper (DETR/DSS, 2000). It has been suggested that local authorities themselves could meet this cost if they were put on the same financial footing as registered social landlords (RSLs) or alternatively the cost of renovation could be met from transfer receipts (without any boost in Exchequer expenditure) if a 10-year target were replaced by one of 14 years (see Moody, 2000).

Currently, RSLs have a financial advantage over local authorities. After incurring the costs of management, maintenance and debt servicing, making provision for future major works and paying a small amount of tax, there is normally a small surplus for the year – after taking into account rent revenue. In contrast, local authorities are liable to substantially higher taxes and therefore have to cut back on repair and modernisation expenditure. The tax paid by local authorities is the 'local contribution to housing benefit' and is collected from the minority of tenants not on benefit, and paid towards the rents of those who are. It saved the Treasury around £1.4 billion in 1998/89 alone (Moody, 1998). Clearly, if local authorities were not liable for tax, savings could finance the £19 billion investment required to renovate stock, and in the medium term – when council houses are brought up to modern standards – rents could be raised to housing association levels, and the revenue obtained would facilitate further improvement to stock without the need for further government subsidisation (Moody, 1998). To ensure that these resources are targeted at those authorities with the greatest need, it would be necessary to redistribute tax savings in their favour.

Alternatively, in order to meet the £19 billion cost of renovation, *either* it would be necessary to boost capital spending by no more than £250 million per annum (assuming that the Exchequer made savings out of the transfer of stock), *or* the target could be reached in 14 years – at a constant level of investment – if the government were willing to forgo savings on the transfer of stock (Moody, 2000).

However, local authorities are concerned that the brakes might have to be applied to large-scale renovation programmes if the number of stock transfers to housing associations and other social landlords decrease. At the end of December 2000 the number of transfers was running ominously below the 200,000 annual target set by the DETR, and predictions of future transfers were unclear.

incentive to continue with its Right to Buy (RTB) and large-scale voluntary transfer (LSVT) policy in the knowledge that the rehabilitation costs of privatised dwellings would increasingly be incurred by the private sector.

Under the Labour government post-1997, much of this policy was retained, although there was less support for RTB. Local authorities, however, were encouraged to transfer their stocks to housing associations and other social landlords, largely to raise funds to facilitate the renovation of the remaining council stock. It was, nevertheless, considered central to Labours policy aims that the Exchequer should also invest heavily in the modernisation and repair of council estates. As an outcome of the *Spending Review* (Treasury, 2000), therefore, the government planned to spend an extra £1.6 billion on the renovation of council estates over the period 2001/02–2003/04, an increase in real terms of 12 per cent per annum. This should ensure that 500,000 homes will be renovated as part of a 10-year programme to eliminate substandard housing, with the number of households living in substandard housing falling by a third by 31 March 2004 (Waugh, 2000).

It was a cause for concern, however, that the government had less control over the condition of housing association dwellings. Whereas the larger 'housing associations may be good at building new properties . . . they appear to have precious little idea about how best to maintain [their] housing stock' (Field, 2001). It might be necessary, therefore, for the government to bring about a transfer of part of the housing association stock and budget from the larger and less efficient associations to smaller and more efficient housing providers in order to facilitate adequate renovation programmes within the sector.

Further reading

Bailey, N. and Robertson, D. with Pawson, H., Lancaster, S. and Jarvis, C. (1997) *The Management of Flats in Multiple Ownership: Learning from Other Countries*, Bristol: Policy Press. An examination of the repair and maintenance of flats in multiple ownership.

Couch, C. (1990) *Urban Renewal Theory and Practice*, Basingstoke: Macmillan. An accessible overview of urban renewal, including a review of housing improvement policy and practice.

Gibson, M. and Langstaff, M. (1982) *Introduction to Urban Renewal*, London: Hutchinson. A comprehensive and detailed analysis of urban renewal including a critical examination of housing improvement policy in the 1970s.

Goodchild, B. (1997) *Housing and the Urban Economy*, Oxford: Blackwell. Includes a study of housing renewal in the 1990s within the context of the urban economy. Possibly a worthy successor to the preceding publication.

Groves, R. and Niner, P. (1998) *A Good Investment? The Impact of Urban Renewal on an Inner City Housing Market*, Bristol: Policy Press. Examines how improvement policy preserves the quality of housing, and maintains and enhances its value, in an area of older two-storey terraced housing with a high level of owner-occupation and a predominantly Asian population.

Groves, R., Morris, J., Murie, A. and Paddick, B. (1999) *Local Maintenance Initiatives for Home Owners: Good Practice for Local Authorities*, York: Joseph Rowntree Foundation/York Publishing Services. Explores the range of initiatives being undertaken by local authorities to promote repair and maintenance by owner-occupiers.

Leather, P. (2000) *Crumbling Castles? Helping Owners to Repair and Maintain Their Homes*, York: Joseph Rowntree Foundation/York Publishing Services. Looks at a wide range of repair and maintenance issues and suggests that, as a basis for effective housing renewal policies, there is a need for a new clear and coherent strategy for the future of the national housing stock.

Leather, P. (2001) 'Housing standards in the private sector', in Cowan, D. and Marsh, A. (2001) *Two Steps Forward: Housing Policy into the New Millennium*, Bristol: Policy Press. Examines improvement in the owner-occupied sector, areas of market failure, and standards in the private rented sector within the context of the Housing Green Paper.

Leather, P. and Morrison, T. (1997) *The State of UK Housing: A Factfile on Dwelling Conditions*, Bristol: Policy Press. Drawing on a wide range of published and unpublished sources, this report brings together national data on disrepair in UK housing.

Leather, P., Littlewood, M. and Munro, M. (1998) *Make Do and Mend: Explaining Home-Owners' Approaches to Repairs and Maintenance*, Bristol: Policy Press. This examines factors other than cost that constrain households from carrying out repairs and improvement work.

McLean, S. (1999) *Repairs for All: How Social Landlords Can Extend Their Repairs Services to Local Home Owners*, York: Joseph Rowntree Foundation/York Publishing Services. Explores the role of social landlords in providing a repairs service to home-owners.

Revell, K. and Leather, P. (2000) *The State of UK Housing: A Factfile on Housing Conditions and Housing Renewal Policies in the UK*, 2nd edn, Bristol: Policy Press. Updates an earlier analysis carried out by Leather and Morrison in 1997 (see above).

Part II **Housing markets and housing tenure**

5 Housing finance

The demand for housing in both the rented and the owner-occupied sectors has in recent years been substantially facilitated by subsidies: mainly rate fund contributions to local authority housing (until they were discontinued in 1989/90), Exchequer subsidies and housing benefit in both the social and the private rented sectors, and – until the start of the new millennium – mortgage interest relief in owner-occupation. Except for housing benefit (which increased dramatically), all other subsidies decreased during the 1990s.

Throughout the 1980s rate fund transfers remained fairly constant, in cash terms, until they were abolished under the Local Government and Housing Act of 1989. But Exchequer subsidies (allocated by the DoE) declined considerably as a result of attempts by Thatcher administrations to curb public expenditure. By 1987, 80 per cent of local authorities received no Exchequer subsidy at all (see Chapter 7). It should be noted at this point that while Exchequer subsidies to local authorities are not paid directly to tenants, tenants nevertheless benefit because the subsidies cover a proportion of building costs and interest charges incurred in the provision of local authority housing and thus help to keep rents below market levels. Clearly, when Exchequer subsidies are reduced or no longer paid, rents can be expected to rise.

Housing benefit, by contrast, is an important form of income support and, since it is made available by the Department of Social Security, it is not officially regarded as an item of housing expenditure. However, over the period 1986/87–1999/2000, housing benefit increased from £3.8 billion to £11.7 billion (Table 5.1), consolidating the shift of emphasis from 'bricks and mortar' subsidies to personal subsidy that began in the early 1980s. With the prospect of continually rising rents, housing benefit was forecast to increase substantially well into the first decade of the

Table 5.1 *Subsidies to housing, Great Britain, 1986/87–1999/2000*

Year	Bricks and mortar subsidies		Demand-side subsidies		
	Rate fund transfer	Exchequer subsidy	Housing benefit	Income support: mortgage interest	Mortgage interest tax relief
	(£m)	(£m)	(£m)	(£m)	(£m)
1986/87	356	512	3,766	351	4,670
1987/88	328	498	3,871	335	4,850
1988/89	342	602	4,059	286	5,400
1989/90	91	716	4,652	353	6,900
1990/91	−5	1,221	5,686	539	7,700
1991/92	−20	919	7,419	925	6,100
1992/93	−26	539	9,018	1,141	5,200
1993/94	−19	126	10,424	1,210	4,300
1994/95	−21	−129	11,143	1,040	3,500
1995/96	−31	−449	11,892	1,016	2,700
1996/97	−46	−538	12,246	867	2,400
1997/98	−47	−627	11,840	660	2,700
1998/99	−65	−819	11,720	648	1,900
1999/2000	−76	−964	11,774	527	1,600

Source: Wilcox (2000)

twenty-first century (DETR/DSS, 2000). Income support for mortgage interest (ISMI) is also allocated by the DSS, and similarly is not an item of housing expenditure. After rising from £351 million in 1986/87 to £1.2 billion in 1993/94 (a consequence of soaring unemployment), it plummeted to £527 million in 1999/2000 with a fall in unemployment and the introduction of alternative means of mortgage protection.

In general, the case for subsidies seems very convincing. For example, Lansley (1982) explained that

> they are needed to ensure that all households can obtain housing of some minimum standard. Without financial assistance to reduce housing costs, a significant proportion of households would be unable to pay for decent housing. Left to itself, the private market would produce both insufficient homes, often of inadequate standard, and produce a very unequal distribution of housing resources. Lower income groups would be concentrated in poor quality and overcrowed conditions, often paying rent accounting for an unacceptably high proportion of income.

However, in practice, housing subsidies have been applied in an illogical way and have had perverse effects. Further, it was mainly housing consumption that was being boosted (through 'demand-side' subsidies), rather than housing production (through 'bricks and mortar' subsidies), with resulting scarcity and escalating rents and house prices.

It is against this backdrop that this chapter:

- examines the role that housing benefit plays in supporting demand in both the rented and the owner-occupied sectors, and looks at housing benefit under Labour;
- analyses the effects of mortgage interest relief on owner-occupation and on the wider economy during the late twentieth century;
- discusses the reform of mortgage interest relief and its eventual abolition in 2000;
- concludes by commenting on the failure to reform housing benefit during the late 1990s and early years of the new millennium.

Housing benefit

By the year 2000 as many as 4.5 million households in Great Britain claimed housing benefit to enable them to pay rent for their accommodation. Although most claimants were of working age, around 40 per cent were aged 60 or over. In Great Britain as a whole, nearly 60 per cent of claimants were council housing tenants, 19 per cent were tenants of RSLs and 22 per cent were tenants of private landlords, whereas in Scotland 75 per cent of claimants lived in council housing and only 11 per cent lived in private rented accommodation. In Wales the tenurial split was broadly similar to that in Great Britain as a whole (DETR/DSS, 2000).

As Table 5.1 indicates, housing benefit grew rapidly in the 1980s and 1990s as a consequence of rising rents and an increase in the number of eligible households, particularly pensioners and one-parent families. However, the 1994 Budget reformed the amount of housing benefit that could be claimed by private tenants, and thus from January 1996 100 per cent benefit was limited to any rent that was at or below the 'local reference rent for the area', rather than, as before, in relation to 100 per cent of reasonable market rents.

This limit, together with lower unemployment and the the introduction of the working families tax credit, contributed to a 30 per cent reduction in the number of private tenants receiving housing benefit and a £6.5 billion saving in housing benefit expenditure, 1996–2001 (Wilcox, 2000). Benefit claimants, however, were eligible for a 'top-up' benefit for that part of the rent between the amount paid and the market rent. However, from October 1997, top-up benefits were paid only up to the level of local reference rents (see Gibb *et al.*, 1999). This system was also applicable to housing association rents, if rent officers considered 'the rent to be unreasonably high or the tenant to be over-accommodated' (SSAC, 1995: 9).

The implications of reform were wide-ranging. Under the new system there was the danger that rents would be unaffordable to vulnerable groups, and to prevent this from happening, local authorities were given discretionary powers to fund higher amounts of housing benefit where circumstances were exceptional. There was also some concern among prospective tenants about how much housing benefit they could claim, and as a result, local authorities were encouraged to set up a pre-tenancy rent determination system to provide information on this matter. Although it was not mandatory for landlords to participate, those who did were legally obliged to charge the rents proposed. Perhaps of most concern, it was clear that the new system was quite clearly inequitable. Since eligible rents for single persons under the age of 25 were set at no higher than the local reference rent for single rooms, the new system 'obliges low-income single young people to move into shared housing and out of larger accommodation' (Gibb *et al.*, 1999: 189), yet the same restriction did not apply to single persons aged 25 or over. Finally, by forcing rents down to a ceiling in an attempt to prevent housing benefit from rising out of control, reforms to the benefit system in the 1990s might – like rent control in previous decades – create a reduction in both the quantity and the quality of private rented housing.

ISMI grew rapidly in the early 1990s, largely as a result of an increase in unemployment, before decreasing in the late 1990s as unemployment fell. However, the decrease in ISMI payments was also attributable to the 1994 Budget, which reduced the eligible mortgage ceiling to £100,000 and lengthened the period before borrowers became eligible for limited and subsequently full income support. From 1995, existing borrowers would only start to receive income support after two months from making a claim, and then receive only 50 per cent assistance for a further 16 weeks, before receiving 100 per cent help subsequently. New borrowers received help only after nine months, and then at 50 per cent for a further 16 weeks, before being eligible for 100 per cent help thereafter. The underlying motive for these changes was to bring about a cutback in public expenditure and to fill the gap by private mortgage protection insurance (PMPI). However, since bad risks were screened out, insurance was not equally available to all households in need, and take-up (by 1999) was as low as 20–40 per cent (Gibb *et al.*, 1999). However, because PMPI was introduced at a time of rising house prices and falling unemployment, many borrowers had little incentive to take out private insurance, and only a downturn in the economic cycle will indicate whether or not private insurance is an adequate substitute for ISMI at the time of greatest need.

Apart from the weaknesses of some of the changes brought in during the mid-1990s, there are a number of inherent problems associated with the housing benefit system:

- The delivery of housing benefit is often complex, confusing and time-consuming, and eligibility rules are too complex. Only around half of all local authorities in England and Wales administer benefits effi-ciently, and claimants are often left with worrying rent arrears and risk eviction.
- The benefit system is subject to fraud and abuse. Unscrupulous landlords might charge high rents for poor quality housing – while tenants might be unconcerned if rent is reimbursed in full, often directly to the landlord. The system also encourages 'upmarketing' since, in order to protect post-housing disposable incomes, it ensures that whenever rents rise, the actual amount of rent paid remains the same. Arguably, this provides the tenants with 'an incentive to trade-up and occupy housing larger than they require since the marginal cost in rent terms is zero' (Gibb *et al.*, 1999). In the private rented sector, however, upmarketing has been substantially reduced since 1996, when the introduction of 'reference rents' in effect placed a ceiling on rents eligible for housing benefit.
- The system creates an 'unemployment trap', since recipients of ISMI might exercise a preference for remaining unemployed and receiving income support, rather than securing employment and meeting the whole of their mortgage payments from their own resources. It is also a major cause of the 'poverty trap', whereby a household incurs a very high marginal rate of income tax when its gross earnings rise just above the Income Support thresholds.

Housing benefit under Labour

Although the Labour government, 1997–2001, believed that it was right to provide housing benefit to enable low-income households to afford reasonable accommodation regardless of marked variations in rent levels across the country, in its Green Paper it recognised most of the problems associated with housing benefit that needed to be addressed. The government thus proposed that it would introduce changes that would 'improve customer service; reduce fraud and error; improve work incentives; and, for the future, explore other options to support housing policy' (DETR/DSS, 2000: 106). More specifically, it proposed that:

- In respect of customer service, the government planned to improve the delivery of housing benefit by making sure that local authorities had the tools to administer benefit effectively – in terms of better-integrated information technology, and appropriate adminstrative practice.
- With regard to fraud and error, although work was already under way 'to provide local authorities with the tools to improve their anti-fraud work' (DETR/DSS, 2000: 110), the government planned to introduce tighter and more integrated ways of administration, a better performance regime and robust sanctions against offenders.
- In the case of work incentives, it was considered 'essential that housing benefit actively supports the transition into work' (DETR/DSS, 2000: 110), and to achieve this transition the government planned to draw together the Employment Service and parts of the Benefits Agency in 2001/02 in order to deliver a single integrated advisory service to benefit claimants of working age and employers, although it will not take over the administration of housing benefit. However, little or nothing was set out in the Green Paper on how the poverty trap could be eliminated.

Reflecting on other options to support tenants on low income, the Green Paper examined the principal advantages and disadvantages of systems employed in other countries and other possible reform ideas, but concluded that 'structural change may be something worth pursuing in the longer term, but it needs to be considered very carefully and it is unlikely to be possible before the reforms to the social housing system [as proposed in the Green Paper] . . . have been successfully implemented' (DETR/DSS, 2000: 121).

Mortgage interest relief

By far the largest subsidy historically, mortgage interest relief escalated from £4,670 million in 1986/87 to a peak of £7,700 million in 1990/91 (Table 5.1), but subsequently fell to £1,600 million in 1999/00 owing to lower interest rates and relief being limited to progressively lower rates of tax liability. It was often argued that tax relief was not a subsidy at all; it was more a concession. But mortgage interest relief had all the attributes of a subsidy.

First, the Exchequer had £7,700 million less in revenue in 1990/91 than it would have received had there been no interest relief. Second, the revenue forgone would have had much the same short-term fiscal effects on the housing market and the economy in general as, for example, public expenditure on cash grants of an equivalent amount. Third, the Exchequer had to keep rates of both direct and indirect taxation higher than would otherwise have been necessary to compensate for the amount of relief received by owner-occupiers; and finally, housebuyers had more money to spend (or save) than would otherwise have been the case, and much of this pulled up the price of existing housing. Since fewer than one in five houses purchased are newly built, around 80 per cent of mortgage interest relief had no effect on housebuilding.

Owner-occupiers also benefited from not having to pay the full cost of acquiring an asset – in effect receiving a further subsidy. Since 1962, owner-occupiers have not had to pay Schedule A tax on 'imputed rent income' (the free use of accommodation equivalent in value to rent), and are also normally exempt from capital gains tax – exemptions which, even in the 1980s, were probably worth as much as £10 billion and £5 billion per annum respectively. In addition, council tenants buying their own homes received discounts of over £1 billion per annum throughout much of the 1980s.

Effects of subsidising owner-occupation

In terms of subsidisation, owner-occupation was undoubtedly the most privileged sector throughout the latter part of the twentieth century. King and Atkinson (1980) commented that assistance to the owner-occupier had led to 'patent inequalities and to a bizarre pattern of incentives', and Goss and Lansley (1981) thought that 'it would be difficult to devise a more unjust, wasteful and inefficient subsidy system' than the one that currently prevailed. Reform – or so it seemed – was long overdue.

Subsidies had become increasingly inequitable and had exacerbated inflation. They were inefficient and wasteful; adversely affected productive industry; eroded the tax base and impeded household mobility; distorted tenure preferences and rural values; did not provide effective protection for the poor; and did little to facilitate repairs and maintenance. Each of these effects will be examined in turn.

Inequitable assistance

Historically, mortgage interest relief increased in relation to the rate of tax. For example, in the early 1980s, when the mortgage rate was 10 per cent (gross), the basic rate taxpayer paid only 7 per cent net (the basic rate of tax being 30 per cent), but a taxpayer with a marginal rate of tax of 60 per cent would have paid only 4 per cent net, a negative cost of purchase below the rate of inflation and the rate of increase in average house prices. But interest relief was doubly regressive since the larger the income, the greater the possibility of a larger mortgage and the greater the tax relief (on mortgages up to £30,000 after April 1983). Even though the Budget of 1988 lowered rates of income tax to 40 and 25 per cent, the distribution of mortgage relief remained clearly inequitable. Table 5.2 shows, for example, that in 1990/91 the 6.4 per cent highest-earning mortgagors received 10.9 per cent of total mortgage interest relief, whereas the 5.2 lowest-earning mortgagors received only 3.5 per cent of total relief.

Table 5.2 *Distribution of mortgage interest relief, 1990/91*

Annual income	Mortgagors receiving interest	Average relief per mortgagor	Total cost of relief	
(£)	(%)	(£)	(£m)	(%)
Under 5,000	5.2	550	270	3.5
5,000–9,999	11.2	610	640	8.3
10,000–14,999	21.9	740	1,520	19.7
15,000–19,999	22.4	800	1,670	21.8
20,000–24,999	16.3	810	1,240	16.1
25,000–29,999	9.0	880	750	9.7
30,000–39,999	7.6	1,090	770	10.0
40,000 and over	6.4	1,400	840	10.9
All ranges	100.0	820	7,700	100.0

Source: Inland Revenue

Although there were reductions in the basic rate of tax in recent years, during periods of inflation (for example in 1980/81 and in the late 1980s) the negative cost of house purchase was often substantial. Because of this, owner-occupiers (blessed with average incomes twice those of council tenants and with substantially larger subsidies) were able to trade up and accumulate wealth, mainly within the second-hand market, a process aided and abetted by the building societies, which normally allocated only a small proportion of their mortgage loans for the purchase of newly completed housing.

However, with several Budgets from 1988 onwards reducing the lower rates of income tax (in stages from 25 to 10 per cent), the distribution of mortgage relief became less inequitable. Table 5.3 shows, for example, that whereas in 1990/91 the highest-earning mortgagors received two and a half times that received by the lowest-earning mortgagors, in 1998/99 mortgagors in the top income band received only 33 per cent more relief than that received by mortgagors in the lowest band. While mortgagors in the middle income bands received the greatest amount of total tax relief throughout the 1990s, mortgagors in the lowest band increased their share of total tax relief from 3.5 per cent in 1990/91 to 7.3 per cent in 1998/99.

Table 5.3 Distribution of mortgage interest tax relief, 1990/91–1998/99

Income band	Average tax relief (£ per annum)			Total tax relief by income band (%)		
(£)	1990/91	1994/95	1998/99	1990/91	1994/95	1998/99
0–5,000	550	290	150	3.5	8.9	7.3
5–10,000	610	290	150	8.3	10.3	7.3
10–15,000	740	330	160	19.7	22.0	15.7
15–20,000	800	360	180	21.7	22.0	18.8
20–25,000	810	380	180	16.1	14.9	16.2
25–30,000	880	390	190	9.7	8.0	11.5
30–40,000	1,090	410	200	10.0	7.4	12.6
40,000+	1,400	400	200	10.9	6.6	10.5
Basic rate of relief (%)	25	20	10	25	20	10

Source: Wilcox (2000)

Inflation of house prices

House prices were pulled up by mortgage tax relief, tax-exempt imputed rent income and tax-exempt capital gain – effective demand being up to £22,000 million greater by 1990/91 than it would have been otherwise. Yet such inducements to buy had relatively little effect on housebuilding, owing largely to the relatively inelastic supply of new houses. Instead, the second-hand housing market (accounting for well over 90 per cent of all transactions) became increasingly inflated, particularly in the late 1980s.

The house price boom of 1985–89 had damaging effects on the United Kingdom economy. Owner-occupiers inheriting property, relocating or trading down withdrew over £20 billion of equity by 1988, fuelling a high level of inflationary consumption and imports. The consumer boom was also facilitated by home-owners taking out loans against the value of their housing assets. But although this might have seemed feasible, since house values exceeded mortgage debts by £750 million in 1988, a growing number of home-owners faced severe financial difficulty as the level of consumer financial liability increased as a ratio of income from 0.55 in 1982 to 1.12 in 1989 and interest rates rose rapidly in 1989/90 (Maclennan *et al.*, 1991). By 1991, 183,610 households were more than six months in arrears with their mortgages, and 75,540 homes were repossessed.

Owing in part to marked regional disparities in income during the boom years, there were distinct regional variations in house prices and subsidies to the home-owner (the South East, for example, received a disproportionate distribution of mortgage interest relief and exemption from capital taxation). Inter-regional labour mobility, in consequence, was severely impeded, resulting in labour scarcity and inflationary wage demands in the economically over-heated regions, and continuing unemployment in the slow-growth area.

Inefficient or wasteful use of resources

Since the Second World War a very substantial amount of financial resources has flowed into housing, but into housing demand rather than housing supply. Whereas share prices increased fivefold and retail prices went up nine times between 1945 and 1980, house prices increased by a factor of 16. House prices would not have risen so dramatically had the supply of funds for house purchase not increased as a proportion of the

GDP. Whereas in 1956 net loans for house purchase equalled only 1.0 per cent of GDP at factor price, and net personal savings amounted to 4.1 per cent, by 1978 the respective percentages had risen to 4.0 and 11.3 respectively, showing a more than proportionate increase in loans for house purchase – an increase dependent upon a sizeable growth in personal savings.

The *Report of the Royal Commission on the Distribution of Incomes and Wealth* (the Diamond Report, 1976) indicated that there had been a major change in the means by which wealth was being accumulated. Whereas in 1960, 23 per cent of personal wealth was held in dwellings and land and the same percentage in company shares and securities (a ratio of 1:1), by 1974 real property accounted for 47 per cent and shares and securities only 11 per cent (a ratio of over 4:1). Kilroy (1980a) showed that in the United Kingdom 90 per cent of net personal savings flowed into the three channels which receive the most favourable tax treatment – housing, life insurance and pension funds – whereas in the United States the proportion was about 60 per cent. The rapid growth of home-ownership is therefore associated with the drift of savings into less productive real property (most of it consisting of buildings constructed in the past) rather than into more productive forms of investment, not least new housebuilding. Mortgages, and remortgaging, are thus substantial 'capital guzzlers' (Kilroy, 1980b). The diversion of financial resources away from the primary circuit of capital should be a prime cause of concern.

In recent years annual gross capital formation in housing has scarcely increased, the proportion of GDP spent on housing capital formation in the United Kingdom being very low by international standards. As the late Victorian housing 'bulge' is deteriorating at the same time as demand is rising, far too little is being spent on replacing, improving or repairing the housing stock – the net increase in stock being very inadequate compared with increases in most other industrial countries.

Scarcity was compounded by widespread under-occupation. Because of tax exemptions, many households occupied larger and more expensive houses than they really required while investment in the basic stock was jeopardised by insufficient available resources. Kilroy (1981a) introduced the term 'housing sector borrowing requirement' (HSBR) to denote the total amount of money borrowed for investment in housing. Most of the HSBR by 1981 was spent on existing housing – on exchange rather than construction – and in total was as much as the public sector borrowing requirement (PSBR), approximately £8,251 million, indicating very clearly the magnitude of 'unproductive' resource allocation. From the mid-1980s to 1992 the HSBR remained substantially greater than the PSBR, inflating demand rather than stimulating supply.

But it was not just a question of the housebuilding industry requiring subsidies. Kilroy (1981b) argued that subsidies to housebuyers inflate not only the price of houses but also the price of land, to the detriment of most builders who do not hold land banks and to the benefit of land hoarders and speculators.

Effects on industry

It may be difficult to attribute Britain's poor industrial performance in recent years to the system of housing finance applied over the past few decades, but the Deputy Governor of the Bank of England, Mr 'Kit' McMahon (1981) contrasted the building societies' success in quadrupling their deposits to £41,111 million (1971–80) with the difficulties experienced by companies trying to raise finance, especially for long-term projects. Regarding the clearing banks diverting funds into housing, he warned that 'greater competition in housing finance may be thought to give the authorities some cause for concern'.

It was very common for proponents of a free market economy to argue against high levels of public expenditure on the grounds that investment in the more productive private sector is crowded out, and that monetary growth should be curtailed by cuts in public spending in order to counter inflation and release resources for investment in privately owned industry. However, Kilroy (1980a) argued that it was housing (probably more than public expenditure) which had an adverse effect on industry:

> the personal fortunes of . . . [the majority] of Britain's population who live in their own houses depend directly on distortions in the financial markets and the tax system which are now threatening to damage the economy on a scale that the government cannot afford to ignore.

He went on to state that if new institutional investment was attracted into mortgages, there was a clear danger that housing would supersede government policy in crowding out industrial investment and fuelling monetary growth. Similar views were put forward by the National Economic Development Council in 1975, the Meade Committee and the London Clearing Banks in 1978 and Professor Douglas Hague (Economic Adviser to Mrs Thatcher) in 1979.

Effects on taxation

The tax base was being continually eroded by mortgage and life insurance tax exemption, enabling the circumvention of high marginal rates of tax. This resulted in other taxes such as value added tax or even income tax being higher than would otherwise be necessary.

Effects on household mobility

It is generally thought that security of tenure in much of the private rented sector, and allocation procedure and subsidised rents in the public rented sector, both reduce mobility. In owner-occupied housing the effect of subsidies is similar. There is relatively little trading down because of loss of subsidy, and much more trading up in the same area. Whereas immobility in the rented sectors may be associated with unemployment, mortgage tax relief shackles middle-income groups
to their employers (and locations) just as much as pension arrangements do.

Distortion of tenure preference

Kilroy (1979) estimated that under current subsidies the average owner-occupier would be £7,500 better off than the average council tenant after 25 years, a figure surpassed by Mr Michael Heseltine (1979), Shadow Minister for the Environment, who stated that they would be £14,000 better off. The financial scales were thus weighted heavily in favour of owner-occupation and, together with the many other current advantages of owner-occupation, persuaded the young and relatively well-off to opt for it, relegating council housing to an increasingly residual and welfare role.

Distortion of rural values

Mortgage tax relief enabled commuters increasingly to bid up the price of country cottages within travelling distance from work, pricing out of the market rural households who either have to seek council housing, continue to live in tied accommodation or migrate into the towns and cities to be faced possibly with unemployment and homelessness.

Lack of protection for the poor

Since subsidies to owner-occupiers boosted demand and prices, it was largely a myth that they helped first-time buyers by making home-ownership easy and inexpensive. Even when low-income buyers could afford to buy, there was the risk of foreclosure – especially during unemployment – if repayments were in arrears, while the housing benefit scheme introduced in 1982–83, though it benefited tenants, did not assist the owner-occupier. The system closed the door to many would-be buyers, and since it crowded out resource spending it may have directly helped to cause homelessness.

Effects on repair and maintenance

Although mortgage interest relief helped to boost house prices, this by itself did not ensure that home-owners would undertake necessary repairs and maintenance. Many owners were asset-rich but income-poor and were not willing to incur what they may have perceived to be an avoidable cost – particularly if they did not intend to sell their properties in the foreseeable future and realise some of their capital. But nearly 50 per cent of owner-occupiers owned their homes outright, most of them having paid off mortgages over 20–25 years. Not only did outright owners fail to receive interest relief, but many lived in the oldest houses in the poorest condition and might not have qualified for renovation or minor works grants under the provisions of the Local Government and Housing Act of 1989.

Reform of housing finance

There have been many proposals for reform over the past three decades, but a major opportunity for change was lost in the 1970s when a Labour government failed to adopt radical measures to deal with the problems of housing finance. At the beginning of the decade, Labour were enthusiastic about reform. Mr Anthony Crosland (1971b), Shadow Environment Secretary, seemed to be stating the view of the Labour Party when he argued:

> If our object is not simply to eliminate the worst housing, but more generally to reduce inequality, then the present distribution of

> financial aid is strikingly perverse . . . it is highly inegalitarian in its
> effects. It brings no help to those on the lowest incomes, who cannot
> afford to buy in any case; generally it goes to a group which is better
> off, on average, than either council or private tenants; and within this
> group it gives the most relief to those on the highest incomes.

He stated that what housing policy needed was 'above all, a reform of
housing finance'. In 1974, after becoming Secretary of State for the
Environment, Anthony Crosland set up a Housing Policy Review to deal
with the main shortcomings in what he described as the 'dog's breakfast'
of the prevailing system of financial support. The review, proclaimed Mr
Crosland (1974), was intended to be a 'searching and far reaching
inquiry into housing finance. The aim . . . [was] to get back to the
fundamentals, to get beyond a housing policy of "ad hoc"-ery and crisis
management.' Focusing on three issues – the inconsistencies between
what different consumers of housing paid and the aid they received, the
divisive social implications and the inefficient use of resources – he
promised that he would not spare 'the slaughter of sacred cows' in order
to produce solutions.

The Green Paper *Housing Policy: A Consultative Document* (DoE,
1977a), published after Mr Crosland had left the Environment
Department, was, however, a very conservative document. Despite
showing the considerable inequalities in housing subsidies, it left
mortgage interest relief alone on the following grounds: the construction
industry depended upon it; tax relief gives help when it is most needed
(in the early years of a mortgage when interest payments normally
account for 95–100 per cent of repayments); the high amount of relief
which could be claimed by mortgagors on high incomes and with large
mortgages was not inequitable in view of progressive income tax; and
changes might be politically hazardous in view of an increasing number
of households becoming owner-occupiers and with house prices steadily
rising. The Green Paper did not wish to upset 'the household budgets
of millions of families', but it seemed to have discounted the fact that
reduced tax relief is an essential political counterpart to increased council
house rents. The Labour government (despite recommendations for
reform by the party and annual conference) therefore rejected any
significant change in the system of housing finance in the owner-
occupied sector, and Mr Peter Shore, Anthony Crosland's successor,
argued that he had left the system of tax relief for owner-occupiers
untouched, since 'our system of personal taxation is highly progressive
. . . [and] tax relief is virtually the only significant offset against high

marginal rates of tax. It is an incentive to a socially desirable form of expenditure' (Shore, 1977) – a far cry from the sentiments of Crosland.

The owner-occupied sector

It was clear that the Green Paper had rejected a radical reform of the prevailing system of tax relief on mortgage interest because, in the view of Kilroy (1978), 'the case for it was damned by equating fundamental with sudden', even though most of the evidence submitted to the review had been to suggest gradual reform over a long transitional period.

During the years that followed, detailed proposals were submitted for phasing out mortgage interest relief, lowering the size of mortgages eligible for relief and limiting relief to the first mortgage only (Kilroy, 1978), and restricting mortgage interest relief to basic-rate taxpayers only, with relief being phased out over 15 years (Goss and Lansley, 1981). Other recommendations broadly involved leaving mortgage interest relief unchanged, but taxing imputed rent income and phasing in capital gains tax on an owner's only or main house (Grey et al., 1980; King and Atkinson, 1980). Shelter (1982a) even proposed removing the (then) current limits on mortgage interest relief, but taxing imputed rent income, partly to encourage expenditure on repairs and maintenance which could be set against tax.

Some of these proposals were crystallised in the *Inquiry into British Housing* (NFHA, 1985), which recommended *either* a phased reduction in the eligibility ceiling for relief from £30,000 to zero, *or* the withdrawal of mortgage interest relief above the basic rate of tax followed by a progressive annual reduction over 10 years to complete abolition.

The RICS (1985) likewise called for a reduction in the £30,000 eligibility ceiling – by £3,000 per annum over 10 years. More radically, the Archbishop of Canterbury's Commission on Urban Priority Areas (1985) recommended the diversion of mortgage interest relief (amounting to £4,750 million in 1985/86) to house the homeless, improve housing in poor condition and finance the building of more council estates.

In the late 1980s, during the house price boom, further proposals were put forward to reform housing finance. Muellbauer (1988) argued that mortgage interest relief should be discontinued since it had contributed to soaring prices – putting house purchase beyond the reach of the many first-time buyers it was intended to help. He further suggested imposing capital gains tax (or a sales tax) on all dwellings, and (with poll tax replacing domestic rates) either a 2–3 per cent annual tax on the market value of dwellings (as indicated by local house price indices) or a site value rate which would release building land for development and (by stimulating the supply of new housing) slow down the rate of house price inflation. At the outset of the house price slump, Muellbauer (1990) still advocated terminating mortgage interest relief, and proposed that it should be restricted to the first 10 years of a mortgage – which (according to current estimates) would have saved the Treasury £4 billion per annum, the equivalent to 2p off income tax.

The effects of the complete abolition of mortgage interest relief were examined by Wilcox and Pearce (1991). Not only would there be substantial savings to the Treasury (£7,700 in 1991/92), but abolition would bring down house prices (by up to 9 per cent) and be deflationary, with beneficial (though marginal) effects on the balance of payments. Abolition would not only permit a reduction in income tax but also allow interest rates to fall by 1½ per cent. To counter any reflationary effect of these reductions, the money supply would need to be held constant. On balance, most owners would be no worse off, while the average buyer earning £15,000 a year and borrowing £40,000 or more would be better off.

Reversing its earlier toleration of mortgage interest relief, Shelter (1991) likewise called for interest relief to be abolished, but instead of producing savings to the Treasury it should, in the view of Shelter, be reallocated to help the homeless – notwithstanding the hundreds of thousands of households behind with their mortgage repayments and many facing repossession.

Compared to the 1985 *Inquiry* (NFHA, 1985), the *Inquiry into British Housing, Second Report* (JRF, 1991) was more forthright in recommending a phased withdrawal of mortgage interest relief (which it believed to be both inefficient and inequitable), rather than allowing relief to 'wither on the vine', with its real value diminishing with inflation.

Public policy and reform

Perhaps in anticipation of the Conservatives return to office, the independent Meade Committee (1978) advocated that it was necessary to restore the size of the tax base. This had been gradually eroded by relief and exemptions, the most significant of which related to mortgage interest and capital gains or transfers. After the 1979 General Election, Professor Douglas Hague (then Economic Adviser to Prime Minister Thatcher) hinted that mortgage tax relief might be phased out to obviate criticism of proposed expenditure cuts and rising council rents (McLoughlin, 1979), but this was strongly denied by Mrs Thatcher. The opposite occurred eventually. The mortgage tax relief ceiling was raised to £30,000 (by the Finance Act of 1983), and within this limit relief remained unchanged. Although the ceiling had been set in 1974 (and in real terms would have risen to £76,000 by 1983), its higher level provoked hostile criticism. Mr Neil McIntosh, Shelter's director, said the measure would siphon off funds urgently needed to cope with the disrepair of the housing stock and would be a waste of resources; Mr Peter McGurk, the Institute of Housing's director, likewise suggested that the money would be better spent on giving tax relief for house repair and maintenance, and scrapping value added tax on building work and materials; and the *Economist* (1983b) stated that 'to raise the mortgage subsidy [was] bad economics – and class politics of the worst kind'. This was partly because council house rents had risen by 110 per cent between January 1979 and October 1982 (compared to a 60 per cent increase in retail prices), and council tenants' subsidies had fallen from £1,595 million in 1979/80 to £970 million in 1982/83, while mortgage tax relief had risen to £1,910 million. Home-owners with mortgages of £30,000 had already benefited from a reduction in interest rates from 15 to 10 per cent, 1982/83, saving them £73 per month (equal to a reduction of 27 per cent in their repayments). Although the Building Societies Association may have persuaded Mrs Thatcher to raise the threshold to encourage trading up and filtration, and to stimulate demand in the construction industry (in the upper price markets), in total less than 5 per cent of mortgages were for sums in excess of £25,000 in 1983, although in the Home Counties, a Conservative stronghold, probably over 50 per cent were above this figure.

In the early 1980s the Labour Party was firmly committed to reform. In *Labour's Programme 1982* (Labour Party, 1982a), it stated that 'it would tackle the regressiveness of mortgage tax relief . . . for example by restricting eligibility to the standard rate of tax.' Mr Gerald Kaufman (1983), Shadow Minister for the Environment, stated that limiting tax relief to the basic rate would save taxpayers £170 million per annum, and the proposal was incorporated into the party's 1983 manifesto. At the 1987 General Election, Labour's policy was still confined to the abolition of mortgage interest relief above the basic rate of tax – a policy stance shared by the Alliance. The Conservatives, however, pledged that they would continue the current (regressive) system of relief. However, in the first budget of Mrs Thatcher's third administration, the basic rate of income tax was reduced from 27 to 25 per cent and the upper level of tax was consolidated at 40 per cent. The £30,000 eligibility ceiling, moreover, was not raised, as some commentators had predicted, but relief was no longer available to assist existing residents improve their properties, and multiple tax relief was ended (from August 1988). The total amount of mortgage interest relief was thus destined to fall in due course, and since £30,000 was rapidly diminishing as a proportion of average house prices (particularly in the South East), it seemed likely that the impact of mortgage interest relief on the home-ownership market was very much in decline. Nevertheless, in the early 1990s it still remained the largest housing subsidy and continued to distribute the greatest amount to those households least in need, while simultaneously housing subsidies to local authorities were being slashed, rate fund transfers were being strained (and scheduled to be abolished) and housing benefits to young people were being cut.

In the 1991 Budget it was announced that mortgage interest relief would be limited to the basic rate of tax (25 per cent) and discontinued at the higher rate (40 per cent) – a measure which reduced distortions in the housing market, curbed further excessive increases in house prices and reduced the waste of public resources. Further reductions in mortgage interest relief were set out in the Budgets of April 1992 and November 1993. Relief was consequently reduced to 20 per cent from April 1994 and lowered again to 15 per cent from April 1995. It was evident that the government was attempting to prevent a further house price boom occurring in the late 1990s, and avoid a repetition of over-borrowing and large-scale repossession towards the end of the decade. During the period of the 1997–2000 Labour government, mortgate interest relief was lowered to 10 per cent from April 1998 and abolished in April 2000, saving around £1,600 billion per annum.

Conclusion

With the phasing out and eventual abolition of mortgage interest relief, housing benefit became the dominant, though a highly problematic, means of subsidising households in relation to their housing costs. However, despite the government's intentions, the reform of the housing benefit system failed to materialise during the 1997–2001 parliament. The benefit system continued to be severely marred by inadequate administrative funding and intolerable delays in processing hundreds of thousands of claims – including delays affecting housing associations in respect of temporary housing for the homeless (*Roof*, 2001a; *Housing Today*, 2000). To many observers in 2001, the issue of housing benefit reform appeared to be shelved for the time being – particularly since the DSS had ruled out many of the key proposals of the Social Security Select Committee's report the previous October. Arguably, the failure to reform housing benefit had become 'the biggest social policy failure of Tony Blair's government' (Kirkwood, 2001: 26).

Further reading

Aughton, H. and Malpass, P. (1998) *Housing Finance: A Basic Guide*, 5th edn, London: Shelter. A good starting point for beginners.

Barr, N. (1993) *The Economics of the Welfare State*, Oxford: Oxford University Press. Includes an economic approach to the analysis of housing policy.

Bartlett, W., Roberts, J. and Le Grand, J. (1998) *A Revolution in Social Policy: Quasi-Market Reforms in the 1990s*, Bristol: Policy Press. Examines the impact of quasi-market reforms on the provision of public services, including social housing.

Ford, J. and Wilcox, S. (1994) *Affordable Housing, Low Incomes and the Flexible Labour Market*, Research Report no. 22, London: National Federation of Housing Associations. Examines the relationship between rapid increases in housing association rents in the early 1990s and the deepening poverty trap faced by low-income households.

Ford, J., Kempson, E. and England, J. (1996) *Into Work? The Impact of Housing Costs and the Benefit System on People's Decision to Work*, York: Joseph Rowntree Foundation/York Publishing Services. Examines whether, in the context of recent changes in housing, social security arrangements and the labour market, households – for economic reasons – are more likely to take work rather than remain on benefit.

Forrest, R. and Murie, A. (1991) *Selling the Welfare State*, London: Routledge. Includes an examination of housing finance within the context of the welfare state.

Gibb, K. (1995) *Housing Benefit: The Future*, London: National Federation of Housing Associations. Examines alternatives to the current system of housing benefit.

Gibb, K., Munro, M. and Satsangi, M. (1999) *Housing Finance in the UK*, 2nd edn, Basingstoke: Macmillan. A comprehensive and accessible examination of housing finance across the United Kingdom.

Hills, J. (1991) *Unravelling Housing Finance*, Oxford: Clarendon Press. Possibly the best book on housing finance, although it focuses on the housing finance system in the 1980s and is therefore a little dated.

Hills, J. (2000) *Reinventing Social Housing Finance*, London: Institute of Public Policy Research. Examines the social finance system as a whole and pays particular attention to the housing benefit system.

Holman, A. and Whitehead, C. (1997) *Funding Affordable Social Housing: Capital Grants, Revenue Subsidies and Subsidies to Tenants*, London: National Housing Federation. Includes an exploration of the long-term public expenditure implications of personal allowances as against capital subsidies.

Kemp, P. (ed.) (1986) *The Future of Housing Benefits*, York: Centre for Housing Research, University of York. A useful account of the rationale for change in the housing benefit system.

Kemp, P. (1992) *Housing Benefit. An Appraisal*, London: HMSO. A comprehensive appraisal and discussion of housing benefit.

Kemp, P. (1998) *Housing Benefit: Time for Reform*, York: Joseph Rowntree Foundation/York Publishing Services. An examination of the housing benefit scheme, with a particular emphasis on its relationship with work incentives and the efficiency of the housing system.

Kemp, P. (2000) *'Shopping Incentives' and the Reform of Housing Benefit*, Coventry: Chartered Institute of Housing. An examination of whether there should be an incorporation of 'price incentives' into housing benefit to ensure better value for money.

Kemp, P. and Rugg, J. (1998) *The Single Room Rent: Its Impact on Young People*, York: Centre for Housing Policy, University of York. A study of the impact of housing benefit restrictions on young people in privately rented accommodation.

Kempson, E., Ford, J. and Quilgars, D. (1999) *Unsafe Safety Nets*, York: Centre for Housing Policy, University of York. Provides an evaluation of the effectiveness of both mortgage payment protection insurances (MPPI) and ISMI for those on income support (or, more recently, on Job Seeker's Allowance).

Malpass, P. (1990) *Reshaping Housing Policy: Subsidies, Rents and Residualisation*, London: Routledge. Examines rents and subsidies policy since 1945 and explores the (then) impending new regime for local authority housing finance.

Webb, S. and Wilcox, S. (1991) *Mortgage Benefit*, York: Joseph Rowntree Foundation. A study of alternatives to the then current system of housing benefit.

Wilcox, S. (2000) *Housing Finance Review 2000/2001*, Coventry: Joseph Rowntree Foundation/Chartered Institute of Housing/Council of Mortgage Lenders. A detailed compendium of data and information on housing finance supported by useful analyses. Updated annually.

6 Private rented housing

The supply of private rented accommodation in Great Britain declined from approximately 90 per cent of dwellings in 1914 to only 11 per cent in 1999, compared to around 23 per cent in France, 29 per cent in the United States and 40 per cent in Germany; and, in Great Britain, housebuilding in this sector has been virtually non-existent since the Second World War, unlike most other European countries and North America. Private rented housing in Great Britain includes a high proportion of the oldest and poorest dwellings. Most were built before 1919, and according to the DoE (1985a) there were as many as 334,000 houses in multiple occupation (HMOs) in England and Wales – 60 per cent in serious disrepair, 80 per cent below standard *vis-à-vis* amenities, overcrowding and management, and 81 per cent lacking satisfactory means of escape from fire. The *English House Condition Survey 1991* (DoE, 1993), moreover, indicated that whereas only respectively 5.5, 6.5 and 6.8 per cent of owner-occupied, housing association and local authority dwellings were unfit, as many as 19.9 per cent of the private rented stock were of this condition.

Many types of household still depend upon private lettings to satisfy their housing needs. Those on a low income, the unskilled and the elderly comprise the majority of tenants of unfurnished dwellings; while young couples, single persons, transients and students occupy furnished bedsitters – the older, larger families on low or discontinuous incomes settling for 'last refuge' furnished housing of one sort or another. Within the HMOs, 2.5 million people (81 per cent of whom were single) were condemned to live in exceedingly squalid conditions, and in London in the mid-1980s 40 per cent of all private tenants lived in poor or very poor conditions, often without basic amenities, while one-fifth of London's private tenants were pensioners living alone (GLC, 1986). The young try to move out of this sector as soon as they can because of the low standard of accommodation, but particularly in London the private rented sector, however unsatisfactorily, provides a stepping stone to other forms of tenure.

Taking into account the supply and condition of private rented housing, and the types of households dependent on the tenure, this chapter:

- analyses the reasons why the supply of private rented housing decreased during most of the twentieth century;
- reviews the history of rent policy in both the furnished and the unfurnished sectors up to the 1970s;
- discusses private renting under the Conservatives in the 1980s;
- considers proposals for reform in the 1980s and 1990s;
- concludes by examining the recommendations of the Housing Green Paper of 2000 concerning private rented housing.

Reasons for the decrease in the supply of private rented housing

During the nineteenth century almost all working-class housing was privately rented. Landlords needed to raise about two-thirds of the value of their property on mortgage, and if interest rates increased, landlords passed on the cost as much as possible in higher rents in order to maintain profitability. Gauldie (1974) has shown that rents rose steadily in the period 1780–1918 (even when the general price trend was downwards), and that in the nineteenth century the average working-class family paid 16 per cent of its income in rents in contrast to the 8–9 per cent paid by middle-class families. The majority of private landlords were relatively small capitalists content with a secure return on their capital. An 8 per cent gross return was possible, which compared very favourably with, for example, 3.4 per cent on Consols and less than 3 per cent on gilt-edged securities (in much of the 1980s a return in excess of 10 per cent would have been necessary to compare favourably with returns on government stock). Until the extension of limited liability in the late nineteenth century, investment in joint stock companies was unattractive to those with modest means. But with the development of the stock exchange and building societies, the expansion of government and municipal stock, and increased investment opportunities overseas, private rented property became much less attractive as an investment. Increased public intervention further reduced the attraction of housing investment.

The appalling quality of rented housing led the government (from the Public Health Act of 1848 onwards) to increase its control over housing in an attempt to improve standards. These controls resulted in *either* higher rents (to compensate landlords for improvement costs incurred) *or* a decrease in the supply of accommodation if investment became no longer profitable. In the twentieth century the factors which led to the decline of the private rented sector were numerous:

1 *Slum clearance*
   Since the end of the nineteenth century and particularly since the 1930s many hundreds of thousands of houses have been demolished, with demand being diverted from the private to the public rented sector, or to owner-occupation.
2 *Policies aimed at dealing with overcrowding*
   The Housing Acts of 1961, 1964 and 1969 increased controls over multi-occupation, and consequently reduced the number of private tenants. Landlords often diverted supply to the owner-occupied sector as a response to the constrained rent income, and demand was diverted mainly to the public sector.
3 *Housing rehabilitation*
   Particularly in the early 1970s and again in the late 1980s, housing improvement grants were often taken up by landlords for the purpose of 'gentrifying' property. In Inner London, for example, tenants were displaced, the dwelling improved with the aid of a grant, and the rehabilitated building sold for owner-occupation. The combination of the provisions of the Housing Acts of 1969 and 1980 and the housing price booms of 1971–73 and 1986–90 led to a substantial diversion of supply away from the private rented sector.
4 *The unattractiveness of investing in private rented housing*
   It has already been stated that there was an aversion to investing in this sector towards the end of the nineteenth century. In the twentieth century investment in private rented housing was even less attractive. Greve (1971) recorded that 130,000 houses per annum were built in England and Wales in the period 1901–06, and the number fell to 100,000 (1907–10) and 60,000 (1911–14) – almost all the houses being built to rent. Rising standards, the higher cost of construction and rent control adversely affected production after the First World War. The price of new houses increased fourfold between 1914 and 1920, and economic rents would have been beyond the means of most would-be tenants. The Housing and Town Planning Act of 1919 enabled local authorities to partly make good the deficiency of supply

by giving them powers to provide subsidised housing for the needs of the working class.

5 *The cost of repairs and maintenance*
Inflation since the Second World War has increased these costs and driven many landlords out.

6 *The desire to own one's own house*
Home-ownership became a reality for an increasing proportion of the total population in the 1930s as interest rates were low, and land, materials and labour were cheap. Planning legislation did little to impede extensive residential sprawl.

7 *Subsidies and tax allowances*
The benefit of subsidies to council tenants and tax advantages to building societies and mortgagors put local authority tenants and owner-occupiers in a comparatively privileged position compared with the private tenant and landlord. Only since 1973 have private tenants received rent allowances, yet landlords were not able to set a 'depreciation allowance' against taxation as it was assumed that a property lasts for ever. The lack of an allowance encourages landlords to sell for owner-occupation, and this was apparent even when rents were decontrolled, 1957–65.

The above factors were investigated by the House of Commons Environment Committee (House of Commons, 1982), which reported that decline was mainly due to slum clearance and transfer to owner-occupation. Private rented housing fell short on all the general criteria of investment – the level of risk; liquidity; expected return on capital; and management involvement – this last being very burdensome when compared to large-scale investment in offices, shops and factories, or small-scale investment in insurance policies, unit trusts or building societies. Landlords could not expect gross yields of more than 2.75 per cent in London in the early 1980s, or net yields in excess of 1 per cent after management, maintenance and tax. Even if rents could have been set at a level to produce a competitive yield of, say, 10 per cent (way above the registered rent), they would have been unrealistically high; for example, a landlord would have needed a rent of £3,000 per annum (or £250 per month) on a £30,000 flat. (Mean registered rents of unfurnished tenancies were only £61 per month in England and Wales in 1982, and £81 in Greater London.) The tenure also fell short in that it could not offer tenants secure accommodation at rents they could afford – private tenants receiving on average a rent allowance of only £23 (in 1982) in contrast to owner-occupiers getting on average mortgage assistance of £366.

In London large commercial landlords were withdrawing from the market, selling off their freeholds to insurance companies and friendly societies, and they in turn were selling long leaseholds to owner-occupiers. Hamnett and Randolph (1983) reported that whereas there were 50,000 purpose-built flats in 1,300 blocks in central London in 1966 (owned by leading landlords such as Freshwaters, Key Flats and London City and Westcliffe), by 1982 only 25–32 per cent of these were still let under the Rent Acts. Most of the remainder had been sold off at prices ranging from £30,000 to over £200,000. But even if rent regulations were abolished, a market rent of £6,000 per annum (£130 per week) would only attract short-stay visitors or companies, and intensify the pressure on tenants to move out and buy. Hamnett and Randolph argued that only if mortgage tax relief were to be abolished and house prices fell would private rented accommodation be revitalised, although its long-term future would remain in doubt.

Harloe (1979) pointed out that in the United States (where there is negligible competition from the public housing sector) and in France, West Germany, Denmark and the Netherlands (where there are only mild rent controls, coupled with subsidies), private renting was also in decline and confined mainly to older and less attractive housing. Other forms of housing investment are more efficient, more subsidised and more geared to inflation. It is still argued, however, particularly on the political right, that rent control and regulation has been the principal cause of the decline of the private rented sector. But opponents of rent control are often unaware of the reasons for its introduction and subsequent history.

Up to the time of writing, rent control or regulation has been in force continuously since 1915, although to a varying extent. No government abandoned it completely. Why? Electoral considerations might be offered as an explanation – there are more tenants than landlords – but this reason is not altogether credible. There are also more employees than employers, but this does not guarantee continual government support for the interests of labour over capital. Possibly the main reason why governments control rent is that both labour *and* industrial capital benefit if the landlord's desire to maximise investment return is constrained.

Marx in the 1880s –argued that rent payments involve the transfer of already created value (in the form of wages) to the landlord, and because of this transfer, the landlord was essentially parasitic – he did not produce value but intercepted part of the flow of surplus value (in the form of rent) which otherwise would have been retained by the industrial capitalist or gone to the worker (*Capital*, Vol. 3, 1981 edition). Harvey (1974) called this intercepted increment 'class monopoly rent'. In a relatively free market, he claimed, private landlords (particularly in the inner city) are able to exploit tenants who cannot obtain social housing or afford owner-occupation. By keeping some accommodation vacant, over-letting other dwellings and shirking repairs and maintenance, landlords are able to create artificial scarcity, thereby forcing up rents. But with the imposition of rent control, there is a hastening in the decline of private landlordism, representing a victory of industrial capital over landed property (Massey and Catalano, 1978). Landlords, in effect, are 'sacrificed on the altar of capitalist profitability, and the explanation for rent control is to be found . . . in the requirements of capitalist profitability for a cheap and appeased labour force' (Merrett with Gray, 1982). It is this Marxian interpretation, ironically, which probably explains why Conservative governments have been reluctant to abolish rent control or regulation completely.

The unfurnished sector

Rent control and decontrol, 1915–39

Rent control was first introduced by the Increase of Rent and Mortgage Interest (War Restrictions) Act of 1915. Rents were controlled on property with rateable values of less than £35 in London, £30 in Scotland and £26 elsewhere in the United Kingdom (Table 6.1). Mortgage interest rates were also fixed and building societies had restrictions placed on foreclosure. Apart from protecting tenants during the First World War (when many breadwinners were in the armed forces or strategic industries), the Act was a response by the state to the contradiction between industrial capital's needs for cheap and reliable labour, and the desire of private landlords for market rents. Employers not only welcomed the Act but supported the preceding Glasgow rent strike. The Increase of Rent and Mortgage Interest (Restrictions) Act of 1920 extended controls into peacetime, making adjustments for inflation.

During the inter-war period it was recognised that rent control was a disincentive to invest in new private rented housing and repair and maintain older housing. The Onslow Report 1923, Marley Report 1931 and Ridley Report 1938 therefore recommended decontrol. Consequently

Table 6.1 Legislation controlling rents, 1915–38

Rent Act	Main provisions	Rateable value (£)	London	Scotland	Elsewhere
1915	Rents controlled at 1914 levels	Not exceeding	35	30	26
1920	Rent controls continued	Not exceeding	105	90	78
1923	Decontrol by possession; letting freed from control when tenant left				
1933	(a) Decontrol of houses	Not below	45	45	45
	(b) Decontrol by possession	Not below	45	35	35
	(c) Decontrol on registration of possession	Not below	35	20	20
	(d) No decontrol by possession unless decontrolled 1923 to 1933, and registered	Not below	20	20	20
1938	(a) Decontrol of houses	Not below	35	20	20
	(b) No decontrol by possession or self-contained dwellings	Not exceeding	35	20	20

the Rent and Mortgage Interest (Restrictions) Act of 1923, the Rent and Mortgage Restrictions (Amendment) Act of 1933 and the Increase of Rent and Mortgage Interest (Restrictions) Act of 1938 gradually decontrolled 4.5 million dwellings (although a further 4 million remained controlled at the lower end of the market). By the late 1930s, as a result, investment in the development of medium- and high-rent housing became attractive.

With the outbreak of the Second World War, the Rent and Mortgage Interest Restriction Act of 1939 abolished decontrol by vacant possession and extended rent control to over 10 million dwellings with rateable values of less than £100 in London, £90 in Scotland and £75 elsewhere (Table 6.2). Until 1957 the rents of these properties were frozen at their 1939 level but the general price level had increased by 97 per cent by 1951. The principal economic effects of government intervention in the private rented sector were clear:

1 Rent restriction at the time it was introduced was at the market level, but over a period and with inflation, hypothetical market rents rose further and further above controlled rents and created scarcity ($q_1 - q$ in Figure 6.1).
2 The scarcity of private rented accommodation diverted demand to other sectors of the housing market. Unsatisfied low-income households increased demand for local authority housing or (until the Rent Act of 1974) for uncontrolled furnished accommodation, and higher-income households increased demand for owner-occupied

Table 6.2 *Legislation controlling and regulating rents, 1939–65*

Rent Act	Main provisions	Rateable value (£)			
			London	Scotland	Elsewhere
1939	Rents controlled	Not exceeding	100	90	75
1957	Rent decontrolled	Not below	40	40	30
	Owner-occupied houses partly let				
	New unfurnished dwellings				
	Remaining tenancies had rents fixed				
	at twice their 1939 rateable value				
1965	Rent regulation	Not exceeding	400	200	200
	Rent control continued	Not exceeding	110	80	80

housing. One wonders what proportion of the 66 per cent of households which are now owner-occupiers would have preferred to have been private tenants if satisfactory rented accommodation had been available. Scarcity also rendered many households homeless, especially when inflation widened the scarcity gap, impeded the increase in the supply of local authority housing and put owner-occupation further out of reach of low-income families.

3 During inflation there was a transfer of income from landlords to tenants, as the former's money rent was fixed and fell in real terms, and the latter's money wage or salary normally rose and usually also increased in real terms. Because of rising costs, the landlord's ability and incentive to repair and maintain his property was reduced. Housing consequently deteriorated in quality and whole areas of rented accommodation, overwhelmingly concentrated in the inner cities, degenerated into slums. But as a house was considered to last for ever, landlords could not claim a depreciation allowance to set against taxation, and even payments into a sinking fund to replace the dwelling were not tax deductible.

The process of deterioration is explained in microeconomic terms by Frankena (1975) and Moorhouse (1972). Rented housing, they argue, consists of a combination of services. It is not just accommodation, but includes such items as repairs and maintenance, decoration, and possibly cleaning, lighting and heating – all being supplied at a price which in total constitutes rent. When rents are controlled below their market level, landlords' profits will be reduced or eliminated if they continue to provide services in full. They will consequently reduce the supply of services in an attempt to maintain profitability. Figure 6.2 shows that if a controlled rent (*cr*) is set below the market rent (*r*), then the price per unit of

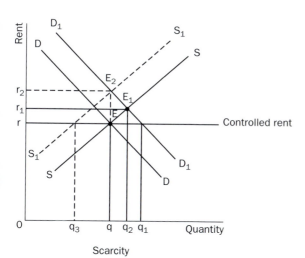

Figure 6.1 *Effect of rent control on the supply of private rented dwellings*

Note: E = initial market equilibrium

E_1 = hypothetical market equilibrium

E_2 = eventual hypothetical market equilibrium after supply decreased from SS to S_1S_1

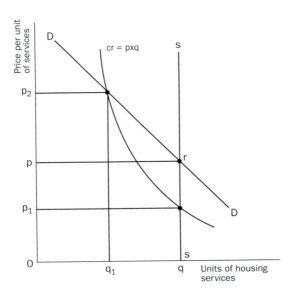

Figure 6.2 Rent control and the decrease of housing services

services falls from p to p_1. The landlord might therefore respond by reducing the provision of services and raising their price. But since this landlord's total revenue must not exceed the controlled rent, the price–quantity combination stays the same – a rectangular hyperbole being traced by the cr curve. If the provision of services is reduced to q_1, the price per unit of services will be the same under rent control as under free market conditions, and any further reduction in services would be unlikely since demand and supply would be in equilibrium. However, it might not be possible for the landlord to cut back on services quite to this equilibrium level since standards might fall to below those permitted by environmental health law.

4 Because of the above factors, investment in the building of private housing to rent has been negligible since 1939, but it was difficult to quantify to what extent rent restriction was responsible in view of the high returns on investment elsewhere – for example, office and shopping development competitively attracting long-term capital.

5 There was an inducement for landlords with vacant possession to sell their properties for owner-occupation. This reduced the supply of rented accommodation (from SS to S_1S_1 in Figure 6.1) and increased the degree of scarcity (to q_1-q_3). Landlords also converted their properties to uncontrolled furnished accommodation.

6 Rent control led to an under-use of the housing stock. Many small households clung to large dwellings, and many high-income tenants benefited from very low rents. Conversely, some large families, often with low incomes, had to settle for small furnished dwellings at high uncontrolled rents. There was thus little relationship between household size, income, housing space, amenities and standards. Sometimes a single room in an unfurnished tenancy was let furnished at a rent higher than the controlled rent for the whole.

7 Rent control produced a number of nefarious results such as key

money, premiums, licences and 'furniture and fittings' payments – all ways of increasing the landlord's revenue and control over tenants without infringing the letter of the law.

8 Rent control may also have impeded the mobility of labour. Even if the householder was unemployed, secure and low-rent accommodation was often preferable to employment opportunities but problematic housing elsewhere.

Rent decontrol, 1957–65

Landlords obviously wanted the 1939 Act to be repealed, and within the Conservative government there was a desire to return to free market rents. The Housing (Repairs and Rent) Act of 1954 was a step in this direction. It allowed controlled rents to rise following repairs being undertaken by the landlord, but it was the Rent Act of 1957 which brought about a significant return to (almost) free market conditions. The Act aimed to decontrol some 5 million dwellings, notably those with rateable values of more than £40 in London and Scotland and £30 elsewhere (Table 6.2), although 3 million were to remain controlled at rents twice their 1939 rateable values.

By 1965, 2.5 million dwellings had been decontrolled, but far from resurrecting the sector, decontrol led to an even greater loss of accommodation than in the previous period of control – 290,000 dwellings per annum (1957–60) compared to 180,000 per annum (1951–56). The landlord's desire to sell off for owner-occupation could be more easily realised after the 1957 Act than before. Tenants of decontrolled dwellings were often faced with either higher rents or notices to quit and harassment if alternative accommodation could not be found, and tenants of controlled accommodation faced even greater pressures to vacate their homes. The plight of the private tenant was eventually acknowledged by the government, and it appointed the Milner Holland Committee in 1963 to investigate particularly the private rented sector in London. Its report paved the way for the repeal of the 1957 Act.

Rent regulation, 1965–70

The Rent Act of 1965 (consolidated by the Rent Act of 1968) was one of the most important pieces of legislation introduced by the 1964–70

Labour government. Rent regulation was to apply to unfurnished dwellings where the rateable value was less than £400 in Greater London and £200 elsewhere (Table 6.2). Those properties which had not been decontrolled by the 1957 Act remained controlled. Regulated rents were 'frozen' at the amount payable in 1965, and for new tenancies, rents were to be equal to the amount payable under the previous regulated tenancy.

Machinery was set up to fix and review rents for regulated tenants. Rents were to be assessed and registered by a rent officer after an application by a tenant, a landlord or both. The rent officer was to objectively assess a 'fair rent', although there was no fixed formula available to enable them to determine what was 'fair'. Officers were to have regard 'to all circumstances (other than personal circumstances) and in particular to the age, character and locality of the dwelling house and its state of repair'. Scarcity value had to be disregarded, therefore the 'fair rent' was to equal the hypothetical market rent which would result if supply and demand were in equilibrium in the area concerned.

During an inflationary period, rent officers had an enormous responsibility to ensure that their assessment of fair rents was as realistic as possible, and this was no easy task as rent levels had to be fixed for three years. There was almost inevitably a lack of consistency, and it was a doubtful advantage that many officers lacked professional qualifications and had to rely on their impartial but non-expert judgement. The Rent (Control of Increases) Act of 1969 ensured that there would be an element of control over the rent officer's power to increase rents. Only one-third of an increase in the fair rent would be payable in each of the three years following an assessment.

The 1965 Act was intended to benefit both tenants and landlords, and to enable the market to function without producing the adverse effects of the 1957 Act. Donnison (1967) explained that

> Unlike rent control, which was designed to freeze a market, thus eventually depriving its prices of any systematic or constructive meaning, rent regulation was designed to recreate a market in which the overall pattern of prices responds to changes in supply and demand, while the local impact of severe and abnormal scarcities is kept within bounds. . . . The first task of those responsible for regulating rents is to bring down some of the highest to a level that is rationally related to those that are freely determined in the open market.

Within the major cities, however, fair rents were being assessed at well below hypothetical market rents. Rent regulation failed to take supply and demand into account. Because of the resulting low returns on investment in unfurnished rented housing, landlords were deterred from continuing to supply accommodation in this sector. On obtaining vacant possession, or by means of harassment and 'winkling' (offering deceptive cash sums to tenants to quit), landlords sold their properties for owner-occupation or to larger landlords, or converted them into uncontrolled furnished dwellings. Outside of the luxury and controlled sectors, private rented housing was becoming almost extinct. The 1965 Act, like the 1939 and 1957 Acts, failed to safeguard this sector for the working classes. Scarcity had resulted from the same demand and supply interrelationship which had characterised controlled rents (Figure 6.1), and unfurnished properties were being diverted to other sectors no less rapidly than during the period 1957–65.

Rent policy, 1970–74

In order to review and report on the working of the 1965 Act and to consider any possible amendments to the legislation, the Francis Committee was constituted in 1969 and its report was published in 1971. The report confirmed the belief that the supply of private rented unfurnished accommodation was drying up, presumably because other forms of investment were more attractive. To help slow down this decline or prevent further deterioration in the quality of stock, the report recommended that the 1.3 million controlled tenancies in England and Wales should become regulated. The Conservative government's White Paper *Fair Deal for Housing* (DoE, 1971) incorporated this recommendation and the Housing Finance Act of 1972 subsequently decontrolled most controlled tenancies bringing them into the fair rent system (Table 6.3).

The second main provision of the 1972 Act was that tenants of private unfurnished accommodation could apply to local authorities for means-tested rent allowances. Since means-testing would be on a unprecedented scale, this provision provoked much criticism. It would have seemed logical for rents to have been assessed at a level at which the majority of tenants could afford to pay without an allowance. But rents were likely to be assessed at a level so high that most tenants were eligible for an allowance. The Shadow Environment Minister, Mr Anthony Crosland

Table 6.3 *Legislation regulating rents, 1969–74*

Legislation	Main provisions	Rateable value (£)		
			London	Elsewhere
Housing Act 1969	Controlled dwellings rehabilitated up to a 12-point standard to be decontrolled and regulated	Not exceeding	400	200
Housing Finance Act 1972	Most controlled tenancies to be decontrolled and regulated	Not exceeding	40	30
1973	Rent regulation extended to higher rateable value properties	Not exceeding	1,500	750
Rent Act 1974	Rent regulation extended to furnished tenancies	Not exceeding	1,500	750

(1971a), referring to the public sector where a comparable system was to operate, forecast that

> An army of bureaucrats will be employed to pay back part of the excessive rent in rebates. There is a risk of a low take-up of rebates with consequent family hardship. And the creation of yet another means-tested benefit aggravates the increasingly serious problem of working class incentives.

The same could have been said about private sector rent allowances. But accepting that these allowances were to be introduced, Mr Crosland argued that their cost

> should fall on the State and not, as the government propose, in large part on . . . local ratepayers. The relief of poverty is a national, not a local, responsibility; and its cost should be borne by central government and the national taxpayer.

The 1972 Act did nothing to prevent house prices from soaring; indeed, by making renting less attractive to tenants it boosted the demand for home-ownership. Landlords were increasingly eager to sell their properties (even their high-rent luxury flats) for owner-occupation. Tenants were consequently being displaced by winkling and harassment. Sums of up to £4,000 were being offered by landlords to tempt tenants to leave, although the average inducement was far less. Lloyd (1972) reported that offers of £500 have become common place to people to leave controlled tenancies for which they have been paying low rents. Lesser sums are offered to higher paying occupants to vacate valuable flats. Such inducements are frequently meaningless to pensioners, widows and the lowly paid who are unable to raise a mortgage or pay market rents elsewhere.

With the severe housing shortage, landlords needed possession to be able to realise substantial capital gains. Short leases were therefore not renewed and flats were left empty for up to 18 months to allow the property to deteriorate so that remaining tenants would leave. Alternatively, high-rent luxury accommodation (much of which had been recently rehabilitated) was being sold leasehold to former tenants, the freehold often being acquired by property speculators. If leaseholders were unacceptable to the new freeholders, they were winkled out.

The 1972 Act was thus clearly not preserving the stock of private rented housing, nor was it having a favourable effect on the quality of rented housing. Whereas the Housing Act of 1969 permitted landlords to decontrol a property if it was rehabilitated up to a 12-point standard (and about 80,000 tenancies were decontrolled in this way), the 1972 Act made decontrol virtually automatic, and not dependent upon the condition of the dwelling, except in the case of very poor housing.

Rent policy, 1974–79

On the return of the Labour government, rents were frozen in March 1974, and imprisonment became a penalty for illegal eviction. The Housing, Rent and Subsidies Act of 1975 replaced the 1972 Act and introduced new measures concerning fair rents. Rents were to be raised in three stages spread over two years, but landlords could apply for a further increase in the third year. In 1973 the upper limit to Rent Act protection was raised to £1,500 in Greater London and £750 elsewhere in England and Wales, while the decline in the number of unfurnished tenancies (due mainly to the diversion of supply to the owner-occupied sector) continued apace.

The furnished sector

General lack of control before 1974

Until 1974 furnished accommodation had not been subject to the same control or regulation as unfurnished rented housing. Although there was rent control on some furnished lettings, if the rateable value was under £400 in London and £200 elsewhere, it was much looser than in the unfurnished sector.

With the introduction of fair rents in the unfurnished sector by the Rent Act of 1965, landlords of unfurnished dwellings were induced to convert to furnished tenure. The Housing Act of 1969 gave a further encouragement to let furnished (rather than unfurnished) after a property had been rehabilitated, with the assistance of an improvement grant matched by at least an equivalent amount of expenditure by the landlord. There was also a tax advantage in letting furnished, as the provision of domestic services such as cleaning and laundering, and breakfast (unlikely to be provided by the landlord in unfurnished accommodation) permitted rent to be taxed as earned rather than unearned income. Also, depreciation allowances in furniture and fittings could be negotiated with the Inland Revenue, or a capital allowance could be made in respect of total cost to be offset against income in the first year of the tenancy.

By the early 1970s the supply of furnished accommodation was increasing at a rapid rate, in contrast to the diminishing supply of unfurnished tenancies. There had been 530,000 households in furnished lettings in England and Wales in 1966, but by 1971 the number had increased to 760,000. Some examples of this increase at a local level are shown in Table 6.4. The swing away from unfurnished accommodation was notable. In Wandsworth, for example, there was a decrease of 9,000 unfurnished tenancies, 1966–71, but a gain of 4,000 furnished tenancies (demolition and sale for owner-occupation resulting in a net loss of rented accommodation), and in Bristol there was a decrease of 1,000 unfurnished tenancies and a gain of 5,000 furnished tenancies (a net gain due to conversion into smaller units).

But because of the scarcity of housing in other sectors, the demand for furnished accommodation also increased. In Inner London in particular, where the proportion of furnished households was generally higher than elsewhere (16 per cent in contrast to 3.7 per cent in England and Wales overall), this gave cause for concern.

Table 6.4 *Households in furnished dwellings, selected boroughs, 1966–71*

Borough	Furnished households as % total households		% increase
	1966	*1971*	*1966–71*
Bristol	5.9	10.6	79.7
Wandsworth[a]	10.0	15.5	55.0
Coventry	2.9	3.9	34.5
Leeds	5.6	7.5	33.9
Hammersmith[a]	17.1	21.7	26.9
Brent[a]	12.7	16.1	26.8
Lambeth[a]	12.1	15.2	25.6
Islington[a]	16.1	19.2	19.3
Camden[a]	22.4	26.1	16.5
Kensington & Chelsea[a]	35.0	38.0	8.6
Westminster[a]	24.5	26.1	4.9

Source: Census, 1966, 1971

Note: [a] London borough

The Francis Committee paid considerable attention to the furnished rented sector. It warned the government that furnished tenancies should not be brought into the fair rent system. Since the supply of unfurnished accommodation had diminished under the Rent Act of 1965, so the supply of furnished dwellings would dry up if rent regulation were extended to this sector. But in contrast, the Greve Report, *Homelessness in London* (Greve, 1971), attributed the increase in homelessness to the increase in furnished accommodation, with its insecurity of tenure. The furnished sector had seen continued eviction of tenants with minimal delay while court orders were obtained and implemented, and tenants were apprehensive about approaching rent tribunals. Greve consequently argued that protection should be extended to this sector, particularly in the case of multi-occupied properties owned by absentee landlords.

The Francis Committee based its misleading argument on the assumption that the Rent Act of 1965 *caused* the decline in the number of unfurnished rented dwellings. Although the number fell by an average of 3 per cent per annum (1965–70), the decline had been at a rate of 3.5 per cent per annum in the years immediately before regulation. Although sub-market rents initially cause shortages, the scarcity gap widens as a result of landlords withdrawing their property from the market. Critics of the Francis Report argued that this would not happen if furnished accommodation were to be subject to rent regulation. They would not wish to divert their properties to the unfurnished sector (since this was already regulated), nor would furnished accommodation be easily sold off, as the committee feared. However, critics did not take into account that the supply of furnished accommodation might decline because landlords faced with rent regulation would prefer to occupy the whole of the property themselves, leave it empty or, given a house price boom, deconvert for owner-occupation.

However, in the early 1970s the furnished sector declined even though free market rents generally prevailed. 'Creeping conversions' were taking place (hotels acquiring adjoining residential property and converting it into extended hotel accommodation) and many furnished tenancies were being converted into 'bed and breakfast' boarding houses providing landlords with higher rents and greater control. Furnished accommodation was getting beyond the means of young families or single persons wishing to live alone – sharing becoming increasingly necessary.

Although furnished tenants qualified for rent allowances under the 1973 Furnished Lettings (Rent Allowances) Act, the claims procedure was complex, and as in the unfurnished sector, the take-up of allowances by those in need was low. Where demand was high, landlords were able to push up rents by an amount equal to the allowance.

Rent policy, 1974–79

As in the unfurnished sector, rents on furnished dwellings were frozen from 8 March 1974 to the end of the year, and landlords were prohibited from evicting tenants in order to be able to charge higher rents to new tenants. The incoming Labour government also extended rent regulation to the furnished sector. The Rent Act of 1974 (consolidated by the Rent Act of 1977) enabled tenants to apply for a fair rent, and security of tenure was granted – both provisions applying to properties with rateable values of up to £1,500 in London and £750 elsewhere. These measures covered 90 per cent of the 764,000 furnished lettings in the United Kingdom.

However, many groups of occupiers did not qualify for full Rent Act security of tenure. They comprised the following:

1 Tenants of resident landlords. But where a dwelling was part of a purpose-built block of flats of at least two storeys, the tenant had full protection if the landlord lived in one of the flats, and a landlord previously not a resident could not reduce his tenants' security by moving in with them after the tenancy had been created.
2 Occupiers of dwellings where a substantial part of the rent was for board and attendance; student and holiday lettings; and licensed premises. In these situations, occupiers (like all lawful occupiers) could be evicted only with a court order.
3 Business tenants under the Landlord and Tenant Act of 1954, and farmworkers under the Rent (Agricultural) Act of 1976.
4 Persons with a 'licence to occupy' – usually where accommodation is not exclusively their own, for example shared flats, or where no rent was paid, for example housing tied to employment.

The 1974 Act was clearly intended to alleviate the growing problems faced by furnished tenants (or would-be tenants), and was a belated response to the recommendations of the Greve Report of 1971. But although it generally benefited existing tenants, the Act may have aggravated the problems faced by people seeking accommodation. The supply of furnished accommodation continued to dry up.

In Inner London, because of the shortage of properties to buy in relation to demand, the price of flats accelerated; for example, a four-bedroom flat in a 'desirable' location might have sold for £65,000 in 1977, but would have fetched £200,000 by 1978, and prices of £50,000–£100,000 were unexceptional for one- or two-bedroom flats in 1978. With values like these, landlords of furnished accommodation would have been tempted to sell even if there had been no 1974 Rent Act. The scarcity of rented accommodation, due partly to properties being sold off, led to soaring rents. Foreign visitors were prepared to pay up to £250 per week for two-bedroom (holiday-let) flats.

It seemed as though the furnished dwellings market had reacted to rent regulation in much the same way as unfurnished lettings had responded to control and regulation, but the 1974 Act, if anything, slowed down the decline in the private rented sector (Table 6.5) by making it more difficult for landlords to gain possession and sell off for owner-occupation. It is significant that the number of dwellings in the private rented sector declined by 10.1 per cent (and the average decline per annum was 3.3 per cent) in 1971–74, whereas the number declined by only 8.8 per cent (and the average decline per annum was only 3.1) in 1975–78.

Table 6.5 *Households renting privately, United Kingdom, 1971–78*

Year	% housing stock	Change from previous year (%)
1971	18.9	–2.7
1972	18.0	–3.5
1973	17.2	–3.4
1974	16.4	–3.5
1975	15.7	–3.3
1976	15.0	–3.1
1977	14.4	–2.6
1978	13.8	–3.3

Source: DETR, *Housing and Construction Statistics*

One considerable advantage of the 1974 Act was that the number of households made homeless due to dispossession by landlords noticeably decreased. Table 6.6 details this reduction in London. But the number of households made homeless through not being able to obtain private rented accommodation probably more than offset this decrease.

The Rent Act of 1974 undoubtedly had an effect upon the market for private rented accommodation, but it is difficult to differentiate the effects of the Act from the effects of soaring property prices which tempted landlords to sell off their housing.

Although there was an annual loss of about 100,000 private rented dwellings in 1975–78, this must be contrasted with the annual loss of 290,000 units in 1957–60 – a period of decontrol and insecurity of tenure.

Table 6.6 *Households accepted as homeless, London, 1974–77*

	Total accepted	Homelessness as a result of repossession by landlords	
		% total	Number
1st half 1974	5,760	28	1,601
1st half 1975	6,280	17	1,090
1st half 1976	6,520	13	820
1st half 1977	5,410	9	510

Source: DETR, *Housing and Construction Statistics*

Private renting under the Conservatives

The Conservative Party in the late 1970s had no intention of removing security of tenure from tenants of absentee landlords, and recognised that rent regulation was necessary. But under the Housing Act of 1980, the Conservative government introduced shorthold tenancies – believing that many properties stood empty because the Rent Acts of 1974 and 1977 had got in the way of landlords and tenants wishing to agree to a lease for a short fixed period. Shortholds are applicable only to new lettings, and (under the 1980 Act) at the end of fixed-term agreements of 1–5 years landlords had the right to regain possession. Shorthold is in effect a form of decontrol. If a landlord decides to evict tenants who do not depart at the end of the agreed term, he must issue an eviction notice which gives the tenants three months' notice of his intention of going to court. The court may then grant an order of possession (normally to be effected within 14 days). During its first year in operation, the 1980 Act required that landlords charge fair rents, and that these were to be registered within 28 days from the start of a tenancy; but after 1981, if landlords and tenants agreed, market rents on new shorthold tenancies were negotiable outside Greater London – regulation remaining in the capital since some 21 per cent of housing was private rented (and 30–40 per cent in some boroughs), compared to 13 per cent nationally in 1981.

The Conservatives thought that supply would also be maintained (or increased) if controlled tenancies became fair rent lettings. The 1980 Act therefore, at a stroke, decontrolled 300,000 dwellings and subjected their tenants to fair rents. To ensure that the income of landlords kept more in line with the rate of inflation, fair rents were to be re-registered every two years instead of three.

The 1980 Act also introduced 'assured' tenancies whereby approved landlords were permitted to let their new dwellings outside of the Rent Acts. Building societies, banks, other finance houses and construction firms could be licensed by the government to build homes for rent.

Some effects of the Housing Act 1980

Shorthold was adopted very slowly. By March 1981 only 320 shorthold tenures had been created in England (with only 19 in Greater London and 1 in Merseyside) and by August 1981 the number had increased to only

3,500. Landlords were clearly not withholding accommodation because of the Rent Acts. The government's survey of empty housing in 1981 showed that of the 550,000 empty private dwellings, only 31,000 were left vacant because of rent legislation. The House of Commons Environment Committee (House of Commons, 1982), moreover, predicted that shorthold would create only 50,000 new tenancies by 1983/84. Landlords still preferred 'phoney' holiday lets (especially in London) to shorthold. Generally, large corporate landlords and many smaller non-residential landlords found that it was more profitable to sell off their properties and invest elsewhere, and the 1980 Act only hastened this process by creating more vacant possession – as happened after the 1957 Act. Although the House of Commons Environment Committee (House of Commons, 1982) pointed out that landlords could expect a return of only 3.5 per cent (gross) with fair rents, if landlords were free to aim for a 'competitive' return of at least 6 per cent on vacant possession then at least 40 per cent of furnished tenants and 50 per cent of unfurnished tenants could not afford these rents without a substantial transfer of income (in the form of subsidy) from the taxpayer to the landlord. Residential landlords, by contrast, were less concerned about rents than security of tenure (even under shorthold) – perhaps fearing the repeal of the 1980 Act – and similarly withdrew accommodation from the market. At the lower end of the market, as the North Islington Housing Rights Project (1982) showed in its study area, many small landlords (especially in HAAs) relied exclusively on rent income; over a half preferred to let to friends or relatives; half were elderly; and ethnic minorities accounted for a third. In the inner cities – marginal owner-occupier areas – rent income was less of a return on investment and more a means of enabling small landlords to buy their own homes. For those tenants living in shorthold property in London, fair rents were often determined unsatisfactorily. Since fair rents operated only from the time they were registered, landlords tended to push them up beforehand towards free market levels so as to benefit ultimately from higher rates of return.

Critics of the 1980 Act argued that shorthold would increase homelessness, undermine security and provoke harassment. Shelter (1979c) predicted that

> Shorthold tenancies [would] destroy the long established principle that a tenant who pays his rent should be secure from arbitrary eviction. More tenants will become homeless, and tenants will be too scared to ask for essential repairs and improvements.

From the start, the Labour Party was opposed to this form of tenure, and Mr Gerald Kaufman (1981), Shadow Environment Minister, confirmed that 'the next Labour government . . . [would] repeal shorthold tenancies and give full Rent Act protection to shorthold tenants'. The withdrawal of the private landlord was all too common in the letting market. The *General Household Survey, 1980* (ONS, 1980) showed that unfurnished private rented accommodation had decreased from 12 to 6 per cent of the total housing stock, 1971–80, while homelessness and council waiting lists had rapidly increased (in London alone, 197,000 households were on waiting lists and 447,000 needed rehousing in 1981). It is often argued that rent control fails to protect low-income households since it encourages the withdrawal of accommodation and leads to homelessness, with adverse effects on family life, child education, labour mobility and earning potential (because of uncertainty and limited means). But the opposite, rent decontrol in the form of shorthold and Rent Act loopholes in the 1980s, may have the same effect – perhaps on an even greater scale. This was demonstrated previously when the Conservatives' Rent Act of 1957 brought about the biggest reduction in private rented accommodation since the Second World War.

The other new form of letting, assured tenancy, also proved unsuccessful. Although 121 financial and construction companies were on the approved list by 1983 (including the Abbey National Building Society, the Prudential and Barratt), they planned to build only 600 dwellings by 1985 – comparatively high rates of interest and the risk of future regulation making this form of investment unattractive.

As in the late 1950s and early 1960s, under decontrol in the 1980s, harassment of tenants was rife – reaching 'epidemic proportions' according to Mr Allan Roberts, MP, introducing his Private Tenants' Rights Bill in 1983 to protect more effectively the rights of private tenants. The Minister of Housing, Mr Ian Gow, announced (in November 1983) that an inquiry was to be set up into the problems of management of privately owned mansion blocks of flats, where there was much concern about abuse, particularly in London, and Westminster City Council seemed disturbed by the incidence of 'harassment by neglect' on the part of large private landlords. The Organisation of Private Tenants (OPT) claimed that harassment had become 'widespread among landlords who want[ed] to make big profits by selling properties with vacant possession'. The OPT warned that 'a new breed of aggressive property speculator' had been buying up tenanted properties with the 'sole intention of removing the tenants in as short a time as possible'

(*Evening Standard*, 1985). Likewise, the Greater London Council (GLC, 1986) reported that over 14,000 households (4 per cent of the capital's tenants) had recently suffered serious harassment.

Through loosening up of the private rented market, a sizeable proportion of low- and average-income tenants experienced a drop in their living standards. Rent escalation was a consequence of the abolition of controlled tenancies by the 1980 Act. Rents increased from as low as £1 per week on average (in 1980) to £6 per week (or over £12 per week in parts of London). Although the government obviously preferred this to municipalisation (notwithstanding the large consequential increase in rent allowances and supplementary benefits – the former rising by £72 million in 1982/83), some form of social ownership would have been beneficial to tenants, since with higher rents there would have been at least a greater likelihood of repairs and maintenance being undertaken. Higher rents became an increasing problem after the Conservatives changed the system of income support. Based on the DoE (1982) consultative paper *Assistance with Housing Costs*, the Social Security and Housing Benefits Acts of 1982 introduced unified housing benefits, administered solely by local authorities and replacing rent allowances (or rebates) and supplementary benefits in respect to rented housing. It was estimated that whereas 1,140,000 households would gain from the scheme, 1,990,000 would lose, and that 124,000 below the 'poverty line' would need topping up with income supplements. Both SHAC (1981) and CHAR (the Campaign for the Single Homeless, 1981) warned that many households in London would lose £3–£4 per week under the Act, and *Labour's Programme 1982* (Labour Party, 1982a) noted that the scheme did not ensure equal treatment for those in and out of work, and still required a local rates contribution towards what was a national income support programme. By mid-1983 one-fifth of all local authorities were unable to cope with the housing benefit scheme, and many thousands of private sector tenants were piling up debts because of consequential rent and rate arrears. Rising rents were compounding these difficulties.

In November 1983 the Department of Health and Social Security (DHSS) announced plans to cut back on housing benefits to save £230 million. Up to 5 million people (public as well as private sector tenants) would have in consequence faced higher rent and rate bills. The Institute of Fiscal Studies calculated that more low-wage families would be brought down to the level of the unemployed, thereby reducing work incentives and increasing the poverty gap. Age Concern, the Child Poverty Action

Box 6.1

Private rented housing in Scotland, Wales and Northern Ireland

Although, historically, the private rented sector was a major provider of housing in Scotland, public policy has reduced its share of the total stock to only 5.7 per cent in 1999. Whereas in 1991 this tenure was spread fairly evenly over rural and urban areas, by 1996 as much as '70 per cent of private rentals were in urban areas' (Earley, 2000: 25). The demand for private rented housing will undoubtedly increase in the future as a consequence of increased labour flexibility, the continuing expansion of higher education leading to later entry into the owner-occupied market, and increased pressures on the social rented sector, but it is far from certain whether there will be a reciprocal increase in supply because of uncertainty about long-term financial returns and the lack of interest from corporate bodies (Earley, 2000).

The private rented sector was also in decline in Wales, decreasing from around 223,000 dwellings in 1961 to 97,000 in 1991. However, because of an increase in demand in the 1990s, the sector began to expand again, reaching 109,000 (or 8.6 per cent of the housing stock) by 1999 (Williams, 2000). Gross yields, which had previously been falling, were now reasonably stable. In the fourth quarter of 1999 yields averaged 9.6 per cent (6.2 per cent net) and thus helped to ensure the success of the government's buy-to-let scheme. For landlords who speculated in house prices, the impact of reduced yields was offset by capital appreciation resulting from the housing boom of the late 1990s (Williams, 2000).

Because of reduced job security and the availability of housing benefit, the private rented sector in Northern Ireland increased in size during the 1990s, but represented only 4.4 per cent of the housing stock in 1999. Growth was attributable both to an increased demand among students for unregulated rental housing, notably around the university towns, and among better-off households at the upper end of the private rented sector, for example in Belfast city centre (Stevens, 2000). As elsewhere across the United Kingdom, the buy-to-let market has attracted investors into this sector.

Group, the Low Pay Unit and SHAC all warned the government of the consequences of cuts in housing benefits. In January 1984 the government partly recanted and reduced the proposed cut to £190 million, the Secretary of State claiming that 'only' 2 million people out of a total of 6.5 million receiving benefits would be worse off. But critics argued that the average pensioner and low-income family would still suffer because of the cuts. Mr Michael Meacher, Labour's social security spokesman, stated that the proposals were still mean and vindictive and that he would be calling for a full-scale review because of administrative confusion and the breakdown of the system throughout the country.

Reform: the 1980s

There were several major sets of recommendations for reforming the private rented sector in the 1980s. First, Lansley (1982) proposed that private tenants should pay the same rent as council tenants for equivalent accommodation. They had been paying more for inferior accommodation and on worse tenancy terms, therefore statutory rent freezes or reductions would necessitate either the payment of subsidies to landlords or the municipalisation of the stock by local authorities.

Second, the Labour Party in *Labour's Programme 1982* (Labour Party, 1982a) pledged that if it soon came to power it would abolish shorthold; freeze rents for one year after coming to office; reform the means-tested 'safety net' of the Housing Benefits Act and increase benefits in real terms; close loopholes in the Rent Acts (for example, holiday let and bed and breakfast evasions would be outlawed); strengthen tenant repair rights; and tighten the law against harassment. More fundamentally, the party was poised to adopt a policy of municipalisation as the only realistic long-term solution. Echoing the sentiment of Harold Wilson, who argued that 'rented housing is not a proper field for private profit' (Wilson, 1964), Mr Gerald Kaufman (1982) declared that with the exception of owner-occupiers letting their properties when they were away for a limited period 'it would be best for landlords to be relieved of a burden they cannot shoulder properly, which gives insufficient rights to tenants, and which can permit appalling exploitation'. However, Labour was not returned to office at the General Elections of 1983, 1987 and 1992 and therefore it was more likely that proposed solutions mooted by inter-party or independent bodies (rather than the political left) would pave the way to reform.

The third recommendation was of this sort. The House of Commons Environment Committee (House of Commons, 1982) suggested that since private tenants got poor value for money compared with council tenants and owner-occupiers, and landlords (currently) did not get an adequate return on their capital invested, there should be tax concessions to landlords on their rental incomes; tax relief for tenants on their rent payments; the establishment of comparable rents across the whole rented sector; and the introduction of subsidies to landlords (similar to specific methods of subsidy overseas). Government would need to decide whether it would assist this sector *either* by increasing public expenditure by shifting funds from elsewhere, *or* by enabling it to realise a rate of return competitive with other forms of investment. Although assured tenancies could be expanded in this respect, rents would have to be astronomically high if they were to attract investors. Labour members of the committee regarded municipalisation as the main solution, but also favoured tenant co-operatives (although these might be practicable only where there were large landlords and unfragmented estates).

Fourth, a report by the NFHA (1985) recommended the introduction of capital value rents in the private rented sector. Rents (linked to inflation) would need to provide a return of 4 per cent on capital value if they were to match current yields in the finance market, and in addition landlords should expect to receive an element for management and maintenance, and an element for service charges – where appropriate. For existing tenants, capital value rents would be phased in, but for new tenants these higher rents would be imposed immediately.

Fifth, the RICS (1985) likewise suggested the phasing in of market rents on existing lettings, the instant introduction of market rents for new developments and the modification of the law on security of tenure in order to encourage owners to let their surplus accommodation.

Partly in response to the majority view of the House of Commons Environment Committee (outlined above), the government in 1983 embarked on a thoroughgoing policy review of the Rent Acts. Mr Ian Gow argued that more private rented housing was required to increase labour mobility; the Rent Acts in his view had reduced the supply of accommodation for those people they were intended to help. Before the review was completed, Mr Patrick Jenkin, Secretary of State for the Environment, declared that the government wanted to free approved landlords who converted or improved dwellings from the current legal ceilings on rents. The government hoped that this would slow down the decline of the sector, and help workers move to new jobs.

All the main political parties agree that the staggering depletion of accommodation in the private rented sector is a cause for serious concern. But over the years, governments have been unable to arrest this decline. No government has been able to achieve simultaneously an adequate return for the landlord and satisfactory rents and protection for the tenant. Few people would mourn the demise of private landlords. Their reputation has been generally poor, but the housing which they own must not be allowed to vanish from the rented sector. It is needed by young single persons, young couples without children, transients, students, the elderly (many of whom may have been lifelong tenants), and those at the bottom of the socio-economic ladder suffering from one or many forms of deprivation. It has been an 'open door' tenure and must remain so, in contrast to 'traditional' council housing (which has been allocated mainly to families with children) or owner-occupied housing (restricted to those with adequate incomes and savings and to specific types of property). The Housing and Planning Act of 1986, in consequence, extended assured tenure to refurbished dwellings. Whereas only 600 assured units had been built, January 1980 – March 1987, in the following six months a further 2,400 (mainly refurbished) assured tenancies were created, largely as a result of the 1986 Act.

Some Conservatives, however, were increasingly pressing for general decontrol – regardless of the impact of a free market in rented housing on the social security responsibilities of the DHSS. Mr Francis Maude MP (1985), for example, argued that 'A lot of people feel that the problems of homelessness could be very largely met by liberalising the Rent Acts by bringing into occupation the large quantities of empty housing that exists in this country' and Mr Ivor Stanbrook MP (1985) questionably claimed that 'The Rent Acts by causing shortages have been responsible for more misery and unhappiness in this country than any other social question'. In its third term in office, the Thatcher government subsequently planned to bring back into use 500,000 properties left empty allegedly because of rent regulation, but it was not prepared to risk complete decontrol.

Based largely on the White Paper *Housing: The Government's Proposals* (DoE, 1987), the Housing Act of 1988 and its Scottish equivalent, the Housing (Scotland) Act 1988, aimed to revive the private rented sector by reducing the minimum period of shorthold (renamed assured shorthold) to only six months and extending assured tenancies to the remainder of new lettings. Assured shorthold lettings were to be at market rents which were to take account of the limited period of contractual security of the tenant – and the tenant could apply during the initial period of the tenancy to a rent assessment committee for the rent to be determined. Assured tenancies, on the other hand, although being relatively secure, were to be at rents freely negotiated between landlord and tenant and therefore were at market levels. Existing regulated tenants would (ostensibly) continue to be protected by the Rents Acts, but, as Shelter (1990a) pointed out, this provision encouraged landlords to harass tenants into either agreeing to higher rents and less security by switching to shorthold or assured tenancy agreements covered by the 1988 Acts, or risking eviction. Clearly, the laws on harassment and illegal eviction have to be tightened, and tenants forced out illegally should be able to claim greater compensation than hitherto.

An underlying aim of the 1988 Acts was to increase the supply of housing to facilitate both geographical and tenure mobility. It was thought that many vacant jobs in London and the rest of the South East could be filled if unemployed labour in the North was able to obtain rented accommodation in areas of job opportunity. But this will not occur as long as the financial advantages of owner-occupation (enhanced by massive tax allowances and exemptions) continue to out-perform the benefits of rented housing – in terms of both investment and occupancy criteria. Landlords, moreover, will have little desire to retain their investments in a sector unpredictably shored up by welfare payments.

Critics of the 1988 Acts argue that since rents would rise dramatically (particularly in areas of housing shortage), so too would the need for housing benefits. In 1987, as many as 60 per cent of private tenants received housing benefit but, with rent increases inevitable, the proportion eligible for benefit would escalate – producing thereby a poverty trap. It would be more advantageous for many tenants to be out of work and in receipt of housing benefits than to have a job and be ineligible for assistance (McKechnie, 1987). Tenants are also adversely affected by housing benefits being paid only in relation to levels of rent assessed by rent officers on the basis of rules laid down by the Secretary of State for the Environment. Normal levels of benefit are, for example, not available for accommodation considered to be too large or too expensive for the needs of the applicant. Above these levels, landlords are free to charge any rent the market can bear, with higher-income tenants inevitably squeezing out those in receipt of housing benefits. Many low-paid employees (often in essential services), nurses for example, have been forced to find alternative and better-paid jobs. Nurses (already leaving the profession at 30,000 per annum by the mid-1980s) were faced in parts of London with weekly rents rising from £25 to an unaffordable £150, notably higher than their starting pay.

Under shorthold arrangements, tenants would have less protection than before while assured tenants would have to pay exorbitant rents for dilapidated and unsafe housing. Even regulated tenancies would be under threat since the right of succession at fair rents would be terminated – inflicting upon those 'inheriting' the tenure the option of market rents or eviction, while local authorities would lose their powers to apply for a fair rent to be registered – to the detriment of many existing tenants.

The impact of the 1988 Acts on the supply of private rented housing was not easily predicted. On the one hand it was thought that the Acts would not have a significant effect on supply since most new lettings in the mid-1980s (90 per cent, according to the DoE) were already at market rather than regulated rents. On the other hand, harassment was rife since landlords still preferred to sell (with vacant possession) rather than let (Kemp, 1987). In the period 1979–87, 550,000 dwellings were lost to the sector (ironically, more than the government hoped would be attracted back into rented use by its 1988 measures). It was clear that since the 1920s, rent decontrol had consistently reduced the size of the private rented stock and it could have been strongly argued that the 1988 Acts would reduce its size still further until it reached a core of about 4–5 per cent of the total stock of housing. At the time of writing, however, the sector was buoyant and showed signs of growth rather than decline.

In addition to rent deregulation, the government used fiscal means in an attempt to increase the supply of private rented accommodation. The Budget of March 1988 extended the Business Expansion Scheme (BES) to individuals who invested up to £40,000 per annum in approved unquoted property companies building or acquiring housing for rent under assured tenancy arrangements. (Eligible properties had to have a maximum value of £125,000 in London and £85,000 elsewhere.) Under BES provisions, individuals qualified for tax relief on their investment, and companies (in 1988) were able to raise up to £5 million each under the scheme. However, while the BES might to some extent have revitalised the private rented sector this might have occurred only over the short term and have benefited investors and landlords more than tenants since rents would have to be high to provide a competitive return. Since tax relief ceased in 1993, many individuals have subsequently contemplated withdrawing their investments while landlords have tended to seek capital gains – both necessitating the sell-off of properties for owner-occupation. The BES, however, was clearly intended to kick-start a revival in private renting, but only 16,000 new homes were supplied over four years. The scheme was clearly an expensive use of public money, costing two-thirds as much per home as for a housing association dwelling, yet producing only short-let accommodation at much higher rents (Best, 1992).

An alternative form of fiscal assistance (which could have brought back into use many of the 500,000 empty properties allegedly kept off the market by rent regulation) would have involved a reduction in tax on rental incomes, larger allowances for repairs and maintenance, and the introduction of depreciation allowances on rented property. Unlike the provisions of the 1980, 1986 and 1988 Acts, these measures – proposed by the House of Commons Environment Committee (House of Commons, 1982) – not only would have revived the private rented sector but could have stabilised rents and reduced dependency on housing benefits. The 'subsidisation' of supply rather than demand would not necessarily have imposed any greater burden on the Exchequer, and could have been compatible with the declared policy aim of curbing inflation. Failure to adopt this approach (and the short life of the BES inducements) might have suggested that the government was not serious, at least in the 1980s, in its intention of reviving the private rented sector – the promotion of owner-occupation being its overriding aim.

Perhaps the only encouraging development in the private rented sector in the late 1980s was the increased involvement of building societies as developers and landlords. Building societies might have become aware of the substantial increase in the number of inherited owner-occupied properties – worth about £9 billion by 1990 (Pitcher, 1987). The result of this inheritance, or so it could have been argued, was that there would be a considerable reduction in the demand for mortgages. The extension of owner-occupation, moreover, could not go on for ever (possibly stabilising at about 70 per cent of the total housing stock by the year 2000). Building societies would therefore need alternative investment opportunities, and one such area could be rented housing. In 1987 Nationwide-Anglia thus set up a company, 'Quality Street', to utilise £600 million to develop (and subsequently manage) 40,000 homes by 1992 – mainly in London, the North and Scotland. The Halifax Building Society was also moving into the rented market with a portfolio of 4,000 units in its programme of acquisitions (Levinson, 1988).

Although to be welcomed, such institutional developments probably had little impact on the private rented sector as a whole and thus failed to arrest its further decline. However, by 1993 there were still over 2.2 million private rented dwellings in occupation in Great Britain as a whole, and an additional 764,000 empty units in the private sector in England alone (JRF, 1994). To satisfy needs, new ways had to be found to maintain or increase the supply of private rented sector housing.

Table 6.7 *Contraction and expansion of the private rented sector, Great Britain, 1982–98*

Year	Number of dwellings		Change from previous year
	(000)	*(%)*	*(%)*
1982	2,318	10.9	
1983	2,304	10.7	–0.60
1984	2,209	10.6	–4.12
1985	2,258	10.3	+2.22
1986	2,205	10.0	–2.35
1987	2,139	9.6	–2.99
1988	2,077	9.2	–2.90
1989	2,069	9.1	–0.39
1990	2,123	9.3	+2.61
1991	2,219	9.6	+4.52
1992	2,286	9.8	+3.02
1993	2,333	9.9	+2.06
1994	2,395	10.1	+2.65
1995	2,476	10.4	+3.38
1996	2,536	10.6	+2.42
1997	2,566	10.6	+1.18
1998	2,581	10.6	+0.58

Source: DETR, *Housing and Construction Statistics*

Proposals for reform: the 1990s

The decline of the private rented sector decelerated in the late 1980s and early 1990s and the sector subsequently expanded (Table 6.7) mainly owing to a diversion of both demand and supply from the owner-occupied market. Many potential housebuyers in 1989–91 were deterred from acquiring a mortgage at a rate of interest of between 11.50 and 15.40 per cent to purchase a property which was rapidly falling in value, and instead either postponed their purchase or rented private accommodation where outgoings were often less and where capital losses were zero. (The option to rent social housing was increasingly constrained by public expenditure cuts and by the privatisation of much of the local authority stock.) There was a similar lack of confidence in the housing market among would-be sellers who often preferred to let their properties (normally under shorthold agreements) rather than sell at deflated prices. Where house prices dropped most, in London and the rest of the South East, there was a flood of properties to rent as owners found it increasingly difficult to sell – with rents consequently falling. It is unlikely, however, that local increases in the supply of private rented accommodation will survive a sustained increase in prices in the owner-occupied sector – other things being equal.

But both the outcome of independent investigation and government thinking suggested that significant changes should occur in the way in which fiscal policy and landlord and tenant law have an impact on the private rented housing market. The *Inquiry into British Housing* (JRF, 1991) proposed that while landlords at the upper end of the market should be outside of rent regulation and be ineligible for tax privileges, landlords of property below a certain value should expect no more than a 4 per cent net rate of return, as was recommended by the earlier Inquiry (NFHA, 1985). With regard to new lettings, landlords could be offered a choice of either tax exemption or a 100 per cent capital allowance.

It was becoming increasingly clear that more and more private rented accommodation was required. The JRF (1991) pointed out that this was due in large part to the growing problem of homelessness, the increase in the number of repossessions in the owner-occupied sector, and (at a time of high unemployment) the need for greater job mobility. Along the lines of the 1985 *Inquiry*, the Foundation suggested that the supply of affordable private rented housing would be increased to meet needs only if landlords received acceptable returns ensured through the provision of subsidies – in the form of either tax concessions or grants. Tax allowances would obviously not only benefit landlords, but could (in addition to the provisions of the 1988 Act) stimulate investment from pension funds, banks and building societies. If, as the Halifax Building Society proposed, landlords were offered the same tax relief as owner-occupiers and rental income was tax exempt, this would put private rented and owner-occupied housing on the same fiscal footing, and if tax relief was linked to mortgages/loans temporarily up to £60,000, investment in the private rented sector would, arguably, boom (Kaletsky, 1992). Tax concessions, moreover, would have less impact on the PSBR than direct grants, and would not be in direct competition with the funding of social housing.

At the General Election of April 1992, the Conservative government proposed extending, nationwide, a pilot scheme it had introduced the previous year to bring back some of the 590,000 empty private houses into use. Under the HAMA (Housing Associations as Managing Agents) programme, £1.25 million was available (via the Housing Corporation) to help housing associations act as intermediaries between private landlords and tenants – whereby the associations would undertake management, choose tenants, collect rents, deal with day-to-day problems and (in some cases) improve the properties, with costs incurred being deducted from rents. Lettings would be on an assured shorthold basis – normally for a year and underpinned by housing benefits. A further £5.5 million was allocated through local authorities in Wales to support schemes in the Principality which involved housing associations working directly with landlords. The Conservatives also proposed introducing a 'Rent a Room' scheme under which home-owners would be able to let rooms to lodgers without having to pay tax on the rent they received. The scheme, introduced in June 1992, allowed owners to let a room for up to £65 per week (or £3,250 per annum) before becoming eligible for tax – an incentive, or so it was claimed, in areas of greatest need, for example in London. The scheme would initially cost the Exchequer between £2 and £3 million per annum but the sum would increase if more people rented out rooms.

By 1994, the Major government was reported to be considering relying on private lettings (as an alternative to new housebuilding) as a means of satisfying the needs of the homeless and those on council waiting lists (Shelter, 1994), but it was faced with a problem of its own making in the private rented sector. In 1993/94 housing benefits in total had risen to £8.8 billion since in general there was no specific rental ceiling above which benefit would not be paid. Although the rise in private sector rents and hence the increase in benefit costs were an outcome of the Housing Act of 1988, the DSS in 1994 seemed intent on capping benefit in the private sector to prevent it rising out of control and leaving local authorities to decide whether to pay any difference between a tenant's rent and the benefit received. Tenants would therefore either have to spend a disproportionate amount of their weekly income on rent or move from one place to another in search of the cheapest housing. By August 1994, however, it seemed apparent that the reform of housing benefits would be postponed until the University of York (having been commissioned by the DSS) had produced its research report on the containment of housing benefits in a number of other countries, for example New Zealand, Sweden, Germany and the Netherlands.

The Labour Party, at the 1992 election, had suggested more positive ways of reviving the private rented sector and limiting the payment of housing benefits. Landlords, in the view of Labour, should be free either to let at market rents with tenants not being eligible for housing benefits, or to choose to let at regulated rents with tenants being able to claim housing benefits (and, as an incentive to let at this lower level, landlords would be eligible for tax relief or direct subsidies). Clearly, Labour's long opposition to private renting was over for the time being. It was recognised that there was a need to revive the sector to around 10 to 15 per cent of the housing stock and that this could only be brought about by changing the type and distribution of subsidies within the housing finance system (Merrett, 1991).

Conclusion

In line with Labour's policy shift towards private renting, it was clear that after its return to power in 1997 it had no intention either of changing the structure of assured and assured shorthold tenancies (previously introduced by Conservative administrations) or of reintroducing rent controls in the deregulated market. Instead, it aimed to reinvigorate the sector by attempting to 'raise the standards of reputable private landlords, encourage new investment and tackle problems at the bottom end of the sector' (DETR/DSS, 2000). In its Housing Green Paper, Labour thus proposed to:

- assist well-intentioned landlords to enhance their expertise and standards, and strengthen their position in the market through letting schemes, best practice guidance and voluntary accreditation;
- examine whether tax changes could help to make the sector work better and render investment in rented housing more attractive (unlike owner-occupiers, private landlords were subject to capital gains tax but – unlike most other investors – were not eligible for depreciation allowances);
- ameliorate specific problems of poor condition and exploitation by a minority of bad landlords (in clearly identified areas of low demand) by means of discretionary powers conferred on local authorities to license privately rented homes (in addition to licensing houses in multiple occupation – a 1997 manifesto pledge); and
- deter unscrupulous landlords from profiting from housing benefit while neglecting their responsibilities – for example, by placing conditions on the receipt of benefit or limiting payments of benefit direct to landlords (see Chapter 20).

In these ways, the government hoped that by retaining the many good and well-intentioned landlords (and by encouraging them to raise their standards further), persuading reputable investors to increase the supply of decent rented homes, and making the worst landlords perform better (or get out of the rental market altogether), it could 'secure a larger, better quality, [and] better-managed private rented sector' (DETR/DSS, 2000).

Only time will tell whether the Green Paper's proposals will be fully implemented. In the meantime, it is possible that the private rented sector will expand and its problems will increase rather than decrease as the number of bad landlords – as well as the number of good landlords – escalates. In the late 1990s, and into the new century, expansion was facilitated by 100,000 buy-to-let mortgages and investment of between £6 billion and £8 billion (Insley, 2001). Many landlords were thus able to build up large property portfolios, and with falling interest rates market rents enjoyed high rates of return. During the same period, and particularly in areas where the demand for owner-occupied housing was low, inherited properties were sold – often by auction – at knock-down prices to private landlords, who then let to housing benefit tenants – again ensuring high rates of return on capital (Field, 2001). At the time of writing there were few signs that these practices were diminishing.

Further reading

Bailey, N. (1996) *The Deregulated Private Rented Sector in Four Scottish Cities, 1987–1994*, Scottish Homes Research Report No. 30, Edinburgh: Scottish Homes. An in-depth and comprehensive discussion of the impact of deregulation on private rented housing in urban Scotland.

Barnes, Y. (1996) *Beyond HITs*, London: Savills. Investigates whether – despite potentially attractive rates of return – unresolved practical issues and restrictions and limitations might have limited the success of housing investment trusts.

Black, J. and Stafford, D. (1988) *Housing Finance and Policy*, London: Routledge. Using demand and supply theory, chapter 5 examines some of the perceived consequences of rent control.

Crook, A.D.H. and Hughes, J. (1995) *The Supply of Privately Rented Homes: Today and Tomorrow*, York: Joseph Rowntree Foundation. Provides important insights into ways in which privately rented housing is supplied in the United Kingdom, outlines some of the difficulties of increasing provision in the future, and suggests what policy changes could be made to ensure an expansion of the sector in the future.

Crook, A.D.H. and Kemp, P. (1993) 'Reviving the private rented sector?', in Maclennan, D. and Gibb, K. (eds) *Housing Finance and Subsidies in Britain*, Aldershot: Avebury. Presents a clear and analytical review of ways in which the stock of private rented housing can be increased.

Crook, A.D.H. and Kemp, P. (1999) *Financial Institutions and Private Rented Housing*, York: Joseph Rowntree Foundation/York Publishing Services. A useful report that suggests that in order to attract large-scale private investment into the private rented sector, the government will need to introduce a more attractive medium than housing investment trusts.

Crook, A.D.H., Kemp, P., Anderson, I. and Bowman, S. (1991) *Tax Incentives and the Revival of Private Renting*, York: Cloister Press. An in-depth investigation of the Business Expansion Scheme in its first two years of operation.

Doling, J. and Davies, M. (1984) *The Public Control of Privately Rented Housing*, Aldershot: Gower. A thorough account of the development of public intervention in the private rented sector, produced during a period of deregulation.

Gibb, K. (1990) *The Problem of Private Renting*, CHR Discussion paper No. 20, Glasgow: Centre for Housing Research and Urban Studies, University of Glasgow. Presents a wide-ranging discussion of private renting.

Hamnett, C. and Randolph, B. (1988) *Cities, Housing and Profits*, London: Hutchinson. A well-researched and interesting study of the break-up of flat empires and the decline of private renting in the capital.

Harloe, M. (1985) *Private Rented Housing in the United States and Europe*, London: Croom Helm. A thorough and very readable account of the private rented sector. However, although useful in its time, it has become a little dated in terms of policy.

Kemp, P. (ed.) (1988) *The Private Provision of Rented Housing: Current Trends and Future Prospects*, Aldershot: Gower. Provides a comprehensive examination of the decline of the private rented sector, deregulation and proposals for reform, although inevitably a little dated.

Kemp, P., Crook, A.D.H. and Hughes, J./DoE (1996) *Private Renting at the Crossroads*, London: Coopers Lybrand. An assessment of whether or not housing investment trusts would succeed in channelling capital into housing for rent.

Maclennan, D. (1982) *Housing Economics*, Harlow: Longman. Contains a clear analysis of the economics of rent control.

Oaks, C. and McKee, E. (1997) *City-Centre Apartments for Single People at Affordable Rents: The Requirements and Preferences of Potential Occupiers*, York: Joseph Rowntree Foundation/York Publishing Services. An illuminating study of consumer views of market renting in Leeds, reporting discussions with 'middle-income' single-person households and young childless couples.

Rugg, J. (1996) *Opening Doors: Helping People on Low Income Secure Private Rented Accommodation*, York: Centre for Housing Policy, University of York. Presents an evaluation of 161 schemes that provide low-income households with deposits and rents in advance to enable them to find reasonable accommodation in the private rented sector.

Whitehead, C.M.E. and Kleinman, M.P. (1986) *Private Rented Housing in the 1980s and 1990s*, Cambridge: Department of Land Economy, University of Cambridge. A detailed analysis of the sector, set within the context of 1980s Conservative housing policy.

Whitehead, C.M.E. and Kleinman, M.P. (1988) 'The prospects for private rented housing in the 1990s', in Kemp, P. (ed.) *The Private Provision of Rented Housing: Current Trends and Future Prospects*, Aldershot: Gower. Presents a pessimistically illuminating prediction of the future long-term demand for this tenure.

Whitehead, C.M.E. and Kleinman, M. (1990) 'The viability of private rented housing' in Ermisch, J. (ed.) *Housing and the National Economy*, Aldershot: Gower. A clear analysis of the factors that militate against a revival of the private rented sector.

7 Local authority housing

When the Conservatives won the 1979 General Election, local authorities were the largest landlords in the country, accounting for 32 per cent of the housing stock of Great Britain, compared to 13 per cent in 1947 and less than 2 per cent in 1913. From the late nineteenth century until the 1970s, the state recognised that the private sector was unable to supply enough housing at reasonable prices or rents to satisfy total need. Taking on this responsibility, governments throughout most of this period financially supported the extension of council housing, with owner-occupation receiving comparatively little assistance until the 1970s. However, under three Thatcher governments, local authorities' share of the total stock fell dramatically to only 21 per cent by 1991, and under the Major and Blair administrations to as little as 16 per cent by 2000. It is within this context that this chapter:

- examines the function of council housing in the twentieth century, and the changing responsibilities of local authorities in the supply of social rented housing;
- analyses public sector finance, particularly with regard to the control of capital expenditure, rent and subsidy policy, and the anomalies and inequities of finance;
- considers the shortage of council housing and the problem of empty council houses;
- explores the negative and positive aspects of council tenure;
- reviews party politics and the devolution of housing management; and
- concludes by contemplating the role of local authority housing in the early years of the new millennium.

Function of council housing and the changing role of local authorities in the supply of social rented housing

Over the years the function of council housing has been interpreted in two contrasting ways. On the one hand there was the view that the public sector should supply housing for 'general needs' – to satisfy the demand from households (irrespective of their income) to rent rather than buy, through either choice or necessity; and on the other hand there was the belief that council housing should fulfil a 'welfare role', assisting particularly those households unable to afford or find any other sort of accommodation. This second view was reinforced after 1988 when local authorities forfeited their role as the principal providers of new social housing, becoming instead 'enablers' of housing provision.

Taking the lead from David Lloyd George's pledge to provide 'homes for heroes' after the First World War, and the Liberal-dominated coalition's Housing and Town Planning Act of 1919, Labour governments have continually held the belief that council housing should fundamentally satisfy general needs. The Housing Acts of 1924, 1946 and 1949, and the Housing Rent and Subsidies Act of 1975 all aimed to achieve this objective. Council housing, in the view of Labour, was not intended solely for the poor, the underprivileged or the population of 'traditional working-class areas'. Neither was it intended to sustain the existence of a two-class nation. No one held this view more strongly than Aneurin Bevan, Minister of Health (and Minister responsible for housing) in the Attlee post-war government. His 1949 Act incorporated this view into legislation. It removed the 'ridiculous inhibition' restricting local authorities to the provision of houses for the 'working class'. Instead they could attempt to meet the varied needs of the whole community:

> 'We should try', he said, 'to introduce in our modern villages and towns what was always the lovely feature of English and Welsh villages, where the doctor, the grocer, the butcher and the . . . labourer all lived in the same street. I believe that is essential for the full life of a citizen . . . to see the living tapestry of a mixed community.'
>
> (Michael Foot, 1973)

By the late 1970s, with the adoption of monetarism by the Callaghan administration and in response to the Green Paper *Housing Policy: A Consultative Document* (DoE, 1977a), Labour reluctantly abandoned this aim. Council housing became a victim of public expenditure cuts, and public sector housebuilding fell from 173,800 starts in 1975 to 80,100 in 1979, its lowest level since the 1930s and insufficient to satisfy even welfare needs. Improvements also fell – from a peak of 188,000 in 1973 (under the Heath Conservative government) to 111,000 in 1979 – and in addition many councils slashed their repairs and maintenance expenditure, causing delays in getting repairs done, and frustration to tenants.

Conservative governments (or Conservative-dominated coalitions) had, however, for long been generally opposed to the provision of council housing to satisfy general needs. With regard to housebuilding, slum clearance and redevelopment, the Housing Acts of 1923, 1933, 1935 and 1938, and the Housing Subsidies Act of 1956 were essentially concerned with extending or at least maintaining council housing as a welfare service; and while the Housing Act of 1961 made available general needs subsidies, this provision was not so much an indication of a change in the Conservatives' inherent view of council housing but more a reflection of the government's aim to use housebuilding as a means of reflating the economy. The Housing Finance Act of 1972 reaffirmed Conservative housing philosophy. By raising council house rents to the 'fair rent' level of the private sector (and by introducing a system of means-tested rent rebates), the 1972 Act was an attempt to shift demand (from all but the poorest tenants) away from council housing to owner-occupation. In 1979, moreover, the incoming Thatcher government began an ideological onslaught on local authority housing, turning it even more into a 'residual' tenure (see Chapter 8) – by 1990 the number of local authority starts plummeted to only 8,600 (Figure 7.1). Under the Major government, output further slumped to only 2,000 starts in 1993, while the relatively high level of improvements (315,228 in the local authority sector in 1990) did little to ease the lot of the 1.5 million households on council waiting lists. Thus, over many decades, Conservative policy has been broadly consistent. It has been manifestly antagonistic to the sector in aggregate and to general needs provisions in particular. It was only in the 1950s (and under Labour legislation, the 1949 Act) that a Conservative government perceived council housing as a priority. In the period 1951–55, Harold Macmillan, the Housing Minister, ensured that as many as 940,000 public sector dwellings were constructed (compared to only 312,000 completions in the private sector) – a genuine attempt to satisfy both welfare *and* general needs.

Under the Housing Act of 1988, however, local authorities became 'enablers' rather than providers of social housing. In effect, local authorities began to undertake a 'range of activities that makes possible, encourages or facilitates the provision of social housing opportunities by

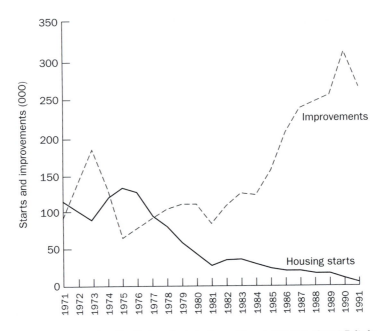

Figure 7.1 *Local authority housing starts and improvements, Great Britain, 1971–91*

bodies other than the local authority itself' (Bramley, 1993), notably the housing associations, private developers, financial institutions and local community trusts. But while local authorities soon ceased to be the main providers of new social housing (by 1993 their 2,000 housing starts compared very unfavourably with the housing associations' 41,300 starts), they remained 'responsible for ensuring, so far as resources permit, [that] the needs for new housing are met, by the private sector alone where possible, with public sector subsidy where necessary' (DoE, 1989). Since the Housing Corporation, however, became the main conduit through which government funds flowed into the provision of new social housing, it seemed a little odd that the local authority became 'the key strategic agent in bringing about the provision of social housing' (Bramley, 1993).

Public sector finance

In contrast to individual borrowers, who normally obtain a 25- or 30-year mortgage for house purchase, local authorities borrow for as long as 60 years – mainly to keep their annual loan charges (interest payments) to the minimum. Loans obtained from the council's Loans Fund (which contains money raised from individuals, institutions and the Public Works Loans Board) require loan sanction from the government, and facilitate spending on housing and associated land, sewers and roads – all items being shown on a council's housing capital account. Until 1976 there were few restrictions on the amount of housing built by local authorities, although the cost of each scheme was controlled by the cost yardstick – loan sanction not being granted if councils failed to comply with this constraint.

On completion of a scheme, there is normally no further capital expenditure, but current expenditure is incurred on management, maintenance and loan interest, and this is met out of rents, annual government subsidies and, in some areas (until 1989), rates – all of these items also being itemised on a council's housing revenue account. Until 1980 there was a 'no-profit rule'.

From the inception of council housing until the 1930s, local authorities set rents for each scheme as it was completed. Rents were calculated on the basis of the difference between loan charges on capital expenditure plus the cost of maintenance and management, *and* the subsidy (and possibly a rate contribution). But the result was different rents for similar dwellings. Rent pooling was therefore introduced whereby the rent income for the whole of the council stock would be based on total loan charges plus total maintenance and management costs *less* subsidies (with or without rate aid). On the basis of this total sum, individual dwellings were assessed subsequently in relation to location, size and amenities, with rents geared to gross values calculated for rating purposes. For historic reasons, rents were very different from one authority to another, depending on the age of the housing stock and the size of the current building programmes.

Rent rebates (introduced by the 1972 Act) and supplementary benefits were being claimed by over 40 per cent of local authority tenants by the late 1970s, although there were official estimates which showed that 30 per cent of tenants entitled to claim failed to do so.

The housing policy review

In 1975/76 the average subsidy to council tenants was £195, and house purchasers received an average of £185 in the form of tax relief on mortgage interest and option mortgage subsidy. But whereas house purchasers received assistance as their incomes rose (Chapter 6), tenants normally received more if their incomes were small. In 1974 Anthony Crosland (Secretary of State for the Environment), realising that housing subsidies were failing the test of equitability and resource effectiveness, set up a committee to review housing finance. Mr Crosland (1975) was particularly concerned about the growing polarisation between owner-occupiers and council tenants, warning that 'a deep divide has opened up between the two'.

The resulting Housing Green Paper (DoE, 1977a), however, drew attention to two intrinsic advantages which local authority housing has over owner-occupation. The first is that past investments are held at their fixed historic cost, which averaged £3,000 per dwelling in 1976 (although 75 per cent of council houses were built post-war), and the Exchequer subsidy on interest charges declines in real terms. The existing stock therefore offered cheap housing to the end of the century or beyond. In contrast, the cost of owner-occupied housing and interest payments generally increase by virtue of revaluation and exchange. Even without a subsidy the average break-even rent – given pooled interest costs and making allowance for management and maintenance – would have been £11 per week in 1976, and with a subsidy of, say, 33 per cent (equal to the standard relief rate for owner-occupiers) on just the interest charge, the rent would have been £8.50–£9.00. The second advantage was that within a local authority area cross-subsidisation could take place – 'surplus' rents from areas of older or cheaper housing could offset 'deficit' rents from areas of newer and more costly housing. Although inflation put an increased strain on cost pooling, especially as a higher proportion of the housing stock was modern, over time the financial self-sufficiency of local authority housing would reassert itself as it was generally free from the process of revaluation and exchange. The review was opposed to inter-authority cross-subsidisation since it would have breached the 'no-profit rule', but it proposed that areas with rising costs would have to meet these by raising rents in line with incomes and claiming a 'deficiency subsidy' – although there would be no automatic entitlement. Even 'falling-cost' and 'stable-cost' authorities might be in no better position, since the former would need to sell off their older houses to avoid making profits, and the latter might need to trim their capital expenditure since it would be difficult to justify on the basis of local need or housing stress. Rising-cost authorities which need to build large numbers of houses would be under greater government scrutiny than ever before.

Control of capital expenditure

Following the publication of the 1977 Green Paper, the Labour government began to limit the number of houses that local authorities could build – a stance compatible with the party's adoption of monetarism in 1976/77 and its more recent adherence to prudent monetary policies. Although loan sanctions – since 1977 – are no longer required, each local authority has annually to submit a Housing Investment Programme (HIP) (currently to the DETR), comprising both a Housing Strategy Statement and a Bid for Resources. The latter document is in two parts. The first part contains an assessment of local housing needs (in respect of new council building, the provision of improvement grants, lending to housing associations, lending for home-ownership, and private building for sale), while the second sets out the financial plans of the strategy. Following HIP submissions each July, and subsequent discussions between local authorities and the DETR, allocations – currently based on competitive bidding between authorities – are announced in December (Gibb *et al.*, 1999).

Whereas in the late 1970s there were block allocations of borrowing permission, the Local Government and Planning Act of 1980 replaced this form of control by one based on block allocations of capital expenditure. However, under the Local Government and Housing Act of 1989, there was a reversion to the control of borrowing. HIP allocations were thenceforth controlled by means of an *annual capital guideline* (ACG) for the following year. In addition, supplementary credit approval (SCA) – granted on the basis of competitive bids from local authorities – is normally 'top-sliced' from the ACG to permit borrowing for Estate Action and HAT activity, and in some years to facilitate the Estate Renewal Challenge Fund – depending upon policy objectives. In addition to credit approvals, HIP allocations include specified capital grants; for example, for housing defects, renovation of private dwellings, area improvement and slum clearance, and injections from the Estate Renewal Challenge Fund (Table 7.1).

In addition to borrowing, a substantial amount of local authority capital provision is 'self-financed' from revenue contributions and receipts from the sale of houses and land for capital expenditure. But whereas before 1989, 20 per cent of sales receipts could be used annually for capital purposes (the remaining 80 per cent 'cascading' to the following year), under the 1989 Act 25 per cent could be used for capital purposes but 75 per cent had to be used to repay debt – despite local authorities having

Table 7.1 *Local authority capital provision for housing, 1990/91–2000/01 (£ million)*

Central government support to local authorities	1990–91	1995–96	2000–01[a]
Annual capital guidelines	1,384	820	1,820
Supplementary credit approvals:			
Estate Action	180	316	64
Housing Action Trusts	—	93	88
Specific capital grants	311	323	72
Local authority 'self-financed'[b]	1,245	1,138	900
Total capital provision	3,120	2,690	2,944

Source: DETR

Notes: [a] Plans.
    [b] Revenue contributions and cash released from the disposal of assets.

enormous problems of repair and modernisation, and despite the fact that the housing debt represented only a tiny fraction of the current use value of the local authority housing stock (Aughton and Malpass, 1991).

During the 1990s, however, capital provisions were reduced and reached a trough of £2,484 million in 1997/98, but with the release of accumulated receipts – an outcome of the 1998 *Comprehensive Spending Review* (Treasury, 1998) – capital provision rose sharply in 2000/01.

Changes in rent and subsidy policy

The Housing Act 1980 introduced a new rent subsidy consisting of a 'base amount' (equal to the total subsidy paid in the previous year) *plus* a 'housing cost differential' (representing the increase in the total reckonable housing costs over those for the previous year) *less* the 'local contribution differential' (the amount a government expects the local authority to pay towards housing through increased rents or rate fund contributions). In principle, the local contribution differential gave the local authority the choice between increasing rents or increasing the rate fund contribution, but since it was the government's intention to reduce Exchequer subsidies, the DoE had powers to specify the target rate of annual rent increase. This resulted in rent increases considerably in excess of the amount that could be fully met from rate contributions.

Housing subsidies to local authorities, in cash terms, thus decreased from £1,423 million to only £520 million between 1980/81 and 1988/89. By

1987, 80 per cent of local authorities did not receive any housing subsidy at all (although tenants continued to receive rent rebates), and rents increased by 39 per cent between 1982/83 and 1988/89.

Under the Housing Act of 1980, surpluses could be made on the housing revenue account (HRA) and carried forward, in contrast to deficits, which had to be offset by rate fund contributions. Surpluses enabled local authorities to keep rates down, and by 1981/82 50 local authorities outside London were transferring money from rent surpluses to the rate fund, to the benefit of ratepayers. By 1987/88 the number of councils in surplus had increased to 126, and in 1987/88 these authorities (largely in suburban and rural areas) were able to transfer surpluses of £61 million to their general rate fund. Carvel (1982a) considered the implications of surpluses on housing revenue accounts. He argued, first, that councils would be charging unreasonably high rents and failing to maintain a fair balance between ratepayers and rent payers. Second, if rents rose so substantially, then a large proportion of tenants would require rent rebates or supplementary benefits (now housing benefits) but ratepayers would become the real beneficiaries since they would be indirectly subsidised. Third, although national rent pooling could ensure that surpluses would be used to cross-subsidise deficit local authorities (for example in the inner cities), this would not be possible if surpluses seeped out of the system at a local level to benefit ratepayers. Fourth, rents would rise above the level of mortgage repayments, encouraging tenants to buy rather than continue renting. Finally, although local authorities might have a preference for 'break-even' rents (since they regard local authority housing as a no-profit social service), they would be forced to subjugate local autonomy to centralist control.

Carvel (1982b) illustrated the effects of surplus rents by referring to East Cambridgeshire. He reported that in this district, surplus rents financed services to the extent of £730,000 in 1981/82 (compared to a contribution from rates of £447,000). He asked, 'Is this a fair balance between ratepayers and tenants?', and pointed out that high rents necessitate more supplementary benefits to help tenants pay them – a transfer of funds from the rest of the economy to the ratepayers of profit-making authorities being necessary. Carvel asked rhetorically, 'Is it a good use of public money to increase income support for East Cambridgeshire tenants so that East Cambridgeshire ratepayers can pay lower rates?'

Over the period 1979/80–1987/88, change in rent policy enabled the government to cut subsidies from £1,667 to £464 million – a decrease of

72 per cent; reduce the number of local authorities receiving subsidies from 367 to only 88; raise the level of rents from an average of £6.48 to £17.20 per week – an increase of 165 per cent (whereas gross earnings increased by only 125 per cent); and eliminate any shortfall by requiring local authorities to raise rate fund contributions from £306 to £499 million (an increase of 63 per cent).

The dramatic increases in rent (in excess of 60 per cent in real terms in the above period) thus resulted in an increased number of authorities showing surpluses on their housing revenue accounts, and contrary to the recommendation of the 1977 Green Paper, rents also increased more quickly than gross earnings – council tenants paying on average 8.0 per cent of their earnings in rent by 1987/88 compared to only 6.6 per cent in 1979/80. Consequently, most local authorities had no need to draw on their rate fund at all, London accounting for over 90 per cent of all rate fund contributions. Although the English districts met only 1 per cent of their total housing expenditure from rates, the London boroughs met at least 20 per cent of their housing costs from the rate fund in the mid-1980s.

The government had a further mechanism for ensuring that rents would increase significantly. The Local Government, Planning and Land Act of 1980 gave the Secretary of State powers, for the first time, to decide how much local authorities could spend per capita. If a local authority's average rent fell below the regional average, a penalty would be deducted from its rate support grant, but even if the average rent was likely to rise above the regional average, the local authority might still be penalised if it attempted to limit the full extent of this increase by drawing on the rate fund.

Soaring rents were reflected by the number of council tenants receiving rent rebates/supplementary benefits/housing benefits. The number increased from 2.1 to 2.6 million, 1979/80–1987/88, or from 40 to 66 per cent of the total – hardly a trend indicative of the success of public sector housing policy.

It was argued, however, that rents were still below their economic level – put at about £28 per week by Mr Nicholas Ridley (1986), Secretary of State for the Environment. Mr Ridley suggested that many people in local authority housing could afford economic rents (actual rents rose from only 6.4 per cent of average earnings in 1979/80 to 8.3 per cent in 1985/86), and that low rents were a cause of the dereliction of much of the local authority stock.

Under the Local Government and Housing Act of 1989, therefore, rent determination was reformed. The DoE would henceforth assess the total value of all the local authority housing stock in the country, with the value of each authority's stock in the country expressed as a percentage of the total national value. The government would then decide what the total national increase in rents should be for the following year and would calculate how much each local authority's share should be. In effect, the government would use the capital value of each authority's stock not to determine rents – as was proposed in the *Inquiry into British Housing* (NFHA, 1985), but to determine 'guideline' increases – regardless of current rent levels. Local authority rents consequently rose from an average of £20.70 in 1989/90 to a guideline £29.81 in 1992/93, a 50 per cent increase, greatly in excess of the rate of inflation. There were, however, substantial regional variations in rent, and large differences between guideline rents and actual rents (Table 7.2) – many local authorities having to exceed guidelines owing to financial stringency.

Table 7.2 *Local authority rents, England, 1991/92–1992/93*

	1992/93		Increase 1991/92–1992/93	
	Guideline rent (£)	Actual rent (£)	Guideline rent (%)	Actual rent (%)
Greater London	35	53	6	12
Rest of South East	33	43	6	8
East Anglia	30	40	6	8
South West	29	38	5	7
West Midlands	27	35	3	6
North West	25	33	3	5
East Midlands	25	31	3	5
North	24	29	3	4
Yorkshire & Humberside	23	27	3	4

Source: DoE, *Annual Report*, 1993

A new *housing revenue account subsidy* had been introduced under the 1989 Act to replace the former housing subsidy and to include rent rebates (previously paid as a form of income support by the DSS). But although the increase in rents towards market levels brought about a reduction in the housing element of the HRA subsidy from £1,356 million to £1,027 million between 1990/91 and 1992/93, it simultaneously led to an increase in the *rent rebate element* from £2,304 million to £2,982 million over the same period.

Unlike the situation before the 1989 Act came into force, the housing revenue account was now 'ring-fenced'. No longer would rent surpluses be used to subsidise local taxation (instead they would help finance the costs of housing benefit paid to two-thirds of all council tenants), while, conversely, local tax revenue would no longer be used to subsidise rents.

Anomalies and inequities of housing finance

Like the building of most owner-occupied housing and most acquisitions of existing private housing, council housing is financed by loans. But whereas private houses change hands every seven or eight years on average, council houses normally remain in one ownership. Loan debts therefore do not increase (and are paid off over a period of up to 60 years), and loan charges are based on the original debt. Since a significant proportion of the housing stock is old (built in the 1950s or before), the average loan debt was only £3,000 in the late 1980s, whereas the average value of council houses exceeded £20,000. Tenants should have benefited, since their incomes had more than kept pace with inflation, while loan debts and loan charges had not increased with inflation (except for a rise in interest rates). However, in the 1980s the benefit of the declining real cost of loan charges was more than eroded by the reduction in subsidies, and increasingly, profit rents were being charged. In the owner-occupied sector, however, debts normally rise with each successive mortgage, and mortgagors enjoyed tax relief on mortgage loans up to prescribed limits.

The second, and perhaps more serious, anomaly concerns the definition of 'spending'. Whereas an owner-occupier obtaining a loan of, say, £20,000 and repaying £3,000 per annum would have spent £3,000 in the first year (the £20,000 being an increase in his liabilities), a local authority undertaking the same cash transactions would have spent (according to the Treasury) £23,000 (see Chapter 2). Apart from this 'exaggerating' the level of public expenditure, the consequences are serious. Local authorities are forced to keep capital 'expenditure' down to a minimum in order to stay within their budget if they are to meet unavoidable increases in revenue expenditure (for example in rent rebates or, since 1990/91, housing benefits). Thus between 1980 and 1998, local authority capital spending decreased from 26.2 per cent to only 8.5 per cent of total local authority housing expenditure (Table 7.3).

Table 7.3 *Local authority capital and revenue expenditure on housing, United Kingdom, 1980–98*

	1980		1990		1998	
	£m	%	£m	%	£m	%
Net capital expenditure[a]	1,692	26.2	769	9.1	941	8.5
Revenue expenditure	4,758	73.8	7,669	90.9	10,114	91.5
Total housing expenditure	6,450		8,438		11,055	

Source: Wilcox (2000)

Note: [a] Gross expenditure – capital receipts.

It has long been suggested that changes ought to be made to the presentation of HRAs (especially on the expenditure side) to avoid both misunderstanding and unnecessary cuts in spending. First, costs include expenditure on road maintenance, open space and street lighting, but in owner-occupied areas these are costs borne directly by the rate fund and are not regarded as costs specific to owner-occupation as a tenure. Costs of preparing and developing sites are also included in the account, but private developers would not undertake these activities in many urban areas without a subsidy. Second, HRAs include debt charges on housing schemes in the process of development before there is any yield of offsetting rent (Table 7.4).

To satisfy long-term general needs, reforms to the system of financing local authority housing are clearly necessary. Not only should the presentation of housing revenue accounts be changed, but councils should be free to determine their own spending, according to local needs; likewise, they should be free to determine rents. A system of subsidies,

Table 7.4 *Expenditure and income of housing revenue accounts, England, 2000/01*

Expenditure	£m	%	Income	£m	%
Gross rebates	4,196.0	39	Gross rent from		
Repairs	2,115.6	20	dwellings	6,943.9	65
Supervision and management	1,992.1	18	Housing subsidy	3,086.6	29
Debt charges for capital	1,977.6	18	Other rents	178.9	2
Revenue to capital	285.2	3	Interest income	92.2	1
Transfers	75.4	1	Other income	353.8	3
Other expenditure	163.1	2			
Total	10,805.0			10,655.5	

Source: DETR, *Housing and Construction Statistics*

both within and between authorities, is also required. Overall, equity with owner-occupation is desirable if for no other reason than to offer households a realistic choice between renting and buying (see Chapter 5).

The inequities of housing finance are probably an even greater cause of concern than financial anomalies. By 1982/83 average rents amounting to 8.8 per cent of average gross earnings were higher than at any time since the Second World War, after being as low as 6.4 per cent in 1979/80. However, many rents were well above the average, and council tenants received lower than average gross earnings, therefore the proportion of earnings paid in rent was normally far in excess of 8.8 per cent. By contrast, a mortgagor starting off paying, for example, 25 per cent of his income on repayments would pay less than 9 per cent after 10 years with inflation. But inequity is compounded by fiscal assistance. Whereas the average Exchequer subsidy to local authority tenants was £104 in 1985/86, average mortgage interest relief to owner-occupiers was over £550, and the housebuyer also received exemption from Schedule A tax on imputed rent income, capital gains tax and (to a great extent) capital transfer tax (see Chapter 5). By the mid-1980s average net mortgage repayments were in many cases lower than rents on similar properties. Relatively high rents clearly encouraged tenants to buy – a declared aim of Conservative policy.

It was increasingly suggested, however, that rents should increase still further. The *Inquiry into British Housing* (NFHA, 1985) recommended that rents on public (and private) housing should be calculated to give a real rate of return of about 4 per cent on capital value (plus an element to cover maintenance and management). This would necessitate an increase in rent of some 14 per cent in real terms, while surpluses accruing to some local authorities could be redistributed to others by a levy. It should be borne in mind that the *Inquiry* also proposed the abolition of mortgage interest relief, and thus the cost of renting would be brought into line with the new cost of buying similar housing. Alternatively, of course, it could be argued that it would be more sensible to stabilise rents, restore Exchequer subsidies to local authority housing (more than halved, 1979–85), and leave mortgage interest relief intact. However, while the *Inquiry* had recommended that subsidies should be related to individual needs rather than to housing (a view shared by the Secretary of State), housing benefits in the late 1980s were already being paid to two-thirds of the 5 million tenants of public sector housing, therefore any substantial increase in rent, other things being equal, would escalate and extend welfare payments.

Box 7.1

Public sector rental housing in Scotland, Wales and Northern Ireland

Although the public rental sector in Scotland in 1999 contained as much as 25.1 per cent of the country's housing stock (compared to only 16.3 per cent in the United Kingdom as a whole), its size has decreased substantially in recent years as a result of the Right to Buy (RTB) and the transfer of Scottish Homes stock (Earley, 2000). Nevertheless, the number of local authority dwellings remains considerable (around 2.3 million in 1999), and the principal cause for concern is not the loss of affordable housing but the condition of the stock. As in England, access to extra resources to facilitate repair and improvement can be secured by 'transferring stock, working in partnership with housing associations and other providers and or taking advantage of the Private Finance Initiative through which authorities can lease property to the private sector (who then fund refurbishment)' (Earley, 2000: 26) (see also Chapter 8).

In Wales the local authority sector – containing 15.6 per cent of the country's housing stock in 1999 – has also declined in recent years as a 'product of the Right to Buy, voluntary sales and to a much lesser extent through the transfer of stock' (Williams, 2000: 33). In total, the number of local authority dwellings in the Principality decreased by well over a third between 1980 and 1999.

In Northern Ireland, public sector housing is the responsibility of the Northern Ireland Housing Executive rather than the local authorities. However – as in the rest of the United Kingdom – RTB sales and the transfer of new housing development to housing associations reduced the size of the sector from 31.3 to 21 per cent of the province's stock between 1989 and 1999 (Stevens, 2000). Nevertheless, the public sector still remains large in relative terms, and, within the United Kingdom, is surpassed only by that of Scotland.

By 1987 it seemed probable that rents would soon rise dramatically. The AMA (1987) predicted that under new legislation the government planned to make HRAs self-financing. No longer were local authorities to be permitted to subsidise rents from rate contributions. Since 80 per cent of local authorities in 1987 did not receive any Exchequer subsidy, the loss of rate subsidy would inevitably push up rents in many areas – by as much as 200 per cent, according to the AMA. It was ironic that 125 local authorities in England and Wales (largely in suburban and rural areas) were already subsidising their general rate fund from surpluses on HRAs (£62 million being transferred in 1986/87). While the White Paper *Housing: The Government's Proposals* (DoE, 1987) stated that rents would be 'kept at an affordable level', it was difficult to see how this would be possible following the introduction of new measures to prohibit the subsidisation of rents from rates but not vice versa.

Access

The shortage of council housing is manifested in two ways: first, by the increasing size of the waiting list, and second, by the growing number of homeless households. In both respects, access is becoming increasingly difficult. By 1994, waiting lists in England and Wales exceeded 1.5 million; there were 141,800 households 'accepted' as homeless, with countless more unregistered, especially among the young. From the late 1970s to the early 1990s, local authorities were obliged to provide permanent housing (where possible) to the statutory homeless but, owing to the shrinkage of the council stock, others on the waiting list (often with comparable needs) had to wait longer and longer for a council house or flat. Controversially, in 1994 the government proposed introducing a single 'streamlined' waiting list as the basis for allocating housing to those with the best claim to it, possibly resulting in many of the homeless not getting permanent homes.

Whether or not local authority housing should be accessible to all those households who need or choose to rent, allocation criteria should be updated to reflect anticipated changes in the composition of households in the foreseeable future. The provisions of the Housing (Homeless Persons) Act of 1977, in particular, were clearly in need of overhaul from the time the Act entered the statute book. The Labour Party (1982a) pledged that if it regained office it would extend the number of 'priority groups' for whom councils are obliged to secure accommodation; limit loopholes in the 1977 Act by clearly defining such circumstances as 'intentional homelessness', and homelessness itself; and strengthen the rights and improve the condition of homeless people in temporary accommodation. The Housing Act of 1985 and Housing (Scotland) Act of 1987, however, tightened the criteria whereby homeless households would be 'accepted' by local authorities and subsequently housed in permanent or temporary accommodation, while it increasingly seemed that until there was a massive housebuilding programme in this sector, measures aimed at assisting the homeless would only tinker with the problems of access but not solve them.

Many local authorities faced with a static or dwindling stock of housing have had little choice but to accommodate a proportion of the increasing number of homeless in bed and breakfast hotels. Local authorities in England, for example, were putting up over 11,043 families in bed and breakfast accommodation in 1989 at an average cost per household of £15,500 per annum, in contrast to £8,200 which could have been the annual cost of supplying newly built or recently acquired council housing (NAO, 1989). A less costly way of tackling the problem of homelessness has been the provision of 'short-life housing'. Leasing schemes have been negotiated between local authorities and private owners whereby owners relinquish their empty properties for up to ten years to house otherwise homeless families.

Local authorities provide management services and grant aid for improvement where necessary, while ensuring that owners obtain vacant possession when desired. Grosskurth (1986) showed that it cost the London Borough of Brent £2.8 million per annum (or £5,000 per family) to house the homeless in bed and breakfast accommodation, but the same number of homeless could have been housed under leasing arrangements for as little as £642,000 per annum (or £1,700 per family). Grosskurth, however, pointed out some adverse effects of leasing: first, it could pull up the general level of rents in an area; second, it is only a stopgap measure; third, leasing could be used as a smokescreen to hide the general inadequacies of housing policy; and fourth, there is not an unlimited supply of potential short-life housing. What is clearly needed is a much greater quantity of permanent rented housing. In 1994, however, it seemed likely that the Major government was going to rely more and more on private renting, rather than on new housebuilding in the social sector, as a means of providing housing for those most in need (Shelter, 1994).

Empty council houses

Housing finance needs to be reformed to prevent an increase in the number of empty council houses or the need to demolish 'unlettable' blocks. Despite long waiting lists, many urban authorities, such as the Inner London boroughs, have rendered unusable and boarded up empty council dwellings; Liverpool is either selling off or giving away properties such as the infamous 'piggeries'; and other authorities, such as Glasgow, are pulling down blocks built in the 1960s.

Although empty council housing accounted for less than 5 per cent of the total stock in the late 1980s, it was highly concentrated in the inner urban areas and it was severely vandalised. Donnison (1979) had suggested that the main reason for the abandonment of these areas was that the younger and more prosperous tenants had moved out to the suburbs or beyond; that the growth of service industries in the inner areas had not compensated for the loss of manufacturing jobs; that because of housing improvement in the private sector, the standard of the worst council housing had fallen below that of the better private rented housing; and that within the public sector the 'fortunate, the well organised and the articulate' had gradually gained possession of the better housing – normally with little or no increase in rent. While the demand curve for council housing slopes downwards as distance from the centre of an urban area increases, the supply price generally remains constant with distance. Therefore, among higher-income tenants, there was a great incentive to abandon the inner areas.

Donnison proposed that in the public sector there should be a reassessment of rents and valuation for rating purposes on the basis of matching 'shelter costs' with tenants' preferences. This would involve charging higher rents and rates in the more popular estates (with increased rent and rate rebates and child allowances for the poorer tenants). Consequently, 'there would probably be a smaller outflow of the more demanding and "respectable" tenants from estates low on the ladders of esteem'. Additionally, it would be necessary 'to create better opportunities for work and for earning for the people in deprived and unpopular neighbourhoods' – presumably so that they would be able to pay increasingly unassisted rents and rates towards the cost of maintenance and repairs. But, by implication, the crux of Donnison's main proposal was that higher surpluses could be realised on housing in the more popular estates and these could be used to increase the cross-subsidisation of currently less popular housing in some of the inner urban areas – reducing the possibility of it becoming 'unlettable' or having to be demolished.

With a change from a Labour to a Conservative government in May 1979, the emphasis was placed not on strengthening the role of local authority housing but on reducing its importance. The selling off of council houses again became a main plank in Conservative housing policy, while limited resources prevented local authorities from bringing back into use many of their empty dwellings in need of repair and maintenance. By 1982 the total number of empty houses in England alone had reached 596,000, equivalent to 3.3 per cent of the total stock (Shelter, 1982b). Of this number, 97,000 were council dwellings, of which 24,000 (or 0.5 per cent of the council stock) had been vacant for more than one year. These were located mainly in the inner cities; for example, in the London Boroughs of Hackney, Hammersmith and Fulham, and Islington, respectively 5.6, 3.6 and 3.2 per cent of the council stock had been empty for more than 12 months. By 1992 the total number of empty council dwellings in England still remained high at an unacceptable 83,000.

Empty dwellings had caused comparatively little official concern. Mr Allan Roberts MP had attempted unsuccessfully to obtain support for his Local Authorities (Empty Properties) Bill in 1981, which (had it been passed) would have forced councils to make maximum use of empty property by making it available at little cost for single people, co-operatives, housing groups or the homeless. Shelter (1982b) claimed that in many local authorities no one had the responsibility for doing anything about empty property, and that some councils did not even know what empty properties they owned. This not only caused a loss of rent (about £30 million per annum) and rates, but did nothing to help the 62,420 households currently registered as homeless. Shelter proposed that at the minimum, councils should maintain a complete record of their residential property, both occupied and vacant. By 1988 the government began to consider measures to transfer empty council property to the housing associations.

Negative and positive aspects of council tenure

Since the earlier twentieth century, local authorities have provided a wide range of dwelling types. Much housing was built to a high standard under the generous subsidy provisions of the Housing Acts of 1924 and 1949, while other dwellings were built to a lower standard or by means of prefabricated methods under slum clearance legislation, both before and after the Second World Wars (Malpass and Murie, 1999). However, there has also been great variation in the quality of housing management between authorities, and in the political interpretaion and application of policy at a local level. By the late 1970s, council housing – as a means of providing affordable good-quality rented accommodation – was arguably flawed. According to the *National Dwelling and Housing Survey* (DoE, 1977b), council tenants (compared to owner-occupiers) have less choice over where they live, are more likely to occupy flats and at a high density, have less independence to do what they like with their home and are not able to accumulate wealth.

By the early 1980s it was evident that a significant proportion of the council stock was in poor condition. There were as many as 546,766 council houses in England classified as unfit (3.1 per cent of the total); 1,033,351 council dwellings (or 5.8 per cent) lacking one or more amenities; and 1,725,375 (or 9.7 per cent) needing repairs (Matthews and Leather, 1982). Although council housing provided homes more and more for people with little choice, there was a limit to the extent to which families in need would tolerate poor housing. Matthews and Leather (1982), drawing on HIP information, reported that there were currently 285,000 difficult-to-let council dwellings in England (equivalent to 5.8 per cent of its total council stock), and in the metropolitan districts and Inner London boroughs the proportions were as high as 11.7 and 15.6 per cent respectively.

The condition of housing might not have been the only cause of tenant dissatisfaction. Coleman (1985) suggested that planning design had an impact on human behaviour, and that certain designs were closely associated with a high incidence of 'social malaise' on council estates, whereas the Audit Commission (1986) severely criticised standards of housing management – a view shared by Mr John Patten (1987), Minister for Housing, Urban Affairs and Construction, who attributed serious disrepair and rising rent arrears to the failings of local authorities as providers and managers of housing; while Power (1987), more specifically, drew attention to management being fragmented, bureaucratic and remote from tenants, and argued that local authorities had been overtaken by the scale of their operations.

Whereas council housing had been seen by most governments since 1919 as a solution to housing needs, the Conservatives in the 1980s and 1990s regarded it as part of the problem. In an attempt to reduce the perceived weakness of housing management, the government (under the Local Government Act of 1988) introduced compulsory competitive tendering (CCT) in 1992 to enable selected local authorities to bring in managers with 'appropriate' expertise to deliver services efficiently. Critics were quick to point out that if new management operated only according to narrow financial criteria, and failed to provide a sensitive service in response to personal and social needs, many council estates would continue to be blighted by social malaise, although these fears – to an extent – were allayed since 95 per cent of all contracts put out by the end of 1996 were won in-house (*Roof*, 1997).

While 60 per cent of council tenants paid little or no rent in the 1990s (they were recipients of housing benefit), they were unable to exert any pressure on their landlords to increase the sensitivity of management – with or without CCT (see Power, 2000). Britain is unique in Europe in providing housing for low-income people on such a scale – the outcome of the mass clearance of Victorian terraced housing in the inner cities, and the diminution of housing shortages after two world wars. Since the private sector would have been unable to take on a building programme on the scale required, the government found it necessary to create 'a top-heavy, state-run bureaucratic housing machine' to construct as many as 10,000 large council estates, which from the outset proved difficult to manage (Power, 2000). By the late 1990s around 10 per cent of council homes were hard to let (even in London), and local authorities throughout the urban areas of the UK were demolishing estates that they were unable to let or manage (Power, 2000). Not only are many council estates unguarded, unsupervised and afflicted by disrepair, rubbish, graffiti and litter, but many would-be tenants have to contend with a rule-book system of allocation designed more to cope with the sheer volume of people and property than to satisfy the variable and personalised needs of households in respect of choice of dwelling, location, and mobility within and between authorities (see Power, 2000).

Despite the rapid growth of owner-occupation after the Second World War, council housing still accommodated 16.9 per cent of households in Britain in 1998, and local authorities remain the largest landlords in the country. In the past, council housing was a privileged tenure, and tenants rarely faced eviction or unreasonable rent increases, unlike tenants of uncontrolled private lettings. Council dwellings were generally of a higher standard than private rented flatlets, terraced houses or tenements. Even recently, house condition surveys have shown that council housing compares very favourably with private rented housing. The *English House Condition Survey 1996* (DETR, 1998b), for example, showed that although 6.8 per cent of council dwellings were unfit, as much as 17.9 per cent of private rented housing was in this condition.

Despite its shrinking size, local authority housing is still popular among many of its tenants. It offers households a comparatively high standard of accommodation at rents that in general are lower than those charged for housing association accommodation, and it thus creates less benefit dependency than is evident among housing association tenants. Arguably, tenants 'also value having their concerns being directly addressed by democratically-elected councillors rather than housing association bureaucrats' (Long, 2000), and therefore are often reluctant to support the transfer of housing stock from the local authorities to housing associations.

However, although council housing in Britain was until recently a 'general needs' tenure catering for all income groups, it now – like welfare housing in the United States – disproportionately provides accommodation for the unemployed, the poor or newcomers. It is within this context that

> housing professionals want to help the government shape a new future for council housing which aims for a high quality service, a broader mix of customers, and a real choice for many people who cannot find what they want in the private market.
>
> (Perry, 2000)

Party politics and devolution

Generally, most of the weaknesses of council tenure (for example disrepair, inadequate maintenance and the absence of improvement) are due to lack of financial resources, but even if the quality of council housing markedly increased, the tenure would still remain relatively unattractive to many households unless council estates were managed in a more responsive and democratic way. While most local authorities have only a few thousand houses and manage these fairly satisfactorily, some have vast numbers of houses with enormous management problems – for example, in the early 1990s Glasgow owned 170,000 dwellings, Birmingham 120,000, Manchester 100,000, Southwark 60,000, Newcastle upon Tyne 45,000 and Cardiff 22,000. Authorities with stocks as large as these appeared to have insensitive allocation procedures and cumbersome, inflexible and remote management arrangements (*Economist*, 1986). The Audit Commission (1986) claimed that 'standards of housing management give cause for concern . . . where money is being spent on a growing bureaucracy rather than on better services for tenants', and Donnison (1987) pointed out that while local government might be good at meeting needs, 'it is . . . bad at meeting demands . . . bad at listening to its customers . . . bad at repairing and maintaining people's homes efficiently'.

The Labour Party (1982a), on the one hand, rightly acknowledged that under a system of tight public sector control, council housing 'spearheaded the vast improvement in housing conditions over the last sixty (or more) years', and pointed out that

> By pooling all costs and revenues albeit locally and by allocating housing on the basis of need rather than ability to pay, council

housing has offered millions of working people a standard of
accommodation which they otherwise have found unattainable and for
many of them it has, as a result, broken the link between poverty and
bad housing.

On the other hand, the party in the early 1980s saw no reason why the
basic freedoms associated with owner-occupation (such as security of
tenure and the freedom to repair, maintain, improve and relocate) should
not be fully extended to council tenants. Hilditch (1981), detecting
inconsistencies in Labour policy, questioned whether (given constraints
on supply) it was possible simultaneously to allocate housing according
to the traditional criteria of need *and* to provide some of the basic
freedoms of the owner-occupier. The Labour Party (1982a) recognised
these apparent contradictions and proposed both a £15,000 million local
authority housebuilding programme over three years and tenant
democracy in the public sector.

The need for tenant democracy was voiced particularly by Mr Allan
Roberts MP, who argued that tenants should be liberated from
paternalistic management. In his Tenants (Consultation) Bill of 1982
(which failed to be enacted) he proposed that tenant representatives
should be consulted locally by councils, or regionally and nationally by
central government, when policy changes affecting tenants are being
considered; and tenants should have a right in law to take over the
management of estates (with government funds).

The Labour Party (1981, 1983) believed that the individual rights of
tenants should include the right to security, repairs and improvement;
and collective rights should involve (in addition to the right to
responsible and decentralised management and maintenance) the right to
participation and the right to form housing co-operatives (see Chapter 6).
Mr Allan Roberts MP (1982) pointed out that Labour could justify
neither its massive council housebuilding programme nor its opposition
to council house sales 'unless it [was] prepared in partnership with
tenants to give them the same rights and freedoms as those enjoyed by
owner-occupiers'.

Mr Jeff Rooker MP (1985), Shadow Housing Minister, in setting out
Labour's housing priorities for the late 1980s, stressed the need to
establish a charter of rights for local authority (and private) tenants;
to draw up local housing plans with full consultation; and for local
authorities to encourage the setting up of tenant management
co-operatives. The Labour Party (1985) acknowledged the important role

of management co-operatives, believing them capable of transforming unpopular estates into places where people would want to live. In its 1987 manifesto the party pledged not only to give tenants a legal right to be consulted about rents and repair and improvement programmes but also to give groups of tenants who wished to take over the running of their homes the legal right to set up management co-operatives.

The Social Democratic Party also advocated decentralised housing management, and wished to see council housing transformed into a 'consumer-driven service'. The party believed that this in part could be achieved if an appropriate legislative and financial framework were established to encourage the formation of co-operatives.

For different reasons, the Conservatives went along the same devolutionary route as the other main parties. The Housing Act of 1980 (and later the Tenant's Charter of 1992) increased the security of tenure of council tenants and permitted them to decorate, obtain improvement grants and take in lodgers. Also, the government encouraged tenants to form management co-operatives to decide how available money should be spent on repairs, maintenance and improvement (although councils retained the responsibility for lettings, rent collection and setting rent levels). In 1993, therefore, tenants and leaseholders throughout the Royal Borough of Kensington and Chelsea voted to manage the borough's entire stock of 9,700 dwellings through their own tenant management organisations (TMOs), while in over 70 other boroughs tenants voted to set up TMOs to manage individual estates – almost certainly in all cases to pre-empt the introduction of CCT by private companies. But in most estates, where co-operatives did not evolve, tenants who could not get their landlords readily to carry out repairs obtained the right, under the Housing and Building Control Act of 1984, to call in builders to do the work at the council's expense. This provision, however, was limited to repairs costing only £20–£200 (structural repairs to flats and maisonettes being excluded), and tenants could claim a reimbursement of only 75 per cent of what the repairs would have cost the public landlord to have carried out. Since the scheme was optional for tenants, it did not diminish the landlord's responsibility to undertake repairs if tenants so wished.

The Housing and Planning Act of 1986, however, is more far-reaching. It forces local authorities to examine seriously any tenant-based proposal for devolution, empowers local authorities to devolve management to co-operatives subject both to tenant safeguards and to tenant and government approval, and offers grants for training in housing management.

One of the most effective means of devolution to have emerged in recent years was pioneered by the Priority Estate Project (PEP) of the DoE and subsequently by Estate Action. A highly decentralised form of management was established with a housing officer located in each estate to deal with problems locally – including rent arrears and transfers. Likewise, repairs and maintenance were undertaken by an on-site team instead of a town hall-based direct labour organisation (DLO). While Estate Action made the worst estates more habitable, it is tragic that since this approach to devolution emanated from a centralising government, it was not adopted by many Labour councils – many of which still clung ironically to centralised, unresponsive and high-cost control (Jenkins, 1988).

At the 1992 General Election, whereas the Labour Party pledged that it would offer tenants real rights over their homes (in large part facilitated by devolved management), and the Liberal Democrats stated that they would encourage local authorities to create more tenant co-operatives, the Conservatives – in signalling future policy – stated that they would 'revolutionise' the management of council houses and flats by obliging all local authorities to introduce CCT (where TMOs had not been set up).

Thus, while the Labour Party in general favours devolution, particularly as a means of extending the democratic process, the Conservatives almost certainly view it as a means of both reducing public expenditure and paving the way to the privatisation of estates (see Chapter 8).

Conclusion

Unlike Conservative administrations of the 1980s and 1990s, the Labour government, post-1997, recognised that there was an important role for local authority housing, albeit on a reduced scale. Together with other forms of social housing (such as that provided by housing associations), it could offer households a choice of landlord and choice of tenure, and involve tenants in running estates, as stakeholders. In the view of the Housing Minister, Ms Hilary Armstrong, 'choice and freedom must not be the preserve of the private sector alone' (Armstrong, 1998).

In its Housing Green Paper, Labour subsequently indicated that it was intent on 'promoting a stronger role for local authorities in housing to reflect the variations in circumstances around the country and to enable solutions to be tailored to local conditions' (DETR/DSS, 2000: 10). It proposed that it would:

- encourage all local authorities to assume a strategic view of needs across both the social and the private sectors, and in relation to homelessness;
- encourage authorities to function in partnership with local communities, registered social landlords and other organisations;
- strengthen the strategic responsibilities of authorities that have transferred their housing to registered social landlords; and
- ensure that authorities integrated housing policy with planning and other policies designed for the wider social, economic and environmental well-being of the community.

As far as funding was concerned, the government, in its *Spending Review* (Treasury, 2000), indicated that it planned to increase local authority capital expenditure on housing from £2.31 billion to £2.55 billion over the period 2001/02–2003/04 (a 10.4 per cent boost), of which the major repairs allowance would increase from £1.4 billion to £1.6 billion, and housing-based credit approvals would rise from £705 million to £842 million – clearly prioritising repairs and renovation rather than new housebuilding. The government also planned to boost the housing private finance initiative (PFI), which would provide an extra £600 million of investment into council housing over the two years 2002/03 and 2003/04, and inject an additional £460 million into local authority-owned housing placed under the control of arm's-length management companies (Raynsford, 2000).

Further reading

Aughton, H. and Malpass, P. (1994) *Housing Finance*, London: Shelter. Contains useful introductory information on the financial aspects of local authority housing.

Clapham, D. (1989) *Goodbye Council Housing?* London: Unwin Paperbacks. Speculated on the demise of council housing at a time when this was beginning to seem probable.

Clapham, D. and English, J. (1987) *Public Housing: Current Trends and Future Developments*, London: Croom Helm. An illuminating examination of the prospects for council housing during the years of Conservative government.

Cole, I. and Furbey, R. (1993) *The Eclipse of Council Housing*, London: Routledge. A valuable addition to the literature on the future of council housing.

Coleman, A. (1985) *Utopia on Trial*, London: Hilary Shipman. A very influential critique of the design, use and management of council flats in London, often used as a polemic against public housing in general.

English, J. (ed.) (1982) *The Future of Council Housing*, London: Croom Helm. Explores the issues related to the doubtful future of council housing, and analyses the implications of council house sales, criticises housing policies and management, and argues forcefully for the retention of the sector.

Harloe, M. (1995) *The People's Home*, Oxford: Blackwell. A very substantial comparative study of social housing in Europe and the United States, which is valuable reading for its British content alone.

Hastings, A. and Dean, J. (2000) *Challenging Images: Housing Estates, Stigma and Regeneration*, Bristol: Policy Press. Examines three stigmatised estates undergoing regeneration programmes, and demonstrates how regeneration initiatives can tackle image problems.

Holmans, A. (1987) *Housing Policy in Britain: A History*, London: Croom Helm. Contains a very thorough review of the history of council housing from the 1920s to 1979.

Malpass, P. (1990) *Reshaping Housing Policy*, London: Routledge. A powerful critique of housing policy, focusing on local authority housing finance.

Merrett, S. (1979) *State Housing in Britain*, London: Routledge & Kegan Paul. A basic text that provides a historic perspective of the development of the public housing sector.

Pearl, M. (1997) *Social Housing Management: A Critical Appraisal of Housing Practice*, Basingstoke: Macmillan. A comprehensive examination of social housing management focusing on changing practice in recent years.

Power, A. (1987) *Property before People*, London: Allen & Unwin. Presents a critique of housing management in the public housing sector.

Reynolds, F. (1986) *The Problem Housing Estate*, Aldershot: Gower. Focuses on a high-profile aspect of local authority housing.

Swenarton, M. (1981) *Homes Fit for Heroes*, London: Heinemann. Presents a good account of the key events in the development of council housing in the aftermath of the First World War.

8 Privatisation and stock transfer

The privatisation of housing by Conservative and Labour governments in recent years has taken broadly two forms: first, the selling off of council houses to their tenants; and second, and in part to facilitate rehabilitation, the disposal of much of the council stock to housing associations, trusts and private companies either for renting or for resale. To some observers, the whole political and ideological purpose of privatisation has been to 'dismantle public rented housing and to promote private ownership and management by all available means' (Daniel, 1987). It was as though the government had a pathological hostility towards council housing and believed that once the municipal landlord was removed from the frame most housing problems would vanish.

It is against this backdrop that this chapter:

- considers the origins of council house sales;
- sets out the arguments for and against the 'Right to Buy';
- notes the initial effects of the 'Right to Buy' policy;
- discusses the political aspects of privatisation in the 1980s and 1990s;
- explores the relationship between privatisation and rehabilitation;
- examines policy relating to estate transfer in the 1990s;
- examines residualisation; and
- concludes by reviewing Labour's plans to hasten the demise of council housing.

Council house sales: the origins

Although council house sales were first permitted in 1925, throughout the following half-century privatisation of this sort was rarely a political and economic issue since housebuilding in the public sector usually had priority over disposal. This was so particularly in the late 1940s and early 1950s when under successive Labour and Conservative governments the respective housing ministers, Aneurin Bevan and Harold Macmillan, ensured that the number of council houses built exceeded those constructed in the private sector (a situation not repeated again until 1969 and the mid-1970s). But in the mid-1950s not only was housebuilding in the public sector cut back to the benefit of the private sector, but the selling off of council houses was encouraged. Local authorities were empowered by the Conservatives to sell council houses to sitting tenants under Section 104 of the Housing Act of 1957, which allowed sales either at full market value with no conditions attached, or at 20 per cent below provided that (a) the tenant offered the dwelling back to the council if he or she wished to sell within five years of acquisition, or (b) if the authority did not wish to re-acquire the property, the tenant could sell it on the open market but at a price no higher than he or she paid for it plus the value of any improvements. Table 8.1 shows that council house sales were particularly high during the Conservative period of office, 1970–74, and low during the mid-1970s when the Labour government was able to control sales. Whereas Mrs Margaret Thatcher, as Shadow Environment Minister, immediately before the 1974 General Election pledged that local authority tenants of three years' standing would be able to buy their council houses at one-third less than market value (instead of the prevailing 20 per cent discount), Labour's Circular 70/74 (DoE, 1974) was cautious about council sales and stressed that

Table 8.1 *Local authority dwellings sold for owner-occupation, England and Wales, 1970–78*

1970	6,231
1971	16,851
1972	45,058
1973	33,720
1974	4,153
1975	2,089
1976	4,582
1977	12,019
1978	28,540

Source: DoE, *Housing and Construction Statistics*

> The first duty of a local authority is to ensure an adequate supply of rented dwellings. In areas where there are substantial needs to meet for rented dwellings, as in the large cities, the Secretaries of State consider that it is generally wrong for local authorities to sell council houses.

The substantial increase in sales in the late 1970s was mainly due to the Conservatives regaining control of many urban local authorities in 1977–78. In March 1979 Mr Peter Shore, Secretary of State for the Environment, therefore announced curbs on the sale of council houses.

These restrictions applied to sales of newly completed dwellings, sales of empty houses, tenants' rights to take out options to buy their houses in the future, and sales to tenants of under three years' standing. Restrictions were not imposed on sales to tenants of three or more years' standing, the building of houses by local authorities for sale, and equity sharing schemes.

The 'Right to Buy' controversy

The return of a Conservative government in May 1979 resulted in a major change of policy and an immediate relaxation of controls over the sale of council dwellings. But local government elections (also in May 1979) ironically brought a change in the political control of many urban authorities from Conservative to Labour, which produced resistance to the government's policy of encouraging sales. Shelter (1979b) claimed, in fact, that 60 per cent of all local authorities were opposed to unrestricted sales, and this included 30 per cent of Conservative councils. It was not surprising that the Labour Party Conference of 1979 endorsed the National Executive Committee's statement condemning the Conservative policy of selling council houses, and carried a motion that the next Labour government would repeal the forthcoming 'Right to Buy' (RTB) legislation.

This legislation, the Housing Act of 1980 and the Tenants' Rights Etc. (Scotland) Act 1980, gave council tenants the statutory right to buy the freehold of their house or a 125-year lease on their flat. The Act also allowed tenants to take a two-year option to buy their homes at a fixed price on payment of a £100 deposit. To counter the possibility that a council might delay or impede a sale, the Secretary of State had the right to intervene and complete the sale under Section 23 of the Act. Discounts of 33 per cent were offered to tenants of three years' standing, rising to 50 per cent to those of 20 or more years' standing. Council mortgages of 100 per cent were to be made available, but if the property was resold within five years, the capital gain would be shared between the owner and local authority.

Although Mr Michael Heseltine (1979), Secretary of State for the Environment, stated at the Bill stage that the legislation 'lays the foundations for one of the most important social revolutions of this century', Mr Gerald Kaufman MP (1979), former Environment Minister and later Shadow Minister, said the measures would 'not provide a single new home and [would] deprive many homeless people or families living in tower blocks from getting suitable accommodation'.

There were, however, many detailed arguments for and against the sale of council houses, and these will be examined in turn.

The case for council house sales

1 By the late 1970s it was recognised that although tenants received accommodation at considerably below market rents and in many cases below cost rents, it was not in their long-term interests to remain tenants. Mr Peter Walker MP (1978) explained that over the period 1950–78 whereas a council tenant might have paid, say, £4,750 in rent, might not have owned his or her dwelling, and would have to incur increased rents in the future, an owner-occupier who bought a house for, say, £2,000 in 1950 would have paid £2,750 (after tax relief) on a 25-year mortgage (plus repair and maintenance costs) and would have a realisable asset of at least £12,000 with no further mortgage repayments. Over the short term, however, few people find it practicable to realise gains of this sort, and also, over short-term periods prices might remain stable or fall. But in the long term, even without interest relief, owner-occupation is cheaper at all stages than renting, at least from the end of the second year (Townsend, 1979) – with the virtual certainty of capital gain.

      Referring to discounts made available under the Housing Act of 1980, Forrest (1982) outlined the main advantages of tenants buying their homes and argued that

> for the majority who purchase, the advantages are clear. They are enabled to buy at substantially less than market value an asset which will almost certainly appreciate. In most cases the dwelling will be well constructed and well maintained. . . . Purchasers will also possess an asset which they can pass on to their children.

   In quantifying the monetary advantages in buying, Kilroy (1982) showed that over 35 years (and at 1982 prices) the gross and net outgoings of an average tenant (who continued renting) would be £15,128, but the net payment of an average tenant who bought his £14,000 home (at a discount of say 40 per cent) would be only £3,125 – benefiting him to the extent of £12,003 (Table 8.2).

Table 8.2 *Comparative costs of buying and renting an average council dwelling*

Housing costs over 35 years at constant value	*Tenant who buys: market value £14,000 (outgoings, £)*	*Tenant who rents: rent at £8.64 per week (outgoings, £)*
Purchase price	8,400	—
Mortgage relief	–2,268	—
Rent	—	15,725
Less rebates	—	–2,022
Rates	same	same
Maintenance paid	7,076	1,425
House insurance	417	—
Gross outgoings	13,625	15,128
Less value of estate	–14,000	—
Add renovation liability	3,500	—
Net outgoings	3,125	15,128

Source: Kilroy (1982)

By 1982 the financial advantages of buying had become very clear. During the period 1979–82 council rents had risen by 117 per cent (to an average of £14 per week), but many tenants paid only £7,000 to buy their homes (at 50 per cent discounts), which would have cost them with an option mortgage only £19.83 per week, a small difference in weekly outgoings, a gap which narrowed as a result of the further rise in rents and the lowering of mortgage interest rates from 15 to 10 per cent in 1982. Taking into account the increase in the retail price index of 6 per cent in 1982, the cost of living of council tenants rose consequently by 8 per cent, but the cost of living of mortgagors rose by only 3 per cent. It was therefore increasingly advantageous to buy rather than rent. Tenants, however, had been aware of the financial advantages of owner-occupation for a long time but significant obstacles blocked entry into the owner-occupied sector, for example market price valuation of property and the virtual monopoly of building societies as suppliers of finance. The introduction of discounts and the availability of local authority mortgages (as well as the provision of building society finance) thus offered tenants more of a choice between renting and buying.

2 Owner-occupation offers many advantages other than financial benefits. Indeed, the feeling of independence and security, the freedom to choose location and to decorate as desired, the greater opportunity to obtain a house with a garden, and the recognition that the property

is 'your own' may be more important factors to tenants wishing to buy than, for example, the investment value of the property or the depreciating debt in absolute or real terms. Sales of council houses, it could be argued, help to satisfy these needs as real incomes rise.

But the above arguments can be refuted, as is demonstrated below.

The case against council house sales

1   The argument that by selling off council houses, local authorities would contribute towards a significant cut in public expenditure (to the advantage of the tax- and ratepayer) was extremely fallacious. Mr Peter Walker's proposal of giving away council property to tenants of long standing and selling off the remainder with mortgages of different lengths provoked much criticism. McIntosh *et al.* (1978) argued that Mr Walker, in stating that local authority housing was running at a loss, included in his calculations the cost of rent rebates and supplementary benefits, yet the latter would still have had to be paid to low-income householders if they became owner-occupiers. But if income support was not taken into account, rent revenue less management and maintenance costs would have amounted to £827 million in 1977/78, leaving £1,100 million of debt to be met by subsidies – an amount approximately equal to tax relief to owner mortgagors in that fiscal year. It was also pointed out that Mr Walker ignored the cost of tax relief on mortgage interest, which normally rises every time the property changes hands.

   Shelter (1979b) estimated that £3,000 per house would be the net total loss over 40 years to public spending from selling the average council dwelling, and if the annual disposal reached 250,000 dwellings (the Conservatives' target), then £3,000 million could be lost, 1979–84. The DoE (1979) put the figure even higher. It calculated that losses would range from £7,285 to £8,535 per house (or up to £10.7 billion over five years) (Table 8.3). Kilroy (1980a) argued that there was no doubt whatsoever that the public purse would lose money from sales. First, over the course of time, repayments of local authority mortgages (tied to the money value at which a house is sold) decline in real terms, but net rents normally keep pace with inflation – sales turning what is initially a profit into a loss (Figure 8.1A). Second, subsidies on council houses (fixed in nominal terms according to their original cost) decline in value over time, whereas mortgage relief

Table 8.3 *Expenditure and receipts for an average council house sale*

Expenditure (£)		Receipts (£)		
Rent forgone	−14,460	Disposal price		+7,400
Tax relief	−3,225	Repairs savings	small scale	+1,750
			large scale	+3,000
		Loss: *either*		8,535
		or		7,285
Gross expenditure	£17,685	Gross receipts		£17,685

Source: DoE (1979)

increases every time properties are resold (on average every seven to ten years) (Figure 8.1B). Third, local authorities do not receive any capital repayments if they provide the mortgages (yet the building societies – as alternative suppliers of finance – would need to increase their deposits by as much as one-third to finance, say, 250,000 sales per annum – an unlikely possibility).

Although the DoE (1980), reporting to the Conservative government, claimed that sales would be profitable (by as much as £9,128 million) – in contrast to its earlier calculations – this was on the basis that rents would fall in real terms and be below mortgage repayments over the short and long term (Figure 8.1C). Kilroy (1981c) thought this unlikely because historically rents have kept their 'true value' by fluctuating with the rate of earnings growth and rate of inflation. If they did not, then rented housing would become a wasting investment and be more expensive to replace, or it would be worthwhile for the local authority to sell up and invest in the money market. Kilroy argued:

> [the] Department of Environment's proof that council house sales can be financially profitable, on the grounds that rents are unlikely to keep their value against interest rates and earnings, is based on an imaginary world, which came nearest to materialising in wartime and some of the depression years. . . . According to real past performance, selling now must inevitably represent a serious financial loss.

Kilroy's argument was substantiated by the House of Commons Environment Committee (House of Commons, 1981b), which reported that according to its calculations, there would be a long-term financial loss to the public sector of approximately £12,500 per house sold, but, although its estimate was higher than any other figure to date, the committee predicted that there would be only 350,000 sales, 1980–84,

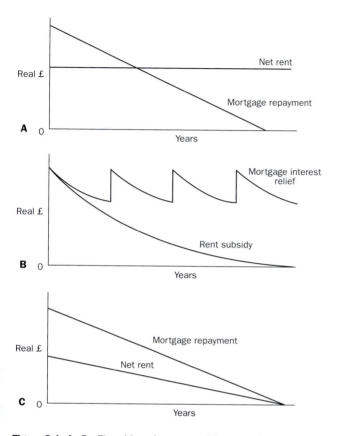

producing a loss of about £4,375 million over 50 years (at 1981 prices).

The long-term loss of about £12,500 per dwelling referred to above is essentially the difference between the net receipts from letting and selling over 35 years (Table 8.4). Kilroy (1982) explained that although the local authority may make short-term gains when selling at a discount, these are turned into long-term losses because average selling prices (of for example £8,400) are far less than rental income (of, say, £10,125 net of repair and maintenance costs at 1982 prices). Local authorities would make a profit on sales only up to 30 years at the very

Figure 8.1 A *Profit and loss from council house sales*
B *Effect of resale on mortgage interest relief, and its relationship with rent subsidy*
C *Continuation of local authority gains over the long term*

most. But if sales took place without discounts, the selling price and rental income would represent the same investment value, of for example £8,400 and £8,103 respectively.

Because of mortgage interest relief and other tax concessions on owner-occupation (see Chapter 10), the Exchequer was also certain to lose from council house sales (although a proportion of the buyers would have qualified for these concessions had they bought in the open market). Kilroy explained that nationally the sale of council houses on a large scale would be detrimental to the total housing stock since the borrowing requirement would increase, while construction programmes and employment would be vulnerable to fluctuations in the building cycle. At the same time, discounts (amounting to £4.6

Table 8.4 *Comparative costs of selling and letting an average council dwelling*

	Public purse receipts over 35 years (£)	
	From a tenant who buys	From a tenant who rents at £8.64 per week
Sale price	8,400	—
Less mortgage rates	–2,268	—
Rent income	—	15,725
Less loan charges	same	same
Less rent rebates	—	–2,022
Less repairs and maintenance	—	–5,600
Gross receipts	6,132	8,103
Add value of house	—	14,000
Less renovation costs	—	–3,500
Net receipts	6,132	18,603

Source: Kilroy (1982)

billion, 1979–85) not only would result in local authorities writing off part of the exchange value of their housing stock but also would facilitate the leakage of productive capital into consumption. Housing investment at best would then be diverted away from the priority of renewing the obsolescent stock to making good the loss of supply of public sector housing, or at worst would be reduced to zero.

2 Despite the advantages to tenants of buying their existing homes (discussed above), in the late 1970s it was doubtful whether the sale of council houses would have satisfied the desire of all tenants wishing to buy. Normally only high-income tenants (those who earned more than £6,000 in 1979) would have been able to buy their homes since prices (despite discounts) would still have been high, especially in the inner cities. Based on a 15-year, 100 per cent local authority mortgage, and assuming a current interest rate of 10.75 per cent, a typical council house would cost a buyer £61 per month initially if the rate of discount was 50 per cent, rising to £97 per month at a discount of 20 per cent. With these outgoings, the vast majority of the then 6 million council tenants (who earned less than £4,000 per annum) would remain tenants.

But even tenants who were able to buy often faced severe difficulty. Knight (1982) foresaw that tenants buying their homes with local authority mortgages faced three pitfalls. First, if they received 25-year mortgages many would be retired before their debt was paid off, possibly buying their house at the expense of adequate food and heating. Second, many mortgages might have been based on the incomes of one or more other members of the family who would still have this responsibility if circumstances changed (for example if they left home). Third, if these members did move away, they would not qualify for mortgages of their own, unless for example they forced the sale of the mortgaged council house. In addition, since RTB encouraged many people to take on burdens they could not afford, mortgage arrears soared in the 1980s (Carvel, 1984a). In 1985, 15 per cent of households having exercised their RTB were behind with their mortgages.

Tenants buying their homes soon became aware of the cost of repairs, and reluctantly postponed expenditure on maintenance. They may also have feared that compared to a normal owner-occupied house they might have difficulty in selling their property, or (groundlessly) if the local authority were to be given 'first option' on resale, they would receive less than the market price.

3 It would be mainly the better-quality dwellings that would be sold off – notably two-storey houses with gardens rather than high-rise flats. Indeed, soon after the 1980 legislation was enacted it was found that there were serious problems involved in selling flats leasehold. The 1980 Act failed to provide a method for dealing with service charges on flats which are part of a block owned by a local authority, either because of an oversight or because it was assumed that not many tenants would wish to become 'isolated' owner-occupiers in a block of tenanted flats.

The best properties have been 'creamed off' by the wealthiest tenants – mainly families with children growing up or grown up, while families with low incomes and young children have been increasingly confined to less desirable and often high-rise property since there have been fewer vacancies for transfer to the more sought-after stock. This has compounded residualisation – a process examined later in this chapter. Allocation has become generally inflexible because, first, as sales proceed, some estates are perceived to be suitable for privatisation and others for continued letting, and second, when council tenants die, buy private houses or rent elsewhere, their dwellings normally become available for new tenants (as many as half of all new tenants have been housed in this way), but if the properties are sold and then become vacant, these houses are no longer used for allocation (unless the local authority buys them back).

In rural areas the depletion of the housing stock poses special problems, since houses bought by their tenants might subsequently be resold as second homes or to commuters, depriving local working-class people of a decent rural home for ever. Although Clause 198 of the 1980 Act gives local authorities in 'designated areas' the first option on the resale of a former council house within ten years of its initial acquisition, since local authorities would be compelled to pay the market price and elect to buy within one month of being notified of the proposed sale, this is not a proper safeguard against the all-time loss of low-income housing.

Both with regard to urban and rural housing, the House of Commons Environment Committee (House of Commons, 1981b) was highly critical of the DoE's 'erroneous claim' that council house sales would not result in any loss of relets – dwellings which become available when a former tenant moves or dies. The committee estimated that over the following 30 years, 78,000 relets would be lost for every 100,000 dwellings sold, with substantial effects on the number of new lettings and transfers. Although Mr Michael Heseltine, Secretary of State for the Environment, advised the committee that the need to replace a council house which had been sold would not arise for many years, the committee declared that it was not clear on what grounds his belief was held.

But fewer relets would disadvantage tenants in another way. The selling off of older and surplus-yielding housing in the suburbs would reduce the degree of cross-subsidisation, and there would therefore have to be consequential increase in rents in the inner cities, and particularly in areas of housing developed in the 1960s and subsequently.

Although the privatisation of some of the council stock satisfies the demand for owner-occupation from a relatively small number of council tenants, the costs borne by other council tenants (both monetary and non-monetary) are very substantial.

There have been some other disturbing aspects of council house sales. Too often tenants have been persuaded to buy as a result of aggressive sales policies, usually employed by Conservative-controlled authorities, often with the aid of estate agents. Research has shown that a number of authorities even offered bonuses or commission to housing officers to encourage tenants to buy (Power, 1982). After five years from acquisition, during the 1980s, former council properties could have been resold at market prices, probably with a considerable capital gain to the owner (bearing in mind that the property was first bought at a discount). But since the gain was untaxed it was entirely at the public's expense. The council stock, moreover, was being depleted, while by 1983 there were over 1.4 million households on council waiting lists, 700,000 on transfer lists and 65,000 registered as homeless. Very few of these households expected to be housed in the foreseeable future, and Shelter (1980) predicted that the average length of time on a waiting list would soon be 21 years. Housing shortages were exacerbated by many local authorities keeping dwellings empty prior to selling them off. This led to a threefold increase (to 24,000 in 1982) in the number of council houses which stood empty for more than one year. Perhaps the biggest indictment of the RTB policy was that it was being implemented at a time when housebuilding in the public sector had slumped to its lowest level since the early 1920s. While there were only 299,800 house starts in the public sector, 1980–86, as many as 925,325 local authority and New Town dwellings were sold off (a net depletion of about 625,525 units, excluding demolitions). Although it was originally intended that sales receipts would be reinvested in local authority housebuilding, 60 per cent of receipts were frozen by central government in 1981 on the grounds that if all or most of the receipts were spent, the effect would be inflationary and inconsistent with government attempts to curb public expenditure.

Many of the above disadvantages of sales were cited by the House of Commons Environment Committee (House of Commons, 1981b) and it accused the Secretary of State of 'muddled reasoning, inadequate research and incorrect statistics in support of Government policies for encouraging council house sales'. The committee thought that the government had exaggerated the financial benefits to local authorities of selling homes, failed to recognise the special problems of rural and inner city areas, and had underestimated the likely increase in council waiting lists. It reported that in 50 years the public sector would have made a loss on sales; in 10 years the stock of council houses would have been cut by a third; tenants unable to buy (and prospective tenants) would have greatly reduced chances of being rehoused (or housed); and sales in the suburbs would make inner city problems more difficult while resales in the countryside would price low-income people out of the market without leaving them the option of low-rent accommodation.

Possibly the only circumstances therefore in which it is reasonable for tenants to be permitted to buy their own council homes is when there is a real surplus of council housing locally, both at the present and in the foreseeable future, or where sales are 'an integral part of a comprehensive housing policy that develops and protects the public rented sector and is geared towards equality between tenure' (Power, 1982) – a far cry from Conservative housing policy.

Initial effects of the 'Right to Buy' policy

Under the RTB provisions of the 1980 Act, the number of local authority dwellings initially sold off to their former tenants increased from only 568 in 1980 to 196,430 in 1982 (Table 8.5). There were distinct regional variations; for example, in the period October 1980 – March 1982 sales ranged from only 0.7 per cent of the council stock in Greater London to 3.3 per cent in the South West and 3.5 per cent in the East Midlands (Forrest and Murie, 1982).

In the early 1980s sales began to exceed local authority housing completions, and for the first time the local authority sector began to contract in absolute terms (Table 8.5). This, together with the decline of the private rented sector (see Chapter 6), severely disadvantaged one-parent families, newly-weds, divorcees and the elderly – all of whom required a more rather than less diversified tenure structure. It seemed probable that under current policy the council stock would continue to decline throughout the decade as the demand to buy would inevitably increase. Although a high proportion of buyers were in their forties and fifties and had below-average incomes, such buyers would continue to emerge throughout the 1980s in response to substantial discounts.

The General Election of 1983 and its aftermath

Before 1979, council house sales had been at the discretion of the local authorities. The Labour Party wished to return to this situation by repealing RTB legislation. *Labour's Programme 1982* (Labour Party, 1982a) stated that the party (when returned to office) would instigate the following:

1 relieve public landlords of any statutory obligation to sell;
2 cancel any outstanding options to buy taken out after 1 December 1981 and return deposits;
3 confer upon public landlords the right in perpetuity to repurchase former council dwellings at full market value when they next came on the market;
4 legislate to ensure that all future sales were at full market value.

Table 8.5 *Local authority dwellings sold under Right to Buy legislation, local authority completions and the local authority stock, Great Britain, 1980–98*

	Sales completed	Local authority completions	Local authority housing stock (000)
1980	568	76,997	6,499
1981	79,430	54,888	6,380
1982	196,430	33,200	6,180
1983	138,511	33,805	6,035
1984	100,149	31,593	5,924
1985	92,230	26,085	5,820
1986	89,251	21,547	5,724
1987	103,309	18,789	5,599
1988	160,569	18,997	5,412
1989	181,370	16,452	5,190
1990	126,215	15,686	5,015
1991	73,548	9,651	4,878
1992	62,986	4,141	4,788
1993	60,256	2,076	4,669
1994	65,174	1,869	4,528
1995	49,298	1,457	4,401
1996	45,094	828	4,301
1997	58,006	399	4,184
1998	55,914	375	4,117
Total	1,738,308	368,835	Change 1980–98: −2,382

Source: DETR, *Housing and Construction Statistics* (various)

This approach was reiterated in the party's 1983 election manifesto but almost certainly lost more votes than it gained. After the election, however, the party, while accepting that on grounds of political expediency it might have to come to terms with a mainly owner-occupied society and allow tenants to buy with little restriction, also recognised that it was important to focus attention on 'housing need' in its entirety. Instead of embracing the RTB philosophy of the Conservatives, at least a section of the party felt that it should instead initiate a 'Right to a Home' campaign. Hamnett (1983) argued that this would involve nothing less than a total reform of housing finance, centred on a universal housing allowance (not related to tenure) and linked to both income and the economic price of housing, and operated through the tax system.

The Liberal–Social Democratic Alliance, in contrast, wished to extend RTB provisions. In its 1983 manifesto it proposed that the scheme should be extended to include shared ownership and houses on leasehold land (affecting a further 500,000 tenants). The subsequent Housing and Building Control Act of 1984 of the returned Conservative government not only contained these provisions, but also increased the maximum discount by 1 per cent per year for tenants of between 20 and 30 years' standing, up to a maximum of 60 per cent, and reduced the eligibility period of tenure from three to two years. (The government had hoped to extend RTB provisions to cover 2 million homes specifically built for pensioners, but excluding sheltered housing. This proposal was thrown out by the House of Lords since it was feared that if these dwellings – short in supply – were passed on to heirs, they would be lost to this sector for the foreseeable future.) Further assistance to people buying their council homes was offered by the Housing Defects Act of 1984. Repair grants of 90 per cent became available to remedy defects not apparent at the time of purchase. At a total cost of £250 million deducted from local authority HIPs, it is probable that housebuyers benefited at the expense of less privileged council tenants who might have had to forgo repairs to their own homes.

By 1986 the pace of council house sales was flagging (there were only 89,251 sales in that year compared to 196,430 in 1982). Overall, seven-eighths of council tenants in England and Wales had failed to exercise their RTB and more than 25 per cent of all households (in England and Wales) preferred to remain as council tenants. The DoE was particularly concerned about the small number of flats which had been sold off (little more than 4 per cent, 1980–85). The Housing and Planning Act of 1986 therefore introduced discounts of 44 per cent on the purchase price of flats (for tenants of two years' standing). The repayment rule was reduced from five to three years as an incentive to buy and to assist mobility, but a limit of £25,000 was placed on discounts – a less than helpful constraint in London, where even two-bedroom flats could be valued at more than £100,000.

In the period 1980–87, however, nearly 800,000 local authority tenants purchased their homes under RTB legislation (Table 8.5) – tenure conversions contributing to the increase in the owner-occupied sector from 55 to 62 per cent of the total housing stock, and to the decrease in the public rented stock from 32 to 26 per cent (1980–87). But these changes were not attributable to discounts on sales alone. The government in 1985 further reduced the ability of local authorities to maintain an adequate level of housebuilding (even to meet welfare needs) by cutting the proportion of sales receipts which could be spent on construction from 40 to 20 per cent.

Labour's 'new realism'

It was widely believed (though not necessarily accurately) that Labour's opposition to council house sales was a principal reason for the party's electoral defeats in 1979 and 1983. The RTB provisions had in fact 'succeeded in portraying the Conservative party as champions of choice and rights, and Labour as bureaucratic bullies' (Griffiths and Holmes, 1984). Thus within weeks of the 1983 election, despite the many economic disadvantages of council house sales, Mr Roy Hattersley MP (subsequently Labour's deputy leader) argued for a rethink of the party's stance on the RTB issue.

Sections within the party articulated reasons why Labour had to come to terms with the reality of a mainly owner-occupied society. It was argued that the party needed to take account of the fact that in all capitalist countries there was a decline in rented housing and an increase in home-ownership (Lundquist, 1986) and that many of Labour's middle-class members were already owner-occupiers. Mr Bryan Gould MP (1984) claimed that according to a poll taken in 1983, 90 per cent of adults under the age of 35 hoped to be owner-occupiers within 10 years and that opposition to council house sales was not even socialistic since inadequate housebuilding (rather than RTB) was the principal cause of lengthening waiting lists and increased homelessness. Cowling and Smith (1984), moreover, argued that since 75 per cent of Labour voters had supported sales (in 1979), the party 'should develop a programme of universal owner-occupation'.

There were, however, many within the party who did not wish Labour to capitulate on the RTB issue and fall victim to narrow political expedience. Mr Allan Roberts MP (1984) thought that Cowling and Smith had wrongly assumed that everybody wanted to buy – in West Germany, for example, most people rented (from non-profit housing associations). What was required, he argued, was equality of treatment (both financial and otherwise) to enable households to make a realistic choice between renting and owning. Without 'capitulation', there was nevertheless a shift in Labour policy towards sales. At the party's 1984 conference, while it was accepted that opposition to the 1980 and 1984 Acts should remain, it was agreed that RTB would be acceptable provided that it was at vacant possession value and without discount, and provided that there was an increase in the supply of high-standard (public sector) rented housing – an approach broadly reiterated by Griffiths and Holmes (1984). The 1985 conference adopted the National Executive

Committee's statement *Homes for the Future* (Labour Party, 1985), which supported RTB but stated that capital receipts should be used to replace housing sold off (a practice largely blocked by the current Conservative administration), unless there was a surplus – an approach further approved by the 1986 conference, with the stipulation that RTB should be at market value. However, at the 1987 General Election it was the 1985 statement rather than the 1986 conference resolution that set out the party's policy on RTB. In its manifesto, while the party pledged to maintain RTB, there was no mention of market value being the basis for acquisition, but the manifesto did pledge that local authorities would be required to use the proceeds of sales to invest in new housing.

The rationale underlying Labour's revised policy on RTB was based on the belief that council house sales had presented the Conservatives with an issue which potentially guaranteed them electoral success: 'The Tories had noticed, perhaps before the Left, that tenure, not class, was now the single most reliable indicator of voting inclination' (Griffiths and Holmes, 1984). It was therefore thought that if Labour attempted to expand owner-occupation by embracing RTB, then it would at least neutralise the appeal of Conservative policy. Research undertaken by Johnson (1987), however, showed that owner-occupiers tended to vote Conservative by a ratio of 3:1, whereas council tenants had a preference for Labour by 2.5:1. It was therefore far from certain that if Labour increased the number of owner-occupiers it would be rewarded in the polls by a higher share of the vote. There is, however, little evidence to show that traditional Labour supporters vote Conservative after exercising their RTB, and also little evidence to suggest that RTB had any impact on the results of the 1979 and 1983 General Elections. Williams *et al.* (1987) found that although the Labour vote declined much more sharply among RTB purchasers than non-purchasers over the period 1979–83, the Alliance parties rather than the Conservatives obtained the greatest direct benefit. They also found that because the number of purchasers was very small in relation to the total electorate, the general effect of sales on the distribution of party preferences was very limited. Other issues were clearly of greater significance to voting behaviour. It thus seemed very unlikely that Labour gained any political advantage in shifting its stance towards council house sales – a conclusion partly borne out by the party's third successive electoral defeat in 1987.

The Right to Buy: the end of the road?

Despite the introduction of more generous RTB discounts in 1986, the Conservative government – in the aftermath of the party's election victory in 1987 – again raised the level of discounts to stimulate sales. In 1988 maximum discounts increased to 60 per cent for council house tenants of 15 years (starting at 32 per cent after two years), and discounts rose to 70 per cent for flat-dwellers after 15 years (starting at 44 per cent on qualification). In Greater London flats had accounted for only 36 per cent of total RTB sales in 1986, but, with larger discounts, flat sales made up 69 per cent of total RTB sales by 1991. Maximum discounts were raised from £35,000 to £40,000 in 1988 as a further inducement to buy, to take account of the house price boom and to facilitate purchase in the more expensive areas. With the subsequent house price slump, however, and rising unemployment, council house sales plummeted – tenants probably being more adversely affected by the recession than most other households.

There were large regional variations in RTB sales. Where there were large concentrations of flats and deprived inner-city neighbourhoods, such as in Yorkshire and Humberside, the North West and London, less than 40 per cent of the (1979) municipal housing stock had been sold off by the late 1990s, but in areas where there were far more houses than flats and very few run-down inner-city neighbourhoods, such as in the South East and South West, the council stock had been reduced by two-thirds (Table 8.6).

By the 1990s it was clear that local authority tenants who had exercised their RTB had been the recipients of the largest housing subsidy of all. According to Maclennan *et al.* (1991), RTB subsidies in the period 1980–88 averaged £12,094 nationally, and as much as £21,675 in London. The RTB process had dwarfed all other privatisation schemes. By mid-1992 the Treasury had received as much as £23 billion from the sale of council dwellings (and a further £3.7 billion was held by local authorities as housing capital receipts). In contrast, the privatisation of Britain's largest public corporations in the period 1980–86 raised only £7.3 billion – a 'property-owning democracy' being promoted more effectively than wider share-ownership. Without doubt, 'in social, political and economic terms, the most important element in the privatisation programme of the Thatcher governments [was the] sale of publicly owned dwellings' (Forrest and Murie, 1990).

Table 8.6 *Sales and transfers of local authority and New Town dwellings, regional distribution, 1979–99*

	Sales April 1979–March 1999 (000)				Stock at 1 April 1979 (000)	RTB sales as % of stock on 1 April 1979	Transfers as % of stock on 1 April 1979
	RTB	LSVT	Other	Total			
North East	116	0	5	122	286	41	0
Yorkshire & Humberside	143	8	15	165	427	33	2
North West	157	29	40	227	485	32	6
East Midlands	126	5	16	147	280	45	2
West Midlands	163	46	25	234	381	43	12
East	152	39	42	233	290	52	12
London	214	36	70	320	575	37	6
South East	176	125	51	325	268	66	47
South West	120	63	19	202	184	65	34
England (total)	1,367	352	282	2,001	3,178	43	11
Wales	105	0	7	112	197	53	0
Scotland	288	21	2	310	564	51	4
Northern Ireland	—	—	82	82	134	—	—
UK (total)	1,760	373	372	2,506	4,073	43	9

Source: ONS, *Regional Trends*

With regard to the replacement of the sold-off stock, it was regrettable that 75 per cent of the receipts from the sale of council dwellings in the early 1990s had to be paid to the Treasury to help pay off local authority debt, and only 25 per cent could be spent on housebuilding or rehabilitation (and even this was deducted from the amount that local authorities were permitted to borrow for capital expenditure). All receipts were thus deemed 'negative expenditure' and were regarded first and foremost as a means of lowering the PSBR, rather than as a means of funding housing provision.

RTB policy, moreover, had a significant effect on the quality and variety of local authority dwellings. Houses (and particularly three-bedroom semi-detached properties) in popular suburban locations were sold off without difficulty (Kerr, 1988), thus increasing the proportion of flats (including those with design defects) among the local authority stock. Similarly, while the proportion of two- and three-bedroom houses decreased, the proportion of one-bedroom houses or two- and three-bedroom flats increased; while the local authority stock was increasingly concentrated in the inner cities or less popular suburban areas rather than being spatially more evenly distributed (Forrest and Murie, 1984).

Despite generous discounts, however, many flat-dwellers soon began to question the benefits of home-ownership. Although they might have bought their homes with the aid of a mortgage arranged between their local authority and a mortgage lender, they often found it difficult or impossible to sell – even at very greatly reduced prices. Many building societies were unwilling to lend to prospective buyers if the technical structure of the building was unconventional – many flats, for example, were made of reinforced concrete utilising large panel systems (Boliver, 1992). Flats were also difficult to sell if owner-occupiers accounted for only a proportion of households living in an estate or in an individual block. To compound these problems, many former tenants faced substantial repair bills within a few years of buying; for example, an estimated 70,000 tenants in London needed to incur up to £40,000 each on repairs by the early 1990s. Although local authorities indemnified former tenants for 10 years against structural defects, there was no definition of what constituted a 'structural repair' (Spittles, 1992) – often to the detriment of adequate remedial work.

Under the Housing Defects Act 1984, however, the government had provided funds (allocated via HIPs) to enable local authorities to repair or buy back up to 30,000 defective systems-built dwellings bought by tenants under RTB provisions – a sum of £600 million having been spent on this process over the period 1984–91, rising to over £900 million by 1993. It was clearly hoped that this would not only facilitate the repair of defective privatised systems-built housing but also enable former tenants to sell either back to the local authority or on the open market. With much larger HIP allocations, local authorities would have been able to extend the repair or buy-back option to all former tenants having exercised their RTB but living in housing in disrepair. It had to be recognised, however, that allocations for this purpose might then have to be diverted away from the repair of over 200,000 local authority dwellings in a similarly poor condition, and this undoubtedly would be inequitable.

Some local authorities nevertheless bought back housing for reasons other than the need to undertake repairs. Welwyn and Hatfield District Council, for example, re-acquired some of its former housing stock in the late 1980s to house its homeless. Although up to £80,000 had to be paid to acquire each house (conferring large profits on former tenants who originally bought at discounts of up to £35,000), the depletion of the social rented stock was delayed (if only temporarily), to the benefit of the homeless and other low-income households unable to buy in the open market.

The problem of repairs and the difficulties involved in selling former council houses on the open market may not have been the principal reasons why the number of RTB sales diminished in the early 1990s. Despite generous discounts, many social-sector tenants still found it difficult or impossible to buy, particularly in relatively high-price or low-income regions. A rent to mortgage (RTM) scheme was thus introduced – initially in Scotland in 1989, where owner-occupation was 20 per cent lower than in England. Scottish Homes began to administer a scheme in respect of its 75,000 dwellings, which enabled its tenants either to part-rent and part-buy or to convert weekly rent payments into mortgage repayments, and further pilot schemes were introduced in Basildon, Milton Keynes and rural Wales in 1991. Under the Leasehold Reform, Land and Urban Development Act 1993, the government introduced a nationwide RTM scheme, whereby tenants were able to take a part share in their home by converting their rent payments to mortgage repayments and subsequently either stepping up payments or cashing in their equity if they moved home. Tenants were eligible for a discount on the market price of their homes, ranging – in the case of houses – from 30 per cent (after two years' residence) to 60 per cent (after 30 years), and – in the case of flats – from 30 to 70 per cent (after 20 years). It is uncertain, however, whether RTM will be successful. Only 400 tenants took advantage of RTM under the four pilot schemes referred to above in the period to April 1992, and as many as 3.7 million council tenants out of a total of 5.2 million were excluded from the RTM scheme since they were housing benefit recipients and deemed to lack 'credit-worthiness'. The scheme, moreover, was of little help to elderly tenants wishing to move out of London to be near their children living in the provinces or to tenants in high-price areas, where (even with discounts) their homes were unaffordable.[1]

In a further attempt to facilitate home-ownership and, at the same time, help reduce homelessness, the DoE permitted local authorities to award grants to council tenants wishing to buy in the open market. In 1988, 11 London boroughs offered council tenants cash grants of up to £17,500 to move out and buy elsewhere. Although the grants were worth only half the value of the discounts tenants would have received had they exercised their RTB, they were sufficiently large to have enabled tenants to buy in the provinces. Although the DoE funded 75 per cent of the cost of grant assistance, local authorities in London and the South East were authorised to borrow money to fund the remainder of the cost of buying in the open market. Between 1989/90 and 1990/91, credit approvals to local authorities (where homelessness was most acute) increased from £41 million to £50 million. However, compared to the number of dwellings purchased under the RTB scheme, 126,215 in 1990, the number of assisted acquisitions in the open market was negligible – only 1,453 in the same year.

During their 18 consecutive years in office, the Conservatives had clearly succeeded in boosting sales by increasing discounts, especially on flats; by introducing a rent-to-mortgage scheme; and by helping tenants to buy homes other than the one they occupy.

Right to Buy under Labour

In 1997 the new Labour government inherited a situation in which nearly 1.7 million council houses in Britain had been sold under RTB legislation, and the total number of dwellings in the local authority sector had fallen by about 36 per cent from around 6.5 million in 1980 to about 4.2 million in 1997 (Table 8.5). Home-ownership, meanwhile, had grown from 55 to 67 per cent over the same period, with nearly half the increase attributable to RTB.

However, although RTB had encouraged many relatively affluent former tenants to remain in their neighbourhoods 'helping to create stable, mixed-income communities' (DETR/DSS, 2000: 36), there were many a priori reasons for abolishing RTB. It had been estimated that the cost to the taxpayer amounted to about £10,000 for each sale (or £13 billion in total), many buyers faced financial difficulty in maintaining their homes, and RTB led to a depletion of the more desirable local authority stock in the more attractive areas, leaving local authorities with poorer-quality housing in the least popular estates (DETR/DSS, 2000). Not only were there large shortages of affordable rented housing in the urban areas of the South East and South West, but rural areas across Britain were facing acute shortages of low-cost homes as a result of more than 100,000 council houses being sold off under RTB legislation (Hetherington, 2000b).

In the late 1990s, however, lower-income households could satisfy their demand for home-ownership in the open market more easily than during the the Thatcher years. Houses were more affordable to people in employment than hitherto (except in parts of London and the South East); people wishing to buy could now afford to do so at an earlier age; and owner-occupation was higher in the United Kingdom than in most other countries in northern Europe (Perry, 1998)

Nevertheless, because of its apparent popularity among households, RTB was not abolished by Labour but became subject to modest reform to ensure that sales provided value for money to the taxpayer, and that home-ownership was sustainable. Thus, although tenants remained eligible for discounts of up to 60 per cent on a house and 70 per cent on a flat, the national maximum of £50,000 was replaced in 1999 by nine regional limits ranging from £22,000 in the North East to £38,000 in London and the South East.

In Scotland – under its Housing Act of 2001 – the creation of a single social housing tenancy extended RTB to the tenants of almost all RSLs. Since – at the draft stage of the legislation – it was thought that this would 'pose a threat to both their stock and asset base' (Earley, 2000: 29), it provoked some opposition in principle not only from the Scottish Federation of Housing Associations and private capital but from others anxious to maintain the supply of affordable homes for households unable or unwilling to buy. Thus while the Act enabled existing tenants to exercise an immediate RTB and be eligible for discounts set in the late 1990s, new tenants were restricted from exercising their RTB for up to 10 years and discounts were to be capped. Clearly, the supply of afforable rented homes would be depleted in those areas where there was a preponderance of existing tenants, especially where locations were attractive, but maintained in the less attractive areas and/or where there was likely to be an inflow of new tenants.

Estate privatisation and rehabilitation

Estate privatisation, in its many forms, was motivated by the perceived fiscal need to transfer the responsibility of repairs from government to housing associations and the private sector, and by an awareness that there was a limit to the number of council dwellings that could be sold off under RTB and RTM policy.

By the mid-1980s the government acknowledged that the total cost of repairs and maintenance to council housing amounted to almost £20 billion, but refused to recognise that local authorities could afford to spend any more than, for example, the £1.3 billion they spent on renovating their stock in 1985/86. Yet council housing needed to be renovated. If expenditure was insufficient, the condition of the stock would deteriorate further and the cost of future renovation would be astronomical – with dire social and economic implications. Local authorities, however, currently held at least £6 billion of frozen sales receipts in their capital accounts, and the net value of their housing stock (taking into account an outstanding debt of £25 billion) was as much as £75 billion. They would thus have been able to spend £20 billion on repairs, etc. by using their own capital receipts and raising a second mortgage on their assets (Page, 1987). But the government believed that local authorities were incapable of undertaking repairs efficiently even if resources were fully available, and the Treasury would have been reluctant to see the PSBR rise by up to £20 billion and would have been (needlessly) concerned about the effects of subsequent expenditure on the rate of inflation – the inflationary effects of expenditure on construction work would have been minimal. Clearly, the government wished to shed its direct and costly responsibility for renovating housing, and thus adopted a policy of large-scale estate privatisation.

Some local authorities therefore began to undertake large-scale privatisation programmes, involving the displacement of tenants and subsequent rehabilitation. Until 1986, councils could move tenants out of their homes against their will only if repairs and maintenance were to be effected. But in many instances, dispossession or 'municipal winkling' was quickly followed by the sale of the vacated housing to property companies that subsequently refurbished the dwellings prior to selling them off for upmarket owner-occupation. In the London Borough of Wandsworth, for example, flats in hard-to-let estates were sold after refurbishment for up to £75,000 and soon resold for £100,000. Under the Housing and Planning Act of 1986, however (and in respect of redevelopment approved by the DoE), local authorities were given powers to force recalcitrant tenants to move to alternative accommodation to facilitate estate disposal; to sell off blocks of flats or whole estates with sitting tenants; and to put out tenders to private agents or non-profit trusts. Under the Act, urban regeneration grants (city grants after 1988) became available to developers.

Critics of estate privatisation argued that the 1986 Act was irrelevant to the urban housing problems of the United Kingdom in that the legislation did nothing to alleviate the shortage of affordable rented accommodation at a time when council waiting lists were growing and the number of homeless households was increasing. It was also alleged that privatisation was being blatantly used for political advantage. In Wandsworth, for example, where dozens of tower blocks had been sold off as luxury flats, 'councillors . . . made no secret of their aim of bringing more Conservative voters into the borough', while conversely removing '2,000 tenants from the electoral register' (Linton, 1987). It was thus not surprising that Battersea North (within 'gentrified' Wandsworth) elected a Conservative MP in 1987 – the constituency having had a Labour member almost continuously since 1892. Tenure shift continued apace. In 1994, for example, figures showed that the borough was planning to spend £36 million on preparing council houses for sale while simultaneously spending nearly £7 million on accommodating the homeless in bed and breakfast hotels. By 1994, 7,898 vacated properties out of a total municipal stock of 42,000 (in 1980) had been sold off – only about 3,000 less than the number sold under RTB legislation. Similar policies were allegedly being applied in the City of Westminster, where it was claimed that £21.25 million had been spent selling off vacated council dwellings in the late 1980s (Linton, 1994).

Estate privatisation became increasingly attractive to the government when it became obvious that RTB policy alone could not be relied upon to reduce the size of the council stock to that of a residual tenure. With councils owning 6.6 million council dwellings in 1979 (equivalent to 32 per cent of Britain's housing stock), RTB policy succeeded in reducing the council stock to 5.8 million by 1986 (taking into account housebuilding within the sector – albeit on a small scale). From a peak of 196,430 RTB sales in 1982, sales decreased to only 89,251 in 1986, and although they recovered in subsequent years (owing largely to the introduction of more generous discounts), the government nevertheless became increasingly aware that there was a limit to the number of council dwellings that could be sold off. The best properties had already been sold off, unemployment was rising and employment was insecure, and many who remained within the sector were either too old or too young to contemplate buying (Forrest and Murie, 1990). The problem of repairs and difficulties of resale also deterred many tenants from exercising their RTB. Estate privatisation would clearly speed up the process of tenure shift.

The beginnings of estate transfer

The privatisation of estates was initially motivated by the need to provide the means of renovating up to 1.3 million substandard, run-down or badly managed council estates. The DoE thus launched the Urban Housing Renewal Unit (UHRU) in 1985 (renamed Estate Action in 1986) to provide assistance to local authorities, to draw in new private sector funds and urban regeneration grants, and to supplement existing HIP allocations from the DoE. After rehabilitation, dwellings could either be retained under local authority ownership, transferred to trusts and rented at economic levels, or sold off as low-cost housing. By 1986, 42 local authorities (half of which were Labour controlled) had embarked on 80 estate sale schemes under UHRU, although with only £50 million at its disposal in 1985/86 UHRU could not have been expected to make much impact on a £20 billion repair problem (Milne, 1986).

In subsequent years, however, DoE allocations to Estate Action increased from £75 million in 1987/88 to £373 million in 1994/95, the proportion of schemes involving private finance increased from 20 per cent to 65 per cent, and disposals to the private sector ranged from 3,200 to 4,200 per annum over the same period but representing only a small proportion of renovated stock (Table 8.7).

Estate privatisation was not, however, confined to the inner cities or suburbs. Under the 1986 Act, up to 60,000 houses scheduled for transfer from the remaining New Town development corporations to local authorities by 1992 were privatised – the Secretary of State for the Environment having overseen their transfer to housing associations, building societies and property companies. It was questionable whether such transfers were popular. In advance of Peterborough Development Corporation being wound up in 1988, an opinion poll showed that over 90 per cent of the corporation's tenants expressed a preference to be transferred to the local authority rather than to a housing association (Ardill, 1987).

By 1988 it was recognised that RTB policy was unlikely to remain effective as a long-term means of privatising the council stock, and it might also have been thought that 'municipal winkling' (prior to refurbishment and sale) would have only limited use since it could take place only in a few boroughs or districts where (apart from having an appropriate housing stock) there was a radical political commitment to shift tenure from the public to the private sector. New and more effective legislation was thus thought necessary to extend privatisation further.

Table 8.7 *Estate Action investment and disposals, 1986/87–2001/02*

	Financial provision (£m)	Continuing schemes	New schemes started		Dwellings improved	Disposals to the private sector
			Total	Involving private finance		
1986/87	45	0	138	18	—	3,400
1987/88	75	85	106	22	47,000	3,200
1988/89	140	130	190	14	79,400	4,000
1989/90	190	213	162	30	63,700	4,200
1990/91	180	212	118	41	49,500	2,500
1991/92	268	210	163	77	62,600	3,800
1992/93	348	236	162	107	69,200	3,000
1993/94	357	248	163	105	66,000	3,900
1994/95	373	285	115	57	46,178	4,543
1995/96	316	275	—	—	26,313	914
1996/97	252	197	—	—	20,239	151
1997/98	174	109	—	—	10,997	71
1998/99	96	47	—	—	6,956	84
1999/2000	67	—	—	—	—	—
2000/01	64	—	—	—	—	—
2001/02	39	—	—	—	—	—

Source: Wilcox (2000)

The Housing Act 1988

Based on the White Paper *Housing: The Government's Proposals* (DoE, 1987), the Housing Act of 1988 was in part directed at the problems of some of the largest council estates, which appeared to many to be unmanageable. First, under the Act – and subject to local ballot – the government planned to establish a number of housing action trusts (HATs) to repair or rehabilitate housing estates, to improve management, to produce a greater diversity of tenure, and in general to improve social and living conditions. In July 1988 it was announced that the government would explore the possibilities of setting up HATs (responsible for 24,525 dwellings) in Lambeth, Southwark, Tower Hamlets, Leeds, Sandwell and Sunderland, and introducing in these areas a three-year £125 million renovation programme – scheduled to begin in 1989.

In many respects HATs are similar to urban development corporations (UDCs). In principle, they were to be imposed on run-down estates even if local authorities refused to co-operate; they were to be run by bodies chosen by the Secretary of State rather than by locally elected (and accountable) councillors; and following the transfer of their property to new social or private landlords (after building works had been completed and subject to a further ballot among tenants), they were to be wound up. Tenants wishing to revert to a council tenancy after refurbishment would be able to do so subject to local authority agreement.

Critics suggested that since most of the housing likely to be transferred to HATs would be in poor condition and contain sitting tenants, the market value which local authorities would realise on sale would be ludicrously low in relation to the outstanding debts on the properties. Tenants feared that they would lose security of tenure or be faced with substantial rent increases up to market levels (after repair or rehabilitation) unless compensated by housing benefit; and the needs of the homeless would continue to remain unsatisfied since HATs – although being located in areas of deprivation – have no specific responsibilities to this growing section of the community.

Because of considerable opposition from both tenants (expressed through ballots and soundings) and local authorities, the DoE decided to abandon its plans to set up HATs in its initially selected areas (Karn, 1993). It was ironic, however, that HATs were subsequently established in Liverpool, Waltham Forest, Hull North and Castle Vale (Birmingham) largely as a result of local authority instigation or agreement and backed by supporting ballots in the estates concerned. It was clearly recognised locally that the enormous problems involved in rehabilitating the worst of the municipal housing stock could be overcome, in the current circumstances, only if financial resources were attracted into the relevant estates through the medium of HATs. Consequently, in 1991/92 the HATs in Liverpool, Waltham Forest and Hull North were allocated respectively £190 million, £160 million and £50 million by the DoE (Karn, 1993).

The second way in which the 1988 Act facilitated the disposal of council estates was by means of large-scale voluntary transfers (LSVTs) to housing associations and (in principle) to private landlords. Transfers – requiring consent from the relevant Secretary of State – were already taking place under both the Housing Act of 1985 and the Housing and Planning Act of 1986, but the 1988 Act removed ambiguities in the previous legislation and clarified the process of transfer. By 1994, 32 local authorities had transferred the whole of their stock – amounting to a total of 149,478 dwellings – and nearly 250 other local authorities were considering transfer. The largest transfer (involving 12,400 homes) took place when the London Borough of Bromley transferred its total stock to the new Broomleigh Housing Association (set up by the council itself).

LSVTs increased markedly in the late 1990s, and in 1999/2000 alone, 16 local authorities transferred over 80,000 dwellings to housing associations at a total transfer price of £659 million (Table 8.8), and prices per dwelling ranging from as little as £1,942 to over £14,000.

Table 8.8 *Local authorities transfer of housing stock, England, 1999/2000*

Local authority	Number of dwellings transferred	Total transfer price (£m)	Average price per dwelling (£)
West Lindsey	3,926	30.7	7,814
Boston	4,871	43.4	8,907
Tynedale	3,564	33.9	9,500
Newcastle under Lyme	9,887	54.0	5,462
Restormel	3,577	25.1	7,018
Manchester	600	0.9	1,500
North Devon	3,293	44.3	13,438
Burnley	5,330	22.4	4,205
Manchester	1,655	2.4	1,419
Weymouth & Portland	3,150	31.0	9,983
Huntingdon	6,650	63.4	9,534
Elmbridge Test Valley	4,894	56.9	11,626
Test Valley	5,495	80.0	14,559
Wyre Forest	6,056	41.0	6,762
Manchester	1,033	2.0	1,942
Tameside	16,466	127.5	6,763
Total/average	80,408	658.9	7,992

Source: Wilcox (2000)

To ensure that management is sensitive to the needs of tenants and to prevent the emergence of monopolies, the government was generally opposed to the transfer of more than 10,000 houses to a new landlord. Housing associations fund the cost of transfer by borrowing from the private sector (notably from the larger building societies) and pay back the loans from rents and receipts from future RTB sales.

Like the setting up of HATs and the subsequent disposal of HAT properties, LSVTs could proceed only if the majority of tenants voted in favour of transfer (with abstentions counting controversially as 'yes' votes in some of the earlier ballots). Councils clearly benefit from LSVT. If they retained ownership, they might, in any case, lose control of their housing since the management of estates could be delegated to private firms through the CCT process. Councils nevertheless would have the responsibility of paying housing benefits (redistributing rent revenue from better-off to poorer tenants) through the medium of ring-fenced HRAs. If, however, council estates were transferred to housing associations or to private landlords, it would be the DSS which would have the direct responsibility for paying housing benefits and not the local authority. Central government, nevertheless, benefited from LSVTs since capital receipts act as a check on the PSBR (Grant, 1992; McIntosh and Utley, 1992).

Clearly, with a total of 149,478 dwellings transferred out of the local authority sector by 1994, LSVTs were more than compensating for the reduction in RTB sales after 1989. But it was becoming uncertain whether the private sector could continue to fund transfer on a large scale. Only 20 transfers a year would probably require an additional £1.1 billion per annum, or twice the amount that housing associations need for programmed Housing Corporation mixed-funded schemes (Mullins *et al.*, 1993).

Under the 1988 Act, council tenants are able to exercise 'Tenants' Choice' – the third way in which 1988 legislation aimed at transferring housing out of the local authority sector. Adding to the rights and choices already given to tenants by the 'Tenants' Charter' of the Housing Act of 1980, tenants of council houses individually were able to exercise their right to transfer to another landlord, but tenants of flats must decide collectively. In either case, a tenant could choose to retain the local authority as his or her landlord even if the majority of tenants were willing to transfer. New landlords had to gain the approval of the local authority in accordance with criteria laid down by the Housing Corporation. By the end of 1991, however, not a single property had moved away from local authority ownership (Mullins *et al.*, 1993), although subsequently 'Tenants' Choice' gradually took off. Often, the poor condition of the relevant housing deterred potential new landlords from taking over estates. Local authorities were therefore sometimes obliged to pay 'dowries' to the new landlord to cover the cost of essential repairs – Westminster City Council, for example, paid £17.5 million in 1991 to a new tenant-controlled company resulting from an assessment of the estate's negative value by the District Valuer.

Box 8.1

The transfer of council estates: Birmingham and Glasgow

At the beginning of the new millennium, Birmingham City Council (with around 92,000 dwellings) and Glasgow City Council (with about 93,000 homes) were the largest local authority landlords in Britain, and both were planning to transfer the whole of their stocks to newly established housing trusts, bypassing the many housing associations already active at a community level in both cities. Both Birmingham and Glasgow had substantial housing debts (amounting to £650 million and £1 billion respectively) and enormous repair backlogs, and thus had a very clear financial motive for selling off their stocks.

At the time of writing, the details of neither of the two transfers had been finally agreed. However, in Birmingham the formation of a housing action trust (HAT) at Castle Vale in 1993 was possibly the precursor to the city's single housing sell-off, set to take place in November 2002. With a population of about 11,000, accommodated in 34 high-rise flats, Castle Vale was developed as the city's largest estate in the 1960s, but subsequently it became a host for social deprivation and exclusion associated with a high rate of long-term unemployment, crime, drugs and housing afflicted by condensation, dampness and structural defects (see Smith, 1998). In 1993, 92 per cent of its residents voted in favour of the estate being handed over to an HAT, since direct funding from Westminster was essential for the successful regeneration of the estate on the scale required. Castle Vale thus became only the sixth HAT to be established since enabling legislation was passed in 1988. With an initial Exchequer allocation of £42.5 million, a start has been made on replacing each of Castle Vale's tower blocks with attractive three- and four-storey houses, but by the time the project is completed (possibly in 2005), a total of at least £200 million will have been invested in regeneration, with funds derived from central government, the European Commission and rents. From the outset, residential involvement in the regeneration process has been important. An elected Estate Forum was set up, composed entirely of residents, and this liaises with the HAT over tenants' concerns and community issues – a process eased by four of the HAT trustees being tenants themselves. Clearly, a 'bottom-up' rather than a 'top-down' approach is much in evidence. Around 600–700 tenants were involved in development projects in the late 1990s, from a Job Club (to help the long-term unemployed back to work) to health, education and training programmes and the provision of a youth and community centre (see Smith, 1998). When the regeneration programme has been completed, the residents will have the opportunity to vote on whether to transfer to a new social landlord with access to both public and private funding, or return to local authority management.

In Glasgow, as part of the Scottish Executive's New Housing Partnership programme the city council planned to transfer the whole of its stock to a newly set up body, the Glasgow Housing Association (GHA), with the Executive agreeing to service the city's £1 billion housing debt. As an umbrella organisation, the GHA – as the largest ever housing association – will function under the direction of a management committee of four councillors, one MP, one MSP, five tenants and five independent members. To devolve management and tenant involvement, it was proposed that 14 area housing partnerships (AHPs) would be established in the city, each under the control of a board of tenants, councillors and independent members, and serviced by councils and Scottish Homes. Within each of these partnerships, an unspecified number of local housing organisations would be established to provide housing management under contract from the GHA. These would be controlled by tenants and registered with Scottish Homes (see Forshaw, 2000a). While, in some quarters, being hailed as the most progressive way of securing community involvement, the proposed transfer has been opposed by those who see any diminution in local authority responsibility as tantamount to privatisation, even though Glasgow already has more tenant co-operatives and community-based housing associations than any other city in Britain (Power, 2000). Clearly, a ballot will determine whether or not the transfer will go ahead as planned, but at the time of writing, a ballot had not been held.

In both Birmingham and Glasgow, there is the possibility that the transfer of housing from local authority to housing association ownership will be little more than a vast exercise in replacing one form of gigantism with another. In each city, moreover, private financial institutions might prefer to invest in 'less clumsy, more manageable, bite-sized bodies where innovation can drive a fresh start' (Power, 2000: 7). In these circumstances, only time will tell whether in these cities at least, RSLs are as competent at managing mass housing as their local authority predecessors.

In Scotland the Housing (Scotland) Act of 1988, like its English and Welsh counterpart, encouraged alternatives to public sector provision. The government seemed particularly concerned that 'north of the Border' as much as 43 per cent of the housing stock was council-owned compared to only 27 per cent in Great Britain as a whole, and thus attempted to lower the proportion of public sector housing to the national level, which itself was being reduced. The Scottish Act introduced a tenants' choice scheme (*vis-à-vis* council housing) whereby tenants (individually or collectively) were given the right to transfer to another landlord (following a straight majority vote in favour of a transfer). The scheme was open to all council tenants (with minor exceptions) and to any (non-public) landlord approved by Scottish Homes (a new unified central agency) and to Scottish Homes itself. As in England and Wales, (low) market values were the basis for transfer. There was, however, no provision in the Scottish legislation to introduce HATs, although Scottish Homes performed a similar function in respect of run-down estates. A chief criticism of the Act is that local authorities (as in England and Wales) would become increasingly unable to fulfil their statutory obligations to the homeless.

From the outset, the Institute of Housing was critical of the Conservatives' stand on public housing. It believed that a national study should be undertaken to provide up-to-date information before any decisions were taken to privatise estates; it proposed that local and central government should enter into partnership to discover the best solutions; it welcomed increased involvement of private capital but pointed out that local authorities were prevented from spending the bulk of their revenue from council house sales; it warned that the price to be paid for private involvement would be rents 40 per cent higher than currently prevailing, with a corresponding strain on housing benefits; and it feared that welfare housing for the elderly, the sick and the otherwise disadvantaged would be all that remained of the council stock if Conservative policies finally ran their course (Institute of Housing, 1987).

In the late 1980s and early 1990s, the government proceeded apace with its policy of estate transfer. At the General Election of 1992 it pledged that it would continue Estate Action and HAT programmes to concentrate resources on the worst council estates, and would enable tenants to apply for HATs to take over and improve the worst estates; would continue to encourage LSVTs to housing associations, and would bring management closer to tenants by reducing the limit on the number of properties transferred in a single batch; and – as part of Estate Action – would introduce a new pilot scheme to promote homesteading, whereby local authorities would offer those in housing need the opportunity to restore and improve council dwellings in exchange for lower rents or the option to buy at reduced prices. The Labour Party, in contrast, was generally opposed to the transfer of council estates, although individual Labour-controlled authorities sometimes welcomed the setting up of an HAT in their area or contemplated an LSVT to help attract an adequate inflow of funds to facilitate estate improvement. The Liberal Democrats at the 1992 election, however, openly favoured the right of council tenants to opt for a change of landlord, but only after a fair ballot.

By the early 1990s (if not before), it was possible to see the Conservatives' privatisation policy in perspective. In net terms, the number of council houses had decreased from 6.6 million in 1979 to 4.9 million in 1992, but local authorities still owned 21.3 per cent of the total stock of housing in Britain in 1992, while private landlords owned only 7.5 per cent and housing associations 3.3 per cent. Clearly, the majority of council tenants were satisfied with their landlord and had no wish either to exercise their RTB or to transfer to an alternative landlord. In a Gallup poll for the National Consumer Council in 1988, for example, 51 per cent of balloted council tenants wanted to keep their local authority as landlord, 21 per cent preferred to transfer to a building society, housing association or tenant co-operative, and only 1 per cent would have opted for a private landlord (*Guardian*, 1988a). A Mori poll for the London Research Centre, moreover, revealed that about 94 per cent of council tenants thought their tenure was ideal, whereas only 6 per cent of council tenants would have chosen a housing association as landlord and less than 0.5 per cent would have favoured a private landlord (*Guardian*, 1988b).

Undoubtedly, the popularity of RTB policy helped the Conservatives win the 1979 election, but there was little evidence that the transfer of council estates was electorally attractive. An ideological hostility to local government and the desire to privatise the financing of housing rehabilitation, rather than electoral support, was the underlying rationale of the 1988 Act and subsequent policy.

Residualisation

Political attitudes towards council housing have become increasingly polarised in recent years. The Conservative Party has come to regard council housing as a 'safety net' for the disadvantaged rather than a general needs tenure and to favour the disposal of as much as possible of the public sector stock by means of RTB, the RTM scheme and estate transfer. After 70 years of growth, council housing is in decline, while owner-occupation is more and more regarded as the 'natural' tenure for the majority of households. The Conservatives seem to hold the view that 'council housing is only for those who, whether through poverty or lack of moral fibre, cannot make the grade as owner-occupiers – a second-class sector of second-class people' (Griffiths, 1982).

The Labour Party (1982a) claimed that the Conservative government, in selling off council houses (and reducing subsidies to tenants), was withdrawing from the post-war consensus on the need to ensure an adequate supply of affordable rented housing. Council house sales and estate transfer – mainly involving the best housing and the more affluent tenants – are increasingly residualising the public rented sector.

The quality of the council stock is thus deteriorating. Whereas in the inter-war or post-war periods council housing was a privileged or preferred tenure, in recent years much publicity has been given to substandard, unimproved or difficult-to-let dwellings, design and structural faults, and the unpopularity of high-rise flats. In its conception, design and construction, council housing is often blamed for anti-establishment behaviour and social disintegration. Rising crime rates and the widespread urban riots of 1981 and 1983 with further disturbances in 1986 in Tottenham and in Newcastle upon Tyne in 1991 generated concern about the relationship between environment and behaviour (see Coleman's (1985) detailed analysis of design and social malaise). Housing and urban problems, moreover, 'have been brought nearer the centre of the political agenda through critical reports highlighting the extent of urban deprivation [and] the lack of investment in the physical fabric of dwellings' (Forrest and Murie, 1990).

Whereas a relatively high proportion of intermediate non-manual and skilled manual workers were council tenants until the 1960s, subsequently there was less of a 'social mix' and a degree of polarisation. There has also been an increasing and disproportionate number of council tenants (lone-parent families, the sick and disabled, and the elderly) dependent on supplementary benefit, largely because of increased unemployment – for example over the period 1972–86.

The privatisation of much of the council stock has clearly accentuated the social and economic differences between council tenants and other households. Curtice (1991) suggested that an increasing proportion of council tenants belonged to a so-called 'underclass' dependent upon the state rather than the labour market for most of its income. In 1990 it was evident that council tenants not only received substantially lower incomes than owner-occupiers and other renters, but received the highest welfare payments (Table 8.9).

Table 8.9 *Household income and welfare payments – by tenure, United Kingdom, 1990*

Households with at least one person receiving:	Council tenants (%)	Other renters (%)	Owner-occupiers (%)
Income in the lowest quartile	60	33	16
Income in the highest quartile	6	28	29
Housing benefit	43	34	6
Income support	26	25	6
Unemployment	22	21	13
Family credit	12	3	1
One-parent benefit	11	6	2

Source: Curtice (1991)

By the late 1990s the residualisation of council housing (or more widely, social rented housing) was if anything even more marked, although it is difficult to compare indicators for 1998 with those of 1990. Table 8.10 shows that in terms of income, heads of households in the social rented sector were considerably worse off than those in other sectors, they were less likely to be in employment, and more likely to be elderly. The social rented sector also contained a disproportionately large number of inactive heads of households (other than the retired), and an above-average proportion of single-person households and lone parents.

Over the years, council house estates have undoubtedly become 'stigmatised ghettos of welfare housing for the poor'. While the Conservatives preached the virtues of home-ownership, the Labour Party claimed that Thatcherism was based negatively on the belief that 'only the disabled, elderly and the "poor" who cannot make the grade in private markets should be provided with public housing' (Labour Party, 1982a). Undoubtedly, there is 'little disagreement that council housing serves the most vulnerable and marginal groups in society' (Forrest and Murie, 1990). Whereas in the past, the provision of council housing eroded the connections between low income and poor-quality housing, since the late 1970s the decrease in the quality and size of the council housing stock 'has been paralleled by the growth of an underclass of economically and socially excluded households. They are increasingly concentrated in a downgraded public housing sector' (Forrest and Murie, 1990).

Table 8.10 *Economic and social attributes of households – by tenure, England, 1998*

	Sector		
	Social rented	*Private rented*	*Owner-occupied*
Average gross weekly income of heads of households (£)	160	326	482
% heads of household in employment	31	64	67
% heads of households aged 65 & over	34	14	24
	Social rented		*All sectors*
% heads of households who were inactive (other than being retired)	26		10
% one-person households	41		28
% lone parents	16		7

Source: DETR (2000c)

It is arguable whether or not the above trends are truly indicative of residualisation because of either generality or the lack of tenure specificity. A decrease in the proportion of council dwellings to the level of the early 1950s would not necessarily imply residualisation if the tenure regained its privileged status; conversely, if the tenure expanded to levels found in, for example, Liverpool, the degree of residualisation might increase. Qualitatively, there are very great geographical variations in the council stock, and as the *English House Condition Survey 1981* (DoE, 1983) showed, local authority housing compared very favourably with private rented and even owner-occupied housing in terms of the proportion of dwellings which were unfit, lacking amenities or in serious disrepair. Also, it cannot be universally claimed that a 'one-class' community gets a worse housing service than that provided in an area of greater 'social mix' – a look at inner-city local authorities under different political control would cast doubt on that proposition. As far as supplementary benefits (income support) and low incomes are concerned, these are reflections of the state of the economy – for example job deskilling and high unemployment – rather than anything specific relating to tenure, and the decreased level of investment and subsidy in the public housing sector may have been a result of both monetarism and an attempt to expand owner-occupation rather than residualise council housing. However, regardless of whether or not the indicators of residualisation are adequately precise, it is important to realise that they are only symptoms, and not in themselves determining factors.

The residualisation of council housing is therefore the result of broad economic and social processes and changes in other housing tenure. The deskilling of labour (as a result of deindustrialisation and technological innovation) and high unemployment have coincided with the withdrawal of private landlords from the low-rent market and the increased demand for owner-occupation among the more affluent – aided and abetted by government policy. With nowhere else to go (except perhaps into the small voluntary housing sector), the disadvantaged have become increasingly concentrated in council housing – ironically, at the same time as this sector has been starved of investment funds and subsidies, housebuilding has decreased to very low levels, while rents (and by necessity means-tested benefits) have soared. These trends exemplify the notion that all too often government policy and social services not only fail to influence the development of society, but reflect changes in the economy and the needs of the dominant class (in the case of employment, the managerial and professional class; in housing, the owner-occupier).

But why in the context of housing has lack of government concern become so obvious? Perhaps it is because for the first time since at least the Industrial Revolution, a sizeable proportion of the adult urban population is at best marginal to the productive process and at worst unnecessary. It can therefore no longer be assumed that the state needs council housing to satisfy the requirements of the manual classes.

Residualisation is thus the growing dependency of council tenants on welfare payments – a relationship evolving rapidly at a time when the market is having an increasingly determining effect on the supply of housing for those on or near average incomes, and when the satisfaction of effective demand is usurping the satisfaction of 'needs' as the basis for provision. There is, according to Forrest and Murie (1982), 'a possibility that the future role of council housing will be to provide a minimum residual service in an environment of greater housing inequality'. But residualisation also means segregation. It is socially divisive and perpetuates the 'two nations' syndrome. It is, as Aneurin Bevan (1945), as minister responsible for housing, argued, 'a wholly evil thing from a civilized point of view, condemned by anyone who had paid the slightest attention to civics and eugenics; a monstrous infliction upon the essential psychological and biological one-ness of the community'.

Conservative policy has thus not so much divided the rich from the poor as it has separated the manual classes into those who have bought or aspire to buy their homes, and those who cannot afford or wish to buy – the latter group living in estates which are rapidly becoming ghettos. Council housing is sometimes equated with 'poor law housing', and an increasing proportion of the remaining stock is of low standard, unimproved, high-rise, unpopular and ageing, and there is little likelihood of its being replaced in the near future. It was clearly the aim of Thatcherism to relegate council housing to a transitional role placed historically between the dominance of private rented housing in the nineteenth century and the supremacy of owner-occupation in the late twentieth century. While the Conservatives (argued Forrest and Murie, 1982) believed that 'the sale of publicly owned houses at large discounts . . . would distribute wealth, reduce social divisions and increase mobility and independence', their policy has further residualised council housing and created greater social polarisation and at a substantial economic cost in the form of higher housing benefits and rents. It can only be assumed that Conservative governments decided that this was the price which has to be paid to discard council housing and its role in the reproduction of labour power.

Conclusion

By the end of the 1990s, and regardless of the increasing residualisation of the council stock, Labour was intent on increasing the pace and scale of LSVTs. It was clear that the Blair government had no wish to maintain – let alone resurrect – council housing either as a general needs tenure or even as welfare provision. Table 8.11 shows that by 1999/2000, over 80 local authorities had already transferred more than 400,000 homes to RSLs since the relevant legislation was introduced in 1988. Over £6 billion of private finance had been attracted both to facilitate transfer and for investment in repairs and renovation. Following the approval of 160,000 transfers in 2000/01, the Green Paper proposed that from 2001/02 the government would support the transfer of a further 200,000 or more dwellings each year (subject to tenant approval) to eliminate England's 3 million council housing stock within 15 years.

Table 8.11 *Large-scale voluntary transfers, England, 1988/89–1999/2000*

	No. of councils	No. of dwellings	Total transfer price (£m)	Investment in modernisation, new stock and usable receipts (£m)	Treasury levy (£m)	Net balance (£m)
1988/89	2	11,176	98.4	72.8	—	25.6
1989/90	2	14,405	102.2	93.8	—	8.4
1990/91	11	45,512	41.0	296.6	—	117.8
1991/92	2	10,971	92.1	77.8	—	14.3
1992/93	4	26,325	238.0	88.5	—	149.6
1993/94	9	30,103	270.5	168.9	22.8	78.7
1994/95	10	40,510	406.3	218.1	53.4	135.4
1995/96	11	44,595	477.8	330.5	47.4	107.2
1996/97	5	22,248	192.6	117.9	9.6	69.9
1997/98	6	24,405	259.6	109.6	—	150.1
1998/99	11	56,072	484.1	354.2	—	151.4
1999/2000	14	80,405	658.9	515.4	9.6	199.7
Total	87	406,727	3,321.5	2,444.2	142.8	1,208.1

Source: Wilcox (2000)

Note: In most years the 'net balance' is the transfer price less the sum of investment in modernisation and new stock, and usuable receipts. From 1993/94 to 1996/97, and again in 1999/2000, a 20% Treasury levy was imposed on receipts (net of the housing debt).

Clearly, the government recognised that local authorities were increasingly unable to provide low-cost housing in good condition. There was an enormous backlog of repair, estimated to cost around £20 billion, and there were thousands of abandoned homes on difficult-to-let estates, particularly in the North (Hetherington, 2000b). However, local authorities are increasingly unable to meet the cost of estate modernisation. Under Treasury rules they are unable to use their housing stock (worth around £200 billion) or future rent income as security to borrow in the open market, whereas RSLs can obtain loans from private financial institutions without such restriction. This form of resourcing clearly reduces the need to make calls on the public purse at a time when the government is prioritising expenditure on health, education and transport. Transfers, moreover, are welcomed by many authorities and tenants as the only way of attracting much-needed investment for estate modernisation, and tenants feel that they might benefit from a more sensitive provision of 'street-level' services.

The number of future transfers, however, is uncertain. If the remaining council stock is transferred within 15 years (as signalled by the Green Paper), the number of transfers will soon exceed the number of RTB sales (1.7 million), and RSLs will replace local authorities as the major provider of social housing well before 2015. However, the pace of transfer might be less than predicted. As many as 70 out of 100 Labour housing chairs anticipated that ownership of their homes would remain under local authority control (Weaver, 2000), while a number of proposed transfers could be blocked by tenants (by 2000, 10 per cent of all proposals had been rejected). There was also some opposition to transfers within Parliament. For example, Mr Jeremy Corbyn MP (2000) called on Parliament to halt the transfer programme since transfers to local housing companies were effectively a form of privatisation, and in an early day motion resolved 'to join with council tenants and local authority trade unionists to ensure that further privatisation of council housing is halted'. Clearly, transfer is a one-off operation, and there is only one chance to get it right. There is also the danger that some RSLs might become large, autonomous and monopolistic bodies and act irresponsibly (unless they are under the control of the Housing Corporation as a regulator), while local authorities – with a statutory duty to house the homeless but with a disappearing stock of housing – might find it increasingly difficult to establish working relations with a plethora of new housing providers (Regan, 2000).

The extent to which local authorities transferred their stock to RSLs, 1989–2000, varied considerably across the United Kingdom. Table 8.6

reveals that in general, far more LSVTs took place in the south of England than elsewhere, ranging from 125,000 transfers (or 47 per cent of the council stock) in the South East to only 5,000 in the East Midlands (2 per cent) and none at all in Wales or the North East. In London only a modest 36,000 dwellings were transferred (or 6 per cent of the capital's council stock). Clearly, there was little relationship between the extent to which transfers had taken place and the need for funds to modernise the remainder of the council stock. This might suggest that the principal reason for embarking on large-scale transfer programmes in much of the South was the ideological desire to privatise council estates and secure receipts *per se*, rather than to modernise the remainder of the local authority stock, whereas in much of the North, and parts of London, local authorities either were willing to forgo the opportunity of raising finance through transfers because of their opposition to privatisation, or, quite simply, had they wished to sell, might not have been able to dispose of their stock because of low demand.

Under the Labour government, 1997–2001, transfers tended to get smaller as more and more local housing companies were established. Around 50 large run-down estates – in deprived inner city areas – were transferred to local companies, and since these operated on a small scale, were locally based and strongly community-focused, they were able to 'offer a unique chance to change conditions on the ground' (Power, 2000). To further this process, the Homes Bill 2001 proposed that local authorities would be encouraged to transfer their housing to new 'social' companies and trusts, which – like housing associations – are able to borrow in the open market to push modernisation programmes ahead.

Note

1 Although £414,000 was spent on advertising the RTM scheme, according to official figures it was taken up by only two households in the first six months following its introduction in October 1993.

Further reading

Clapham, D. (1989) *Goodbye Council Housing*, London: Unwin. Speculates on the demise of council housing in its current form.

European Capital (1997) *Private Financial Initiatives in Social Housing*, London: National Housing Federation. A useful analysis of how private financial initiatives facilitate large-scale voluntary transfers.

Evans, R. and Long, D. (2000) 'Estate-based regeneration in England: lessons from Housing Action Trusts', *Housing Studies*, Vol. 15, No. 2. An interim assessment of HATs, suggesting that the trusts have been relatively expensive compared with other methods of injecting new investment into social housing.

Forrest, R. and Murie, A. (1995) *An Unreasonable Act?*, Bristol: School of Advanced Urban Studies, University of Bristol. A clear examination of the issues raised by the conflict between Norwich City Council and central government concerning the sale of council houses.

Forrest, R. and Murie, A. (1988) *Selling the Welfare State*, London: Croom Helm. Provides a comprehensive study of council house sales in a number of local authority areas.

Mullins, D., Niner, P. and Riseborough, M. (1995) *Evaluating Large Scale Voluntary Transfers of Local Authority Housing*, London: Department of the Environment. Presents the findings of commissioned research on the effects of LSVTs.

Murie, A. and Jones, C. (1988) *Reviewing the Right to Buy*, Bristol: Policy Press. An illuminating study bringing together data and information on the effects of the RTB on the social rented sector.

Scottish Office Development Department (1996) *Transfers of Local Authority Housing Stock in Scotland*, Edinburgh: SODD. A report on the recommended procedure for transferring local authority housing to housing associations.

Taylor, M. (1996) *Transferring Housing Stock: Issues, Purposes and Prospects*, Occasional Paper on Housing, No. 10, Housing Policy and Practice Unit, University of Stirling. Key reading on the transfer of public housing in Scotland.

Wilcox, S. (1993) *Local Housing Companies: New Opportunities for Council Housing*, York: Joseph Rowntree Foundation. An important text on the role of local housing companies in reinforcing the social housing stock.

9 Housing associations

Although housing associations owned only 1.2 million dwellings in 1998 (accounting for a mere 5 per cent of the total stock of housing), they have become increasingly important in recent years, partly owing to a shift of emphasis from rehabilitation to their traditional role of building houses, partly as a result of their new role as principal providers of new social housing, but mainly because of LSVTs of housing from local authority ownership. Within this context, this chapter:

- examines the growth of housing associations from their beginnings in the early nineteenth century to the 1990s;
- reviews public policy towards housing associations from the late nineteenth century to the 1970s;
- explores the role of the Housing Corporation and housing associations in the 1980s;
- considers changes in investment and rent policies in the 1990s;
- identifies the ways in which housing associations can intervene in the private rented sector and facilitate shared ownership schemes;
- discusses the political consensus *vis-à-vis* housing associations;
- analyses the recommendations of the Housing Green Paper; and
- concludes by examining the provisions of the *Spending Review* (2000).

Growth of housing associations from their beginnings in the early nineteenth century to the 1990s

Housing associations originate from 1830, when the Labourer's Friendly Society was formed. The society built very few houses but those it did build were better than most low-income dwellings at the time and had proper drainage. Throughout the rest of the century, poor people failed to attract financial backing, and therefore charitable trusts were formed to attempt to show that private enterprise could provide decent housing for the working classes. Bodies such as the Guinness Trust, the Peabody Donation Fund, the Joseph Rowntree Trust (JRT) and the Sutton Dwellings Trust, formed in the nineteenth century, are still active today in supplying general family housing. Nowadays funds come less from charity than from local and central government, and increasingly from the financial institutions. Since 1964, registered associations have been eligible for loans and grants from the Housing Corporation for the purchase, rehabilitation and conversion of old houses or building of new dwellings. Until the early 1970s housing associations were able to charge cost rents to cover the cost of construction and maintenance. But realistic cost rents were becoming too high for low-income tenants, reaching £30–£40 per week for an average association dwelling. The Housing Finance Act of 1972 therefore brought all housing associations into the fair rent system, but Section 5 of the Rent Act of 1968 continued to exempt housing associations from the Act's provisions on security of tenure, although tenants of unregistered associations gained security in 1974. Most association tenants therefore not only had to pay higher rents than council tenants (council rents remained below the fair rent level throughout most of the 1970s), but lacked the security of most private tenants.

Only 38 per cent of housing associations provided housing for general renting in 1986. Almost a similar proportion either managed alms houses or managed sheltered housing for the elderly, while others provided accommodation for lone-parent families, former mental patients, discharged prisoners and other special groups who failed to qualify for local authority housing. Housing association tenants are among the poorest in society – more than 80 per cent had incomes of less than £100 per week in 1987 and 75 per cent were eligible for housing benefit. Many associations are technically still charities under the Charities Act of 1960 and can therefore supplement funds by voluntary donation. Associations are non-profit-making bodies run by voluntary committees, and sometimes they have an advantage over local authorities in that they can rehabilitate a few houses at a time or build small infill schemes which contrast with large council estate development.

By 1997 there were 2,436 housing associations in Great Britain. Most were small and lacked expertise. Under 20 per cent employed full-time staff. Only 30 owned more than 1,000 dwellings and as few as 10 owned more than 8,000 (the largest being the North Housing Association, the

Anchor Housing Trust and the North British Housing Association – each with over 17,000 dwellings in 1997). But because the associations are usually small, they can build up a close relationship with tenants – often closer than local authorities are able to form.

Associations often exert a considerable influence in low-income and stress areas such as north Kensington, where in the early 1980s housing associations owned more houses (approximately 5,000) than the London Borough of Kensington and Chelsea and the GLC combined. Associations fill a useful role in urban areas if local government, often for political reasons, is opposed to meeting housing need by municipal housing development or rehabilitation.

Public policy towards housing associations

Under the Housing Acts of 1885 and 1890, public authorities became the main suppliers of housing for the needy and largely usurped the role of charities and self-help organisations. This responsibility of government was further confirmed by the Housing and Town Planning Act of 1919, and financial facilities and subsidies were extended to the voluntary movement. In 1935, the National Federation of Housing Societies (NFHS) was formed as a co-ordinating body to take over from the Garden Cities and Town Planning Association (established by Ebenezer Howard in 1899) the central functions of the 75 societies affiliated to it, and to promote and advise new societies – responsibilities acknowledged by the Housing Act of 1936.

In 1936, 100 societies were registered with the NFHS and there were a further 126 unregistered societies. By 1950 the number had reached 409 despite local authorities having chief responsibility for post-war housing. The Conservatives favoured extending voluntary housing when they were in office in the 1950s – a policy not opposed by the Labour Party. The government extended the role of the housing associations in an attempt to increase the pace of rehabilitation. This was facilitated by the Housing Act of 1957, which allowed associations to obtain loans for house purchase and conversion from building societies, local authorities and other public lending bodies. Like private owners, they could now apply for improvement grants for property to be converted into rented dwellings after consultation with the appropriate local authority. Housebuilding was encouraged by the Housing Act of 1961, which provided cost-rent societies with £25 million of loans – the scheme being administered by the NFHS. Some 7,000 dwellings were consequently built.

Until 1964 the words 'housing association' and 'housing society' were synonymous, and the NFHS, as well as many of its registered associations, used the word 'society'. The statutory definition of 'housing association' in the Housing Act of 1957 had been a broad one covering associations, societies and bodies of trustees or companies. But after the setting up of the Housing Corporation by the Housing Act of 1964, associations and societies became clearly defined and had separate functions, and the NFHS was renamed the NFHA. The Housing Corporation was to encourage the expansion of housing society co-ownership and cost-rent schemes, and to arrange 100 per cent mortgages for the former. In 1968 option mortgage schemes were introduced by the Labour government, offering low-income co-owners a subsidy instead of tax relief on mortgage interest (Chapter 10). By 1968 there were 527 co-ownership societies – in contrast to 471 cost-rent societies, members of the latter having to cover interest payments on society finance with rent which was not eligible for tax relief or subsidy. But co-ownership schemes attracted predominantly middle-income households, and it became difficult for the Housing Corporation to find low-cost co-ownership schemes to sponsor. This sufficiently concerned the Minister of Housing, Mr Anthony Greenwood, that he consequently made an order prohibiting the Corporation from lending to societies developing housing at a cost in excess of £7,000 a unit. With the price of land escalating in the early 1970s, and the paternalistic attitude of many society management agents, new co-ownership schemes generally became both unviable and unattractive.

The 1970s

The voluntary movement became increasingly synonymous with housing associations – a trend encouraged by the Conservative government in 1970–74. Under Part IV of the Housing Finance Act of 1972, housing associations became eligible for subsidies from the Exchequer. On new construction, the government gave subsidies on a sliding scale to cover the gap between income from rent and 'reckonable expenditure', which included running costs and capital expenditure, and on conversion subsidies of £5 per week payable for 20 years. Associations could borrow from the Housing Corporation under Section 77 of the Act and obtain mortgages from local authorities up to 100 per cent of the cost of acquiring property, including loan costs. In return for the mortgage, associations were usually required to take a number of families from local authority housing lists.

Subsidisation was becoming increasingly necessary as cost rents were rising very rapidly. A house suitable for letting as flats might have been bought for £30,000, and a further £20,000 might have been spent converting it. Even on a long council mortgage, cost rents of £20 per week might have been necessary to cover construction and maintenance costs. The government therefore gave help only if the housing association charged fair rents under the 1972 Act rather than cost rents.

Government support for housing associations was emphasised by the White Paper *Widening the Choice: The Next Steps in Housing* (DoE, 1973a). It proposed

> to widen the range and choice of rented accommodation by the expansion of the voluntary housing movement. . . . The voluntary housing movement can and should play a bigger part in eliminating the worst housing conditions and widening housing choice for everyone.

A further White Paper, *Better Homes: The Next Priorities* (DoE, 1973b) suggested that HAAs should be declared in areas of housing stress and that housing associations (helped by the Housing Corporation and National Building Agency) should play a leading role in acquiring, improving and managing properties in those areas. It was hoped that housing associations and local authorities would work closely together to provide accommodation to replace the dwindling supply of private rented housing.

By 1973 housing associations were nationally producing 15,000 new dwellings per annum and were providing approximately a quarter of a million dwellings – about 1.3 per cent of the total housing stock of Great Britain. But associations had not fully grasped the opportunities offered for the improvement and conversion of existing accommodation under the Housing Act of 1969, and by 1973 they accounted for only 1 per cent of total improvements and conversions – less than they had done a decade earlier.

With soaring house prices in 1971–73, housing associations were increasingly catering for general housing need – as many would-be buyers could neither afford a mortgage nor (in 1973) obtain one when interest rates fell but mortgage rationing was introduced. Council housing was also difficult to obtain. Waiting lists were long and it was not possible to add people to the list if they already had accommodation – however unsatisfactory. Homelessness was increasing, especially in London, and local authorities were building fewer houses.

In 1974 housebuilding in total slumped to its lowest level since the early 1950s, and more people were homeless or on housing waiting lists than ever before. It was recognised that local authorities could not by themselves deal with the problem of housing need. The voluntary housing movement was therefore strengthened. Although introduced by the incoming Labour government, the Housing Act of 1974 had been substantially drafted by its Conservative predecessors, and generally enjoyed bipartisan support. It was intended to encourage the expansion of housing associations under public supervision. The Housing Corporation's powers of lending and control were greatly extended. Previously the corporation had sponsored only co-ownership and cost-rent societies. Neither was an accessible form of tenure or popular during the inflation of the early 1970s. Under the 1974 Act the corporation's main function was to promote housing associations and to intervene in their activities where it appeared that they were being mismanaged. The registration of housing associations was introduced, administered by the Housing Corporation – associations having to be non-profit bodies and registered as charities or under the Industrial and Provident Societies Act of 1965. Only registered associations were able to receive loans from the Housing Corporation and local authorities to facilitate the acquisition of land (when necessary) to meet their own development costs (salaries, fees, etc.), and to pay builders as work proceeded. A housing association grant (HAG), introduced by the 1974 Act, was to be paid on completion of a housing association scheme and allowed the association to pay back 75–80 per cent of the loans it had received from the Housing Corporation and local authority to finance the scheme. The annual loan charges on the remaining debt (together with management and maintenance costs) should then have been recovered from rent income, but if there was a shortfall this could have been offset by a revenue deficit grant (RDG). The Housing Corporation was also given powers to provide dwellings for letting by means of construction, acquisition (by compulsory purchase if necessary), conversion and improvement, and it could borrow up to £750 million for this purpose.

Additional finance became available to the Housing Corporation in 1977. It set up a private company (under the 1974 Act) to borrow £25 million (for seven years at the market rate of interest) from the City to help housing associations adversely affected by public expenditure cuts in 1976. The Housing Corporation's Finance Company had 40 per cent of its shares held by the Housing Corporation and 60 per cent held by the NFHA, the Guinness Trust, the Sutton Trust, the Notting Hill Housing Trust, the Paddington Churches Housing Association and the London and Quadrant Housing Trust. The loan was facilitated by Morgan Grenfell (merchant bankers), approved by the DoE and secured with a mortgage on properties in schemes sponsored by the Housing Corporation. This intermeshing of public and private finance led to further injections of private investment into public sector housing, since there was a guaranteed return irrespective of the rent policy of the government of the day.

Encouraged by the government and the Housing Corporation, housing associations therefore rapidly increased their activity. The type of activity was largely determined by Circular 170/74 (DoE, 1974), which emphasised that housing associations should play an important role in relieving housing stress or homelessness by their operations in GIAs and HAAs; provide housing for those with special needs such as the single and elderly; design schemes to maintain the stock of rented accommodation in areas where there were severe shortages; acquire properties from private landlords who were failing in their duty towards their tenants or property; and make provision for key workers.

With inflation and rising fair rents in the late 1970s, however, HAGs awarded a few years earlier (in respect of a wide range of activity) seemed in retrospect unduly generous. The Housing Act of 1980 therefore required housing associations to keep a grant redemption fund into which surpluses from increased rent income would be paid – with the intention of ultimately transferring surpluses to the DoE.

The Housing Corporation in the 1980s

Until 1979 the Housing Corporation could have been rightly criticised for inadequately scrutinising the activities of housing associations and for not establishing objective criteria for the allocation of funds. But after the Conservatives gained power in 1979, the Corporation imposed its will on housing associations in an unprecedented and damaging manner. Wolmar (1982) believed that the Housing Corporation had become a tool of the DoE and had lost its independence, relinquishing its role of lobbying for and representing the housing associations. Bureaucratic insistence on having to work within diminishing annual cash limits (the permitted gross capital expenditure of the Housing Corporation being cut from £903 million to £831 million in cash terms, 1982/83–1987/88) handicapped housing association improvement schemes, which normally took two or three years to complete. There was a case for less meddling by the DoE in the Housing Corporation's affairs, and less interference by the Corporation in association affairs, while in contrast there was remoteness and communication failure, especially over the RTB provisions of the Housing Act of 1980. Although in total there were about 375,000 tenants of charitable associations, most would not have been affected by this legislation, for example the tenants of the Peabody, Guinness, Rowntree and Bournville trusts. The measures would have applied only to properties built since 1974 with the aid of housing association grants. Provisions were to be made to enable tenants who were old, poor or handicapped to obtain 'family' mortgages (where repayments could be made by several members of the family, or relatives). Charities were to be reimbursed in relation to the amount of capital they had invested in the property, but the sum would probably have been insufficient for reinvestment in housing after government loans had been repaid. Opponents pointed out the retrospective nature of the legislation (charity associations never expecting that they would be privatised when they accepted government grants under the Housing Act of 1974), and drew attention to the fact that it could affect neighbouring households living in similar dwellings differently, where one dwelling was built just before the 1974 Act and the other just after it.

Mr Richard Best (1982a), Director of the National Federation of Housing Associations, argued strongly against privatisation:

> In my view it would drive a coach and horses through charity law as we understand it. Charities have a duty which lasts into perpetuity. They must think of the homeless and disadvantaged of tomorrow. . . . Volunteers have worked tirelessly to produce a small number of precious homes to rent. . . . It is vital to keep these homes for the old, the disadvantaged and the young in the future.

He reiterated (1982b) that the housing association movement is

> not a nationalised industry or a public authority. . . . We have obligations to hold on to our homes as long as possible. . . . As independent bodies, we received . . . housing association grants. . . .

When we took the money we never for a moment suspected that by receiving it we had changed our status from being an independent body to being a public authority.

In the early 1980s, for example, the associations were encouraged to place an emphasis on sheltered housing, but many associations were inexperienced in this field and were aware that a concentration on this area would be at the expense of good-quality, cheap rented housing for general needs. By the mid-1980s the Housing Corporation had shifted the emphasis to inner city housing (which absorbed 83 per cent of housing association funding from the Corporation in 1986), but this reduced the ability of the associations to help deal with the growing housing crisis nationwide. It was clear that the associations were no longer free to respond to housing needs according to their own specific criteria and expertise. The Housing Corporation, in the view of Wolmar, had become a controller rather than a nurturer of the housing association movement, and although an element of 'policing' was necessary, if control was too tight, many associations would fold up, or at least function inefficiently.

Housing associations in the 1980s

Conservative support for the voluntary housing movement waned for a while after 1979 and the privatisation of housing association property became increasingly an issue in the early 1980s, once the government's RTB policy in respect to council housing appeared to be running out of steam. But as early as 1979, Mr John Stanley, Minister of Housing, set up a working party to examine the extension of home-ownership to housing association tenants, as the government was intent on associations selling off their rented stock building and improving for sale in competition with private builders and no longer building for rent.

The Housing Act of 1980 granted housing association tenants the right to buy their homes from their landlords, and mortgages were to be made available by the Housing Corporation. As a result of this provision, a total of 83,800 housing association dwellings were sold off over the period 1980/81–1987/88. (Housing associations registered as charities under the Charities Act of 1960 were also given the right to sell, although tenants were not given the right to buy, an exclusion affecting half of all association households.) In 1992 the Housing and Building Control Bill, Section 2, intended giving the RTB to tenants of charities, but the proposal was subject to severe criticism, the *Guardian* (1982) for example stating that the worse housing decision made by the government in 1982 was 'to introduce legislation to extend the right to buy to cover charitable housing associations to favour people already fortunate enough to live in a good home at the expense of present and future generations of people who need one'. The RTB proposals meant that the housing associations would have concenrated more on building houses for sale and this could have distracted them from their principal purpose of providing houses for people in need. Critics of the Bill even claimed that Section 2 contravened the European Convention on Human Rights because it proposed interference with charity activity, and when the Bill reached the House of Lords the government was forced to delete the offending clause by a vote against of 182 to 96.

The government, however, was determined to extend owner-occupation. Under the Housing and Building Control Act of 1984, tenants of charitable housing associations were offered cash handouts to enable them to buy in the open market. Handouts of up to 50 per cent of the value of a dwelling were granted to tenants of two or more years' standing in respect of an acquisition costing no more than £40,000 in Greater London, £35,000 in the Home Counties or £30,000 elsewhere. To qualify, tenants would have had to have attempted unsuccessfully to negotiate the purchase of their own housing association home. Although the legislation met with the general approval of the NFHA as it would free some housing association dwellings for households unable to buy, there was concern that the cost of the scheme (possibly £95 million in the first three years) would be deducted from HAG funds normally used to provide new dwellings. Also, unlike the other RTB provisions, this facility would require additional public expenditure which would never be recouped, there would be no capital receipts to invest in new housing, the same property might be bought with a handout more than once over a period of time, and no other group of households would receive such assistance simply for previously being a tenant in one place for two or more years. As a result of RTB provisions, a total of 83,800 housing association dwellings were sold off in England alone in the period 1980–88.

However, in the early 1980s government funding of the Housing Corporation remained broadly static while housing association starts and renovations plummeted to respectively 11,566 and 13,770 in 1981. As a result, the Housing Corporation and some housing associations soon began to utilise private sector finance (particularly from building societies and pension funds) to supplement or replace HAGs. The North Housing Association, for example (the largest in the United Kingdom, with 20,000 dwellings), embarked in 1986 on a £112 million building programme using £100 million from the London capital market and the remainder from its own resources; while the Secondary Housing Association for Wales and the Wales and West Housing Association planned in 1986 to undertake extensive residential development – with 70 per cent of the finance being provided by the Halifax Building Society and 30 per cent being allocated by the Housing Corporation. In general, schemes such as these were developed for assured rather than fair rent tenure, and private sector finance was to be supplied on an index-linked or low-start basis (with repayments rising in later years when rental income was increased in line with inflation). These innovations enabled housing associations to borrow more than would previously have been feasible and thus made it possible for the sector to develop viable schemes more effectively than hitherto.

Mixed funding schemes could be of benefit to the homeless if HAGs were on an appropriate scale. Even though the private sector might provide 70 per cent of the cost of developing temporary accommodation (through index-linked mortgages) and rents would consequently be above fair rent levels, local authorities would nevertheless find these lower than the exorbitant rents charged by bed and breakfast hoteliers – often amounting to £16,000 or more per annum for a family of four.

Changes in investment policy in the 1990s

Rather than complementing local authority housing (and notwithstanding the effects of RTB), housing associations have now become the principal providers of new social housing in Great Britain, as prescribed by the White Paper *Housing: The Government's Proposals* (DoE, 1987). Housing association net capital expenditure doubled from £1,157 million to £2,308 million between 1990/91 and 1992/93, while the number of housing starts in the sector increased from 12,924 in 1987 to 41,261 in 1993. Local authority starts, meanwhile, plummeted from 18,883 to only 3,713 over the same period.

Increasingly, up to 50 per cent of housing association capital spending was targeted at the homeless, with the development of new housing for the elderly (a traditional priority) dwindling to almost zero. Also, whereas much emphasis was placed on rehabilitation in the 1980s, its share of capital investment within the housing association sector fell from over 50 per cent in 1988/89 to about 20 per cent in the early 1990s, since it was increasingly recognised that rehabilitation was becoming as expensive as or more expensive than building, for example, high-density modern units – not least because of the economies of scale and lower risks associated with the latter form of housing renewal. In addition, it is possible that housing associations sometimes assumed that tenants increasingly preferred new homes to rehabilitated dwellings; that associations were discouraged by the decline in the declaration of GIAs and HAAs and by the slow pace of RA declaration; and they might have believed that most of the worst properties had already been improved (although this was not borne out by the *English House Condition Surveys* of 1986 and 1991 (DoE, 1988, 1993). Housing associations were also diverting their rehabilitation activity away from the inner cities to suburban areas. The proportion of grant approvals in the UP authorities fell from 68 per cent in 1987/88 to 43 per cent in 1990/91, a principal cause of concern particularly in the North, where housing associations in recent years had been virtually the only source of investment in areas of older housing (Walentowicz, 1992). Clearly, there was a significant shift of emphasis from rehabilitation to new housebuilding, and it was on this that government expenditure would be targeted.

Over the three years 1992/93–1994/95, the government (in 1992) aimed to spend £2 billion per annum on housing association investment, and to produce a total of 153,000 homes – each association setting out its investment plans in an ADP agreed annually by the Secretary of State. But with cuts in the size of the HAG for each completed dwelling, more had to be borrowed from the private sector. Although public funding continued (involving HAGs and loans from the Housing Corporation and local authorities), mixed funding schemes were increasingly undertaken – private finance enabling public funds to be stretched over a much greater volume of housing than hitherto. Whereas in 1989/90, HAGs covered 75 per cent of housing association capital expenditure, the proportion decreased to 67 per cent in 1993/94, 62 per cent in 1994/95 and 58 per cent in 1995/96 – with a reciprocal increase in risk incurred by the financial institutions.

In the November 1993 Budget, as part of a package of cuts aimed at reducing the size of the PSBR, the government announced a £300 million reduction in its funding of the Housing Corporation (for 1994/95), to be followed by a further cut almost as large in the following year. While the government claimed that this would not adversely affect housing association development programmes since construction costs were significantly lower than hitherto, the Institute of Housing predicted that the cut would result in 13,000 fewer new homes being built in a single year, and a shortfall of 30,000 in the number of homes being rehabilitated. Clearly, the government was creating the conditions whereby housing associations would have to depend more and more on private finance if they were satisfactorily to perform their role as providers of social housing.

Nevertheless, via the Housing Corporation (and subject to approved development programmes), the government was still the principal supplier of funds for housebuilding and renovation in the housing association sector. In addition, from 1990/91 the Housing Corporation was empowered to provide grants under the *tenant's incentive scheme* to help housing association tenants move into owner-occupation and thereby release subsidised rented accommodation for households in greater need of low-cost housing; and in 1990/91 and 1991/92 the government allocated £73 million to the housing associations to assist local authorities provide housing for the homeless and to reduce the numbers of households in bed and breakfast accommodation.

Causes for concern

Despite the Housing Act of 1988, which relegated local authorities to an 'enabling' role only, it must be questioned whether or not housing associations can effectively become the principal providers of new-build social housing, notwithstanding 150,000 starts in 1992–95.

There are a number of causes for concern. First, there is a very clear market disequilibrium between the supply of readily available low-cost land (normally in rural areas and in much of the North) and areas suffering the most acute housing need (particularly in the inner cities and in the South East). It is essential, therefore, that local authorities, in both their planning and their strategic housing roles, fully exercise their enabling responsibility by ensuring that land is made available for social housing and that (under Circular 7/91 – DoE, 1991) social housing is included in any new housing development in their area. HAGs, moreover, must be allocated to areas of greatest need and not excessively to areas where land is readily available but needs are less (Babbage, 1992).

Second, with the recession and lower interest rates deflating the cost of construction in the early 1990s, the Housing Corporation (at least twice) reduced its total cost indicators, resulting in the reworking of schemes and delays in development (Babbage, 1992). It was anticipated that further adjustment would be necessary and delay occur if or when inflation recurs.

Third, with the increased emphasis on mixed funding, housebuilding in this sector will depend more and more on the ability and willingness of the financial institutions to invest in development. But housebuilding will not be the only destination of institutional social housing investment, particularly if the banks and building societies become concerned at the diminishing input of public sector funding. Housebuilding will have to compete for funds with the many large-scale voluntary transfers of stock from local authorities under the provisions of the Housing Act of 1988. Clearly, the scarcity of private funds could be the biggest factor constraining social housing development in the future, not least when the economy recovers and a wider range of investment opportunities emerge.

The private funding of housing associations, however, amounted to £16.9 billion at the end of the financial year 1998/99 (£14.6 billion up on 1997/98), with grant-aided housing association programmes attracting two-thirds of this sum and transfers receiving the remainder. Of the £5.6 billion lent to housing associations in 1998/99 to facilitate the transfers of stock from the local authority sector, 64 per cent was provided by banks, 31 per cent by building societies and 5 per cent by bond issue (Forshaw, 2000b). However, according to the National Housing Federation, by the end of the 1990s cumulative loans from the private sector to faciliate mixed-funded programmes were close to being matched by cumulative loans to facilitate transfers (Forshaw, 2000b). It was anticipated that over the period 1999/2000–2002/03, private sector loans of £4.6 billion would be required to help facilitate mixed-funded programmes, and £2.7 billion would be needed to fund the transfer of up to 125,000 per annum by 2000/01.

Fourth, there was growing concern that because housing associations were charged with the responsibility of being the principal (or only) provider of new social housing (and since output more than doubled 1989–92), there was the danger that the associations were creating the slums of the twenty-first century (Warrington, 1994). Financial constraint and economies of scale often resulted in the construction of large but comparatively poor-quality estates, which unlike their older local authority counterparts were not under democratically elected control. There was also concern that the space standards adopted by housing associations to ensure financial viability were too low to provide adequate living areas for families with children.

Changes in rent policy in the 1990s

The cutback in the size of HAGs and increased reliance on private finance (which, by necessity, requires a competitive rate of return) in the 1990s had not only an unfavourable impact on housing standards but an inflationary effect on rents. The consequences of mixed funding schemes could have become even more marked if public funding had been reduced to 50 per cent as was suggested in the White Paper *Housing: The Government's Proposals* (DoE, 1987).

Whereas existing lettings are at fair rents as determined by rent officers, and are subject to rent increases every two years, under the Housing Act of 1988 all new lettings are at assured or assured shorthold tenure, with housing associations setting their own 'affordable rents'. Although affordability is not defined by the DoE, it was interpreted by the NFHA as a rent approximately equal to 20 per cent of the tenant's average net income. However, in order to ensure that private capital is attracted into housing investment in this sector, average rents for new housing rose to £48 per week in 1990/91 (notably in excess of average local authority rents), and by a further 21 per cent in 1991/92, compared to an increase of only 1.8 per cent in the retail price index and 5.8 per cent increase in the average income of new tenants. As a consequence, by 1991/92 rents consumed 29 per cent of the income of new tenants. Clearly, the underlying reason for these hikes in rent was the government's intention to reduce its share of total investment in this sector. In 1994/95 it was reduced from 67 to 62 per cent and in 1995/96 cut to 58 per cent, leaving the associations to fund the rest primarily from private loans. Rents inevitably rose further, by at least a third over the period 1994/95–1996/97 according to forecasts by the NFHA.

But although low-income tenants might be able to afford higher rents (owing to their eligibility for housing benefits, although the extent to which benefits might increase in proportion to rents is uncertain), households with slightly higher incomes will be priced out, or face considerable hardship. With up to 50 per cent of households in new housing association dwellings being previously homeless, and with up to two-thirds of tenants being in receipt of housing benefits, housing association estates are far less likely to offer a social mix than hitherto. The social consequences of estates becoming 'welfare ghettos' are undoubtedly severe.

Although several housing associations have been actively involved in acquiring local authority housing through large-scale voluntary transfers (and some housing associations have been set up by local authorities to facilitate transfer), many local authority tenants have been unwilling to opt for transfer owing to the risk of higher rents. If, however, the associations acquire only 10 per cent of the local authority stock, they will not only double the supply of housing they owned in the late 1980s, but will possibly offer tenants a more sensitive and efficient management service – albeit at higher rents, reflecting higher costs. It was evident, however, that the 'good landlord image' of housing associations among tenants in the late 1980s was a direct result of housing associations spending an average of £150 per dwelling on management while local authorities spent only £100 per dwelling (Maclennan, 1989). When the size of many housing association estates grows, management could become less efficient and less sensitive. Although the government is normally unlikely to consent to the transfer of more than 5,000–10,000 properties (4,000 in Wales) to a single purchaser, some housing associations are developing extensive areas of housing which might become the problem estates of the future. Many associations, therefore (and particularly the smaller ones), might be reluctant to expand their activities if this necessitates the adoption of commercial criteria in relation to investment and management, and at the end of the day produces many of the problems faced by urban local authorities in the 1980s and early 1990s.

New responsibilities

In addition to becoming the main providers of social housing through the transfer of a proportion of local authority stock and the provision of funds for the construction of new dwellings, housing associations have also had the responsibility of intervening in the private housing market.

Under pilot schemes introduced by the DoE in 1991, five housing associations received a total of £300,000 to act as intermediaries between the private owners of empty houses and eventual tenants. Under the HAMA programme, associations were encouraged to act as managing agents, choose tenants, collect rents and deal with day-to-day problems – as an inducement to owners to put their properties on to the shorthold rental market. If the empty properties needed to be rehabilitated, the associations undertook the task, with the costs incurred deducted from rent. Housing associations also received £577 million from the DoE, through the medium of the Housing Corporation, to buy up 16,000 newly built and/or repossessed dwellings, ostensibly to house the homeless. The £577 million, however, was brought forward from the £6 billion budget of the Housing Corporation allocated for the period 1982/83–1984/85, and most of the money was spent acquiring the unsold newly built housing rather than repossessions.

Shared ownership

Housing associations, since 1977, have supplied a small proportion of social housing by means of shared ownership. New housebuilding, or the acquisition of existing housing, is partly financed by a Housing Corporation loan and partly by a building society mortgage secured on long lease, with the occupier incurring a mixture of rent and mortgage repayment. There are also leasehold schemes for the elderly – partly financed by the occupier, who also pays rent. In each case, should the occupier wish to move, the house would be sold at the prevailing market price and the occupier would receive his or her proportion of the proceeds. As a simplified example, a £50,000 house taken by a 50 per cent owner-occupier would require a mortgage of £25,000. The sale value after say five years might be £60,000, and the occupier's share of the proceeds would be £30,000.

Despite the initial popularity of shared ownership, the number of houses built for this purpose declined dramatically in the late 1980s, from 3,439 in 1981 to less than 700 in 1991. In areas of high unemployment and falling property values, rent arrears and mortgage default were both at a high level. Only if home-ownership seemed to be a good investment were buyers likely to be attracted by shared ownership, but, as with home-ownership in general, shared ownership had little appeal during the house price slump.

Box 9.1

Housing associations in rural England

Owing, in part, to the out-migration of comparatively high-income commuters and second-home owners, and in part to the relatively low purchasing power of rural labour, an increasing proportion of housing in the countryside has become unaffordable to the indigenous population – a situation exacerbated by the reduction in local authority housing. Housing associations are therefore playing an important role in providing low-cost housing in rural areas, often working in partnership with local authorities and the private sector.

Through the Housing Corporation, the Conservative government funded the provision of 9,100 housing association dwellings in rural areas in the period 1990–95 (and in addition financed the provision of a further 25,000 low-cost local authority rural dwellings, 1991–93). Similar initiatives were also undertaken in Scotland through the medium of Scottish Homes, and in Wales through joint housing association–local authority programmes.

These provisions, however, did little to reduce the shortage of affordable rural housing. It was estimated that although 80,000 affordable homes were needed in rural England in the period 1990–95, only 17,700 new social units were provided, 1990–97. By the late 1990s, therefore, 40 per cent of new households in rural areas were unable to buy a home in the countryside (Shucksmith, 2001). Aware of the extent of the problem, the newly formed Labour government attempted to ensure that the existing supply of social housing would be retained for people who could not afford to rent or buy in the private sector (DETR/DSS, 2000); restricted the resale of RTB and Right to Acquire housing in villages of fewer than 3,000 people; and required the Housing Corporation – in its ADP – to allocate a prescribed proportion of new approvals (3.4 per cent in 2000) to settlements with a population of less than 3,000 people. However, housing association development in rural areas is subject to approval by parish councils if it is to receive government funding, and much development has not been approved because of the reluctance of comparatively well-off newcomers to accept new building. Although the Rural White Paper (DETR/MAFF, 2000) set a target of 9,000 affordable homes per annum to be built in rural areas, this number of new homes, even if they are completed, will clearly be insufficient to meet local needs.

Because of these shortcomings, the Countryside Agency was particularly concerned about the need to produce blueprints to revitalise about 1,000 run-down villages and provide grants for essential services such as public transport. It proposed that on a village-by-village basis there would be an identification of sites for the development of affordable housing, and that such sites and the houses built on them would be reserved exclusively for local people. In this way it would be possible to recreate balanced communities, rather than villages occupied entirely by 'older country folk and rich incomers' (Brown, 2001).

continued

A further boost to the development of affordable housing in rural areas was announced by the Countryside Agency in June 2001. The Housing Corporation's budget for rural housing was set to double by 2003, enabling it to fund the construction of 1,600 rural homes compared to only 800 in 2000/01. However, since it can take three to five years to plan and build a home in a rural area, some of this extra investment will allow the Countryside Agency to speed up the development process by deploying its Rural Housing Enabler scheme across England to integrate the planning and construction activities of local authorities, housing associations, the Housing Corporation and other interested bodies.

With regard to council housing, the supply of dwellings in rural areas is likely to increase if the government is successful at encouraging local authorities to spend 'windfall' revenue of around £200 million per annum on housebuilding using capital resources derived from the doubling of council tax on second homes (until 2000/01, owners enjoyed a 50 per cent discount, but became liable to the full tax in 2001/02). Affordable home-ownership could also be promoted in certain areas such as National Parks if park authorities were able to compel prospective second-home buyers to apply for planning permission when their intended properties were likely to stay empty for more than (say) six months or a year, and to refuse permission where the proportion of second and holiday homes exceeded (for example) 10 per cent of the local stock – a proposal adopted by the Exmoor National Park Authority in 2001.

A political consensus?

Conservative governments for many years favoured housing association activity, recognising that with the virtual demise of the private landlord, an alternative form of rented tenure was necessary, and in many localities housing association dwellings were regarded as an alternative to council housing. Although housing association tenants were given the 'Right to Buy' under the provisions of the Housing Act of 1980, the government – in its Housing Act of 1988 – subsequently saw a new and important role for housing associations. Associations would take over from the local authorities the responsibility of providing most of the new-build social housing, and acquire an increasing proportion of existing local authority housing through LSVTs and by many other processes.

At the 1992 General Election the Conservatives reaffirmed their intention of spending £2,000 million (through the Housing Corporation) to provide 153,000 housing association dwellings by 1994/95, and pledged that they would use some of this allocation to implement a Do-It-Yourself shared ownership scheme. First-time buyers would thus be able to choose a house in the open market and buy a share of it (normally 50 per cent), with a housing association paying rent on the rest until they wished to increase their equity in it.

By 1994, however, it was reported that the Major government was considering switching the role of housing associations from provision to management (Shelter, 1994), the associations possibly losing their historic role, a role reaffirmed only in 1988. In Wales, moreover, the housing associations were becoming instrumental in increasing the size of the owner-occupied sector through an expansion of low-cost home-ownership schemes rather than the provision of much-needed affordable rented housing – a shift of emphasis administered by the transfer of responsibility for supervising Tai Cymru (the Welsh equivalent of the Housing Corporation) from the housing minister to the Secretary of State for Wales (Hughes, 1994).

Housing associations, in their historic role, have appealed to sections of the Labour Party for many decades, largely because associations have been non-exploitive (in contrast to private rented housing), have often had more enlightened and sensitive management than local authorities, have more flexible allocation policies and there is an element of tenant control. But, historically, Labour never anticipated replacing local authorities (as the main providers of social housing) with housing associations. *Labour's Programme 1976*, for example, stressed that 'the main role of housing associations must be to complement the efforts of local authorities and not to compete against them for scarce land and housing resources' (Labour Party, 1976).

At the 1992 election Labour proposed that (subject to regaining power) it would establish a multi-billion pound 'National Housing Bank' (funded from national and international capital markets) to lend at competitive rates to housing associations (and local authorities) to build more homes to rent, and thus reduce the need for private funding and steeply escalating rents. In addition, housing associations would be allowed to lease or buy empty houses in order to provide accommodation for the homeless, and homes left empty without good reason by any public authority would be transferred to a better social landlord.

By the late 1990s it was clear that Labour not only supported the Conservative view that housing associations should be the principal providers of social housing, but – after its election victory in 1997 – was intent on increasing the pace of LSVTs of stock from municipal ownership. Labour had witnessed the number of housing association starts rising from 12,631 in 1984 to 41,804 in 1993 (before falling to

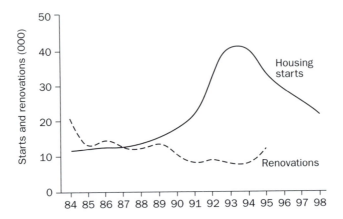

Figure 9.1 *Housing association starts and renovations, 1984–98*
Source: DETR, *Housing and Construction Statistics*

22,263 in 1998) and the number of renovations decreasing from 20,639 in 1984 to around 12,000 in the mid-1990s (see Figure 9.1), and was determined that both new housebuilding and the renovation of stock within an expanded housing association sector should increase substantially in the future.

The DETR and DSS were thus instructed to propose a root and branch reappraisal of the government's policy on social housing in general (including the highly problematic local authority sector), the details of which were incorporated into the Housing Green Paper 2000 – examined below.

Box 9.2

Housing associations in Scotland, Wales and Northern Ireland

As elsewhere in Britain, in Scotland, housing associations rather than local authorities, are now the major providers of new social housing. Whether via new build or rehabilitation, the sector grew substantially during the last 15 years of the twentieth century. By 1999, 5.7 per cent of all dwellings north of the Border were rented from housing associations – the same proportion as in the United Kingdom as a whole. Owing to the historic activity of Scottish Homes in the central belt, about 30 per cent of housing associated stock is concentrated in Glasgow and 10 per cent in Edinburgh (Earley, 2000).

In Wales there was similarly a substantial growth in the provision of housing association dwellings. The sector increased from only 11,000 dwellings in 1981 to 77,000 dwellings in 1999 (or to 4.1 per cent of the total stock), and lost only 1,800 dwellings through Right to Buy sales over this period. However, despite this growth, housing associations have been disadvantaged by cuts in the social housing budget introduced by the Conservatives in the early 1990s but continued under Labour. Output of new rental dwellings thus declined from 2,251 completions in 1998/99 to only 1,752 dwellings in 1999/2000, while the target of 2,300 completions for 2000/01 was cut to 1,600 (Williams, 2000). As an outcome of devolution, the National Assembly for Wales has taken over the regulatory and funding role vested in Tai Cymru in respect of the 94 registered associations in the Principality, and this might ensure that the sector is better equipped to respond to housing need.

In contrast to the rest of the United Kingdom, Northern Ireland has a very small housing association sector (2.6 per cent), partly because the Northern Ireland Housing Executive (NIHE) had until 1998 been the main developer and owner of social housing in the provinces, and partly because there had been an absence of large-scale transfers to housing associations.

The Housing Green Paper, 2000

The government's Green Paper was concerned that 'poor quality, a lack of choice for tenants and changing aspirations have turned much of the social housing sector into a tenure of last resort and have contributed to the phenomenon of low demand' (DETR/DSS, 2000: 55). To reverse this situation, the Green Paper proposed that social housing should be turned into a tenure where the image is no longer one of decline and decay, where tenants have choices, and 'in which the problems of social exclusion, poor quality and poverty of opportunity are confronted and surmounted' (DETR/DSS, 2000: 57).

To this end, the Green Paper set out plans to:

- raise the quality of all social housing to a decent standard by 2010;
- provide new affordable housing;
- improve access to social housing; and
- introduce a fairer system of affordable rents.

Raising the quality of social housing

It was clear that past investment in social housing had not been sustained at adequate levels – leaving a £19 billion repair and modernisation backlog at the turn of the century (DETR/DSS, 2000). Aware of the scale of the problem, the incoming Labour government in the late 1990s began to increase the resources available for investment. Under the Capital Receipts Initiative (CRI) an extra £800 million was invested in improvement in 1997/98 and 1998/99, followed by an additional £3.9 billion to the spending plans for housing for 1999/2000 to 2001/02, in addition to new money for regeneration programmes such as the New Deals for the Communities. There was consequently a significant reduction of the backlog in repairs. Over 300,000 homes in the social rented sector had been improved from 1997/98 to 1999/2000, and it was probable that a further 1.5 million dwellings in the sector would have benefited from new investment by 2001/02.

The Green Paper recognised that it was essential that increased capital spending on improvement should go hand in hand with higher-quality management, more effective investment in stock maintenance (for example through the Best Value regime for housing) and more tenant involvement. It was also acknowledged that the recommendations and targets of the Construction Task Force could help improve the 'performance, quality and value for money of work on the existing stock, and for reducing whole life costs' (DETR/DSS, 2000: 58).

Further measures, however, were considered necessary if the quality of the stock was to be markedly improved. These comprised the following:

Stock transfer

More than 400, 000 homes had already been transferred from around 100 local authorities to registered social landlords since the relevant legislation was introduced in 1988 – attracting over £6 billion of private finance for investment in improvement (and £3 billion of capital receipts for local authorities). The Green Paper proposed that from 2001/02, in order to stimulate further privately funded improvement the government should support the transfer of at least a further 200,000 dwellings each year (subject to tenant approval).

Creation of arm's-length management companies

Arm's-length management companies (where they have been set up under the 1996 Act) have been primarily concerned with the day-to-day management of local authority stock, whereas the relevant local authorities have assumed a strategic rather than a managerial role. However, unless the ownership of housing is transferred to the local housing company (when in effect it would become an RSL), the company is subject to the same controls on capital expenditure as apply to local authorities. The Green Paper therefore proposed that arm's-length management-only companies should be able to retain and use more of their rental income to finance improvement, and that this would be in addition to the credit approvals received from central government through the annual HIP.

The Private Finance Initiative

The Green Paper recognised that the PFI – which provides an alternative to stock transfer – is likely to establish itself as an option that many local authorities may wish to adopt. Under PFI, private sector funds could be attracted into improvement and management, risk could be shared, and investment would not be subject to public expenditure controls.

Providing new affordable housing

Labour was pledged to support the development of new affordable housing 'in line with local needs and priorities . . . [and wanted] to see better links between supply and demand at the local level, higher standards of quality, design and efficiency, and better integration of social and private sector housing' (DETR/DSS, 2000: 70). It recognised that affordable housing must cater for households unable to afford home-ownership and who are likely to need to rent their homes on a long-term basis; households who wish to buy but can afford only properties in the lower price ranges; and people with special needs who require both subsidised accommodation and appropriate support.

Clearly, there are wide regional variations in the need for additional affordable housing. In London and in many southern urban and rural areas, the high demand for housing has inflated house prices and depleted the supply of vacant social housing. However, in many other regions the demand for housing is far less and there is no overall shortage of social housing, but there is a substantial need to refurbish or replace existing dwellings to satisfy local needs. It is within these disparate contexts that the government hoped to ensure that its policies for affordable housing would 'ensure a better mix of housing types and tenures and [would] avoid the residualisation of social housing and its occupants' (DETR/DSS, 2000: 71).

The delivery of affordable housing is facilitated, first, by the provision of social housing grants (SHGs) – through the Housing Corporation's Approved Development Programme (ADP) – to support the development by registered social landlords of housing to let at sub-market rents or for sale on shared or low-cost ownership terms, and second, by the use of local planning authority powers to require an element of affordable housing to be provided in the development of a site under the arrangements set out in PPG 3 (DoE, 1992a) and Circular 6/98 (DETR, 1998c).

By the time the Green Paper was published, the government had agreed with the Housing Corporation that the ADP would

> provide additional affordable housing in areas of economic and demographic growth . . . [and] contribute to the regeneration of deprived neighbourhoods by helping to fund the refurbishment or replacement of existing housing, and fund the provision of new supported housing to meet the needs of a wide variety of vulnerable groups.
>
> (DETR/DSS, 2000: 72–3)

In addition, it was acknowledged that whereas SHG provided help for people to move into homeownership, there was 'now a strong case for a separate low cost homeownership initiative in areas where house prices . . . [were] becoming increasingly unaffordable' (DETR/DSS, 2000: 74). Overall, it was hoped that local authorities, through their development plans, would not only include policies and proposals for satisfying housing needs in their areas, but also – with the aid of PPG 3 – help to create or maintain mixed and balanced communities.

However, it is clearly not enough to provide funds for the provision of social housing if the construction industry is slow to respond effectively and efficiently to changing needs. Although the Housing Corporation has an impressive record of delivering social housing through its ADP, the process has become increasingly complex in recent years. The government, in its Green Paper, therefore declared that it was firmly committed to the implementation of the recommendations of the Egan Report (DETR, 1998a) for improving the speed, quality and cost of construction projects in the social housing sector, and to the promotion of the greater use of new construction techniques, such as prefabrication, through the ADP.

Improving access to social housing

It was the government's intention to enable potential and existing social housing tenants to choose where they wish to live, to create sustainable communities, and to promote the more effective use of the social housing stock.

Clearly, local authorities and other social landlords have traditionally tended to 'allocate' housing to people approximately in accordance with an assessment of those people's needs; people seeking housing within the sector have only a limited opportunity to choose where they can live. The Green Paper therefore proposed that a more customer-centred approach should be adopted that would satisfy the long-term requirements of those who need social housing; empower new and existing tenants to exercise choice to meet their needs; make more efficient use of the social housing stock by widening the scope for lettings and transfers, both nationwide and between local authorities and RSLs; and enable local authorities to build sustainable communities despite wide disparities across local housing markets.

The government clearly recognised that there were a number of ways of encouraging a more flexible and effective use of social housing to satisfy people's needs. At the beginning of the new millennium, most households in the social rented sector whose tenancies existed before the Housing Act of 1988 had secure tenancies, whereas tenants of registered social landlords whose tenancies were established after the 1988 Act had assured tenancies – though similar rights. The government, however, accepted the view of the Chartered Institute of Housing and others that there would be advantages – to both tenants and landlords – in establishing a single form of tenure, and the Green Paper proposed that the 'benefits of, and options for, moving to a new single form of tenure' (DETR/DSS, 2000: 91) would be explored. In line with the Social Exclusion Unit's Policy Action Team, the Green Paper went on to propose that landlords should be permitted to let some of their properties at market rents in areas where there is an excess supply of housing at sub-market rents; that in areas of low demand the existing stock could be used to meet the housing needs of students and medical staff, before new development is considered; and that social housing landlords should be provided with powers to let properties short-term 'to people who do not have a long-term need for social housing, without conferring the rights that are normally given to secure and assured tenants' (DETR/DSS, 2000: 92). Clearly, it was the government's intention, on the one hand, to introduce new arrangements that provide security for long-term tenants, but, on the other hand, to allow landlords increased flexibility to make best use of their stock.

Introducing a fairer system of affordable rents

It was the government's view that 'rents for social housing should be affordable, based on principles that are fair, that provide comparable rents for comparable homes, that inform the choices tenants make about their homes, and that encourage effective management by social landlords' (DETR/DSS, 2000: 103).

Notwithstanding that in April 1999 the average local authority rent in England was as little as £44 per week and the average RSL rent was only £52 per week (in contrast to typical assured private sector rents of around £75 per week in 1998/99), the government, in its Housing Green Paper, was persuaded that there was no case for substantial changes in the average level of social rents, taking into account 'considerations about work incentives, public expenditure and targeting support on those in most need' (DETR/DSS, 2000: 95). However, the government acknowledged that if it was 'to achieve real improvements in social housing . . . [there was a need for] a structure of social rents which tenants see as fair and which complements choice-based letting schemes' (DETR/DSS, 2000: 95). Clearly, there was a need to ensure that in general, rents for properties that are larger, in a better state of repair or in more attractive locations should be higher than those for properties without these attributes, while rents for similar properties in a locality should be broadly the same regardless of whether the properties were local authority owned, or owned by registered social landlords. Likewise, there should be no arbitrary differences in rents for similar properties in neighbouring areas.

The Green Paper thus set out how social rents could be restructured over a 10-year period – each model requiring a weighting for the capital value of the dwelling:

- The first option is based on a 50/50 split between capital values and regional earnings. Although this would result in rents on the highest-valued properties being well in excess of £100 per week, with rents falling in the lowest property areas, this model involves the least disruption.
- The second option is that there should be a weighting of 30 per cent on capital values and 70 per cent on regional earnings. Only a few rents consequently would exceed £100 per week, but rents on the lowest-valued properties would rise by an average of £6 per week. Overall, disparities between local authority and RSL rents would be reduced.
- The third option – derived from an idea by Hills (1999) – is for rents to be based on the cost of management and maintenance, plus an element for capital values. This too would narrow the differential in rents between local authority and RSL housing, since the costs of management and maintenance for both local authorities and RSLs are broadly similar, while the significance of capital values in rent determination would be diminished. However, a disadvantage is that the relationship between quality and rent could be weak in some locations, and the model takes little account of regional variations in affordability.

The Green Paper concluded that the most promising option would be one that took both capital values and regional earnings into account, combined with a system whereby rents would be determined by the cost of managing and maintaining homes. Clearly, if this hybrid option were to be introduced, rents across the two tenures of social housing would be based on the same principles and would, in the long term, converge. Ultimately, not only would rent setting in the two tenures be more equitable, but stock transfers would be easier.

In the view of the government, the convergence of rents (and its effect on transfers) could be hastened since – with an increase in investment (following the CRI and the Spending Reviews) – the condition and quality of the local authority stock will be increased, and this will in itself justify some small annual increases in average local authority rents towards the level of rents charged by RSLs. However, since the Green Paper proposed that the rents charged by RSLs should not rise by more than the retail price index from 2002 (to give local authority rents a chance to catch up), the viability of many housing associations could be in doubt, which would increase the pressure on some associations to merge. Under these circumstances, the Green Paper is right to suggest that it will be necessary to ensure that the SHG is at an appropriate level both to maintain the new-build programmes of RSLs and to prevent rents rising ahead of the retail price index.

Conclusion

To facilitate the government's intention (as set out in the Green Paper) to improve the quality and increase the size of the housing association stock, the *Spending Review* (Treasury, 2000) enabled Housing Corporation spending to escalate from £995 million to £1.5 billion between 2001/02 and 2003/04 – a rise of 47 per cent, whereas local authority capital expenditure was planned to increase by only 10.4 per cent (from £2.31 billion to 2.55 billion) over the same period. A large proportion of this increase was attributable to housing associations being authorised to increase their approved development programmes by 57 per cent over the three years (from £789 million to £1.2 billion). However, since housing associations (together with the financial institutions) need to fund an increasing number of transfers, it is unlikely that the increase in capital spending 'will result in any significant rise in the numbers of new social rented dwellings constructed' (Wilcox, 2000: 80).

The *Spending Review* – with regard to social housing – was concerned not only with investment *per se*, but also with the performance of Best Value and the role of a new Housing Inspectorate. The review, furthermore, announced that a new Community Housing Taskforce would be set up to advance a reformed transfer process and to ensure that new social landlords (established to facilitate transfer) empower tenants and regenerate communities.

Further reading

Chaplin, R., Jones, M., Martin, S., Pryke, M., Royce, C., Whitehead, C.M.E. and Yang, J.H. (1996) *Rents and Risks: Investing in Housing Associations*, York: Joseph Rowntree Foundation/York Publishing Services. Examines the developing relationship between housing associations and private finance, and how existing tenants are subsidising new development.

Chartered Institute of Public Finance and Accountancy (1994) *An Introductory Guide to the Financial Management of Housing Associations*, London: CIPFA. Provides a detailed account of the financial regime and requirements placed on housing associations.

Cole, I., Gidley, G., Richie, C., Simpson, D. and Wishart, B. (1996) *Creating Communities or Welfare Housing?*, Coventry/York: Chartered Institute of Housing/Joseph Rowntree Foundation. A study that examines how housing association residents view their estates.

Cope, H. (1990) *Housing Associations: Policy and Practice*, Basingstoke: Macmillan. Possibly the best introduction to the structure, organisation and role of housing associations.

Hills, J. (1991) *Unravelling Housing Finance: Subsidies, Benefits and Taxation*, Oxford: Oxford University Press. Contains a detailed and useful analysis of the housing association finance system.

Malpass, P. (2000) *Housing Associations and Housing Policy*, Basingstoke: Macmillan. Provides a comprehensive historical perspective of housing associations and related policy.

Nevin, B. (1999) *Local Housing Companies: Progress and Problems*, Coventry: Chartered Institute of Housing. Examines the experiences of local partnerships involved in developing local housing companies.

Page, D. (1993) *Building for Communities: A Study of New Housing Association Estates*, York: Joseph Rowntree Trust. Sheds light on the social considerations involved in developing new housing association estates.

Saw, P., Pryke, M., Royce, C. and Whitehead, C.M.E. (1996) *Private Finance for Social Housing: What Lenders Require and Associations Provide*, Cambridge: Department of Land Economy, University of Cambridge. Provides a useful account of the relationship between housing associations and private finance.

⬤10 Owner-occupation

In 1914 owner-occupation accounted for only 10.6 per cent of the housing stock of the United Kingdom, but largely owing to extensive housebuilding subsequently (except during the First and Second World Wars and their immediate aftermath), the proportion reached 67.6 per cent in 1999 – significantly more than in most other European countries, although in the USA and Australia some 70–75 per cent of dwellings were owner-occupied. In Scotland, however, the proportion of owner-occupation is somewhat less than in the rest of Great Britain – only 62.4 per cent in 1999.

The post-war period in Britain has seen the consolidation of owner-occupation as the most important numerical and therefore most politically sensitive sector of the housing market. Owner-occupation has been encouraged by favourable government policy, the expansion of specialist financial institutions, an investment climate generally favourable to property development, and the difficulty of households finding alternative accommodation. But although 19.4 million houses in all sectors were built in the period 1919–99, only a very small number are added annually; for example, 158,000 owner-occupied houses were started in 1999, equal to 0.9 per cent of the total owner-occupied stock and 0.6 per cent of the total stock – not taking into account demolitions.

Taking account of the scale of owner-occupation in the United Kingdom, and the growth of this sector in recent years, this chapter:

- considers the socio-political view of owner-occupation and the support for home-ownership in recent years;
- discusses the extension of owner-occupation – focusing on leasehold enfranchisement, low-cost home-ownership, protection of home-owners, and the demand for second homes and holiday lets;

- looks at the role of building societies and the economy;
- analyses the house price cycle;
- examines the macroeconomic determinants of the demand for home-ownership; and
- concludes by reviewing the Labour government's approach to the sector in the aftermath of the 1997 General Election.

The socio-political view of owner-occupation

Until at least the 1930s, owner-occupation was by no means considered by most households to be the 'ideal' or 'natural' form of tenure. During the 1920s, except for the council houses produced under the Housing Act of 1923 (the Chamberlain Act), local authority dwellings were 'in every sense the ideal, being better produced at high standard for the better-off members of the working class' (Clarke and Ginsburg, 1975). Owner-occupation began to be popularly attractive only when local authority housing became generally restricted to the displaced families of slum clearance schemes after 1933. The desire for home-ownership was more of a response to a lack of choice than a reaction against renting. The housing policy of the 1930s was, according to Clarke and Ginsburg, 'directly associated with the drive to make the better-off members of the working class into owner-occupiers'.

During the inter-war period the increase in owner-occupation was in large part a combined result of higher real wages, mortgage repayments being lengthened from 15 to 20–25 years, local authorities acting as mortgage guarantors, and the provision of subsidies for private construction (under the Chamberlain Act). Increased car-ownership and the lack of effective planning controls over suburban development meant that cheap land could be used for extensive speculative housebuilding, and there was also a diversion of much private rented housing to the owner-occupied sector since landlords found it more profitable to sell off their property than to remain in the rented sector.

Despite the substantial extension of development control by the Town and Country Planning Act of 1947, owner-occupation (particularly in the suburbs and beyond) has been promoted continuously since the early 1950s. Although the building licensing system (introduced in 1939) was continued after the Second World War, severely restricting private development, licences were abolished in 1952, one year after the Conservative government had replaced Labour, and the Conservatives furthered the expansion of owner-occupation to 1997. Between 1954 and 1957 the Conservative government guaranteed loans to mortgagors in excess of the percentage of valuation that the societies would normally advance; the party fought the 1955 General Election with a pledge that it would create a 'property-owning democracy'; between 1959 and 1962 the Conservative government lent £100 million to building societies to fund the purchase of pre-1919 dwellings; in 1963 it abolished tax on imputed income (Schedule A) from owner-occupied property; in 1971 its White Paper *Fair Deal for Housing* (DoE, 1971) and in 1972 the Housing Finance Act both continued the system of tax relief on mortgage interest; and in the period 1971–74, and from 1977 (at first at a local government level), sales of council housing virtually dominated Conservative policy.

In the 1960s and 1970s Labour governments also attempted to extend home-ownership. The *Housing Programme 1965–70* (MHLG, 1965) outlined Labour's aims for the late 1960s, one of which was 'the stimulation of the planned growth of owner-occupation'. This aim was achieved by encouraging leaseholders to buy their freehold under the Leasehold Reform Act of 1967, by the introduction of the Option Mortgage Scheme in 1968, which made low-income housebuyers eligible for subsidies (equivalent to tax relief on mortgage interest payments), and by the continuation of tax relief to other mortgagors. There was little change of policy during 1974–79, although interest relief was withdrawn in 1974 on mortgages in excess of £25,000. In the same year mortgagors were protected from rising interest rates by government loans of £500 million to building societies to offset a shortfall in funds. Until March 1979 (two months before Labour's defeat in the General Election), the Labour government permitted the sale of council houses – although it did not encourage the practice.

The bulk of the Labour movement has never been opposed in principle to owner-occupation. Democratic socialism draws a distinction between property owned for the purpose of realising a profit, and property which does not carry a cash income with it. The former consists of stocks and shares, rented buildings and land acquired for speculative reasons – all forms of capital; the latter includes the personal ownership of, for example, motor cars, washing machines and owner-occupied houses – all of which could be defined as consumption. If 'profits' are made on the resale of these goods they are usually incidental, and if owner-occupiers sell their houses at a price higher than they paid for them, they may be unable to realise the profit, since they might have to pay the same high price for a comparable alternative property. But by the late 1950s, even though nearly 40 per cent of the electorate were home-owners, Labour had clearly not demonstrated its support for owner-occupation. Abrams and Rose (1959), from their survey findings, reported that with regard to owner-occupation

> it would be hard to over-estimate the importance of housing as a real personal political issue for everyday people and its power to affect party support. Apparently, in the recent past it has affected it in a direction favourable to the Conservative Party.

By 1973, one-third of housebuyers were manual workers (and 40 per cent by 1979), a fact not escaping the Labour Party in formulating its policy towards home-ownership. Miliband (1972) and Coates (1975) suggested that the growing electoral influence of owner-occupiers (who owned over half of the nation's housing stock by the 1970s) went some way to explaining why the Labour Party seemed to be increasingly embracing the Conservatives' idea of a 'property-owning democracy'. But even by 1983, Labour's belief in owner-occupation may not have convinced home-owners. The *Economist* (1983a), referring to a MORI survey, reported that among the 59 per cent of the electorate who were owner-occupiers, 52 per cent voted Conservative at the time of the General Election and 19 per cent voted Labour; while among working-class home-owners, 47 per cent voted Conservative, 26 per cent voted Labour, and 26 per cent voted for the Liberal–Social Democratic Alliance. Both the Conservative *and* Labour parties clearly had a commitment to extend owner-occupation, but they differed about how this should be achieved and the extent to which the sector should be expanded at the expense of other tenures.

Britain's third largest party (historically), the Liberals, has also supported the owner-occupied sector – believing in a property-owning democracy. In the 1970s, the party was concerned with the problems of first-time housebuyers and called for measures to make it easier for young married couples to buy their own homes. This view was inherited subsequently by the Liberal Democrats, who were also in favour of Labour's proposal to phase out mortgage tax relief above the basic rate of taxation (see Chapter 5).

It must not be assumed, however, that the dominance of owner-occupation in Britain is a result of unimpeded market forces, or *laissez-faire* conditions. The situation has arisen since successive governments have equated increased owner-occupation with the advance of capitalism and the promotion of a 'property-owning democracy' (a Conservative election slogan in 1955). To this end, very substantial tax allowances and tax exemptions have been given to owner-occupiers, amounts which in total dwarf Exchequer subsidies and rate contributions to local authority housing. In the absence of this form of state intervention in the owner-occupied sector, a very different pattern of tenure might have evolved.

Government support for home-ownership

A major incentive to owner-occupation in the late twentieth-century was tax relief on mortgage interest payments, an allowance geared to the standard/basic rate of income tax. During the inflation of the 1970s and early 1980s, tax relief meant that the owner-occupier paid in effect a negative rate of interest – as much as 18 per cent in 1975 or 20.1 per cent in 1980. In addition, the doubling of house prices in the 1970s underlined the investment advantages of owner-occupation. If the increase in capital values is included, the effective rate of interest was even more negative. However, with a considerable reduction in the rate of inflation after 1982, net mortgage interest rates were no longer negative – for the first time since the late 1960s – thus reducing the investment potential of owner-occupation, at least in the short term.

Tax relief on mortgages for owner-occupation cost £7,500 million in 1990/91 compared to the £900 million Exchequer and rate subsidy to council housing (the respective amounts in 1972/73 being £802 million and £592 million). It was clearly advantageous to be a mortgagor paying in depreciated pounds (with government assistance) in order to own a capital asset which was appreciating in value throughout most of the 1970s and 1980s ahead of the rate of inflation. In addition, owner-occupiers were free from capital gains tax when they sold (an exemption generally limited to one property), and, from 1963, Schedule A tax (on imputed net income derived from a dwelling by its owner) had not been applied, costing the Treasury initially £48 million in a full year, but nearer £7,000 million per annum (net) by 1982/83. In 1971 stamp duty on house sales was reduced and abolished on mortgages, and in 1973 new houses were zero-rated for value added tax. Owner-occupiers could also qualify for improvement and standard/intermediate grants under the Housing Acts of 1969 and 1974 – financial injections which normally increased the value of their property.

These advantages of owner-occupation increasingly highlighted the considerable inequality of assistance provided for different tenure groups, and the subsidy to housebuyers was itself highly regressive; for example, a basic-rate taxpayer in 1978 paid net mortgage interest of 6.53 per cent, whereas a taxpayer on the highest rate (83 per cent) paid only 1.7 per cent in net interest. This produced some remarkable anomalies: a basic-rate taxpayer buying a £10,000 house with a 100 per cent 25-year mortgage would have to repay £1,081 (gross) in the first year reduced to £759 (net) or £14.60 per week; a single person on the highest tax rate (earning, say, £24,985) buying a £25,000 property would have to repay only £680 (net) (or £13 per week) on a gross mortgage repayment of £2,703.

Box 10.1

Owner-occupation in Scotland, Wales and Northern Ireland

Scotland has the lowest level of owner-occupation of all the countries making up the United Kingdom. Although the proportion of owner-occupied dwellings north of the Border increased substantially from 49.1 per cent of the housing stock in 1989 to 62.4 per cent in 1999, largely as a result of Right to Buy policy, this was still below that in the United Kingdom as a whole (67.6 per cent). Home-ownership aspirations, moreover, were also lower than in other parts of the union. According to the Council of Mortgage Lenders *Annual Finance Survey* (CML, 1999), only 67 per cent of heads of households in Scotland would like to be owner-occupiers by 2009 compared to 74 per cent in Northern Ireland, 79 per cent in England and 81 per cent in Wales – a situation attributable to the 'historically much greater role played by the public sector in Scotland's housing system' (Earley, 2000: 25)

In contrast to the expansion of home-ownership in Scotland, the owner-occupied sector in Wales has grown very slowly in recent years, largely as a result of a downward trend in the annual rate of housebuilding throughout the 1990s. Whereas in 1991 the number of owner-occupied dwellings represented 71 per cent of the housing stock of the Principality, by 1999 the proportion had risen to only 71.6 per cent.

Northern Ireland has the highest rate of owner-occupation in the United Kingdom, rising fairly rapidly from 63.4 per cent in 1989 to 71.9 per cent in 1999, with growth in the latter years being boosted by the sale of Housing Executive properties.

Extension of owner-occupation

Following a steady increase in housebuilding after 1945, the largest number of private sector starts, 247,500, was achieved in 1964 – a reflection of a high level of demand supported by tax relief on mortgage interest. But there were three obstacles to a further expansion of owner-occupation: the leasehold system, the difficulties faced by low-income housebuyers in raising finance, and the problems facing first-time buyers in general.

Leasehold enfranchisement

As early as 1884 a Royal Commission on the Housing of the Working Classes recommended 'leasehold enfranchisement' – the right of a leaseholder to buy the freehold of the land on which his or her house stands. But proposed legislation to implement this right was continually blocked. The Landlord and Tenant Act of 1954 left the leaseholder in a hopelessly weak position compared with the power of the ground landlord. It allowed a household to continue living in a dwelling after the lease had expired but as a tenant paying a market rent and not as an owner-occupier. Normally when ground leases expired, households risked eviction with no compensation for the home which over the years they had saved to buy. The anxiety and insecurity which had been created by the leasehold system had been obvious for decades. The Leasehold Reform Act of 1967 ended this insecurity by giving leaseholders the right to buy the freehold of their houses or to extend their lease for 50 years provided the house had a rateable value not exceeding £400 (in Greater London) or £200 (elsewhere); the lease was originally granted for more than 20 years; the lease was of the whole house; the lease was at a low rent (equivalent to less than two-thirds of the rateable value of the house); and the leaseholder was occupying the house as his or her only or main residence and had been doing so for five years. But a high proportion of leaseholders failed to qualify for these rights. By the early 1980s there were up to 2 million owner-occupiers who were still leaseholders – much as before 1967. They comprised occupiers of flats bought on long leases in the boom development years of the late 1950s and early 1960s; holders of leases of less than 21 years who bought at a premium and paid peppercorn rents (of, say, £100–£2,000 per annum); and leaseholds of old terraced housing in, for example, Birmingham, Cardiff, Liverpool and London, whose 99-year leases would soon run out. All had to face the possibility that either the value of the asset might decrease in real or even absolute terms in the future, or the certainty that they would lose their home when the lease expired. Long before a lease expired, however, the incentive to maintain the property, and its saleability, diminished.

In *Labour's Programme 1982* (Labour Party, 1982a) the party recognised that leaseholds would still be required for flats because of the management responsibilities for common services and maintenance. However, there was a need to regulate service charges, give households the right to extend their leases, and allow occupants of a block to purchase the freehold or head lease collectively, subject to support from a specified majority and safeguards for any tenants involved. But with regard to houses, it was proposed that the creation of new leases should be prohibited, enfranchisement rights should be strengthened by ending periods of qualifying residence and by changing valuation criteria, and local authorities should be given the task of acquiring any freehold not taken up by the occupiers. It was particularly ironic that the Conservative government gave tenants of council houses the right to buy the freehold of their homes, but denied the extension of freehold to leaseholders, a high proportion of whom paid a market price for their homes virtually indistinguishable from freehold value.

Under the Leasehold Reform, Housing and Urban Development Act of 1993, however, 750,000 leasehold flat owners either were able to buy the freehold of their homes (provided their leases were for more than 21 years and two-thirds of leaseholders of a freehold property agreed) or, where they were not eligible to buy, they were given the right to extend their leases by 90 years.

The 1993 Act, however, failed to prevent unscrupulous landlords and managing agents from obstructing the transfer of freehold. Collective enfranchisement clearly did not work, although there was some success in allowing tenants to extend the length of their individual leases to 90 years, even though there were continuing problems of management. Under the Housing Act of 1996, leaseholders' rights were therefore enhanced, and from September 1997 – without tenants having to risk huge legal fees – new Leasehold Valuation Tribunals were empowered to arbitrate in service charge disputes and could determine whether tenants could take over the management of their homes from negligent landlords.

On returning to power in 1997, Labour considered that further reforms were necessary. It therefore aimed to introduce a new form of tenure which would enable properties to be collectively owned and managed by their residents, and to simplify the rules concerning the purchase of freeholds by individual leaseholders. Based on the proposals of the Housing Green Paper (DETR/DSS, 2000), the Commonhold and Leasehold Reform Bill of 2001 proposed to:

- introduce a new form of tenure – *commonhold* – to facilitate the joint ownership of freehold and to create management companies to run communal areas;
- make it easier for a leaseholder to buy the freehold of his/her home (for this purpose, the minimum proportion of leaseholds in a block that must take part in the scheme would be reduced from two-thirds to a half);
- make it harder for landlords to obstruct this process; and
- enable all leaseholds in a block to take over the management of the block even if they did not buy the freehold.

Since commonhold would be unlimited in time (unlike a lease), it would give each resident in a block an interest similar to freehold. Although commonhold would be available for new schemes from the outset, it would also be possible for the new tenure to be created in respect of existing properties if all parties agree. Because of a shortage of parliamentary time, the Bill failed to get enacted in the 2000/01 session – with the Queen's Assent being delayed until after the 2001 General Election.

Low-cost home-ownership

Although, on average, home-owners are better off than tenants (with only a small proportion of owners having low incomes), the owner-occupied sector in 2000 contained around half of all poor households since it was by far the largest tenure, while owner-occupiers comprised 57 per cent of all households within the lowest income decile (Burrows and Wilcox, 2000). It should also be recognised that while most low-income households aspire to own their homes, in recent years fewer low-income households have been moving into home-ownership than have moved out of it. Over the period 1995–97, for example, the owner-occupied sector attracted a further 50,000 low-income households, but failed to retain around 80,000 poorer households – in many cases owing to repossession (Burrows and Wilcox, 2000).

Since at least the 1980s, in an attempt to expand the owner-occupied sector governments have attempted to ease the financial burden of home-ownership among lower-income households. Apart from introducing the Right to Buy, whereby council tenants could buy their home at a discount (see Chapter 8), governments – through local authorities and the Housing Corporation – facilitated a succession of low-cost schemes to meet the needs of relatively low-income households. Such schemes included:

- a conventional shared ownership scheme;
- a cash incentive scheme;
- a 'do-it-yourself' shared ownership scheme;
- a 'home-buy' scheme;
- a 'starter home initiative' scheme; and
- a 'new homesteading' arrangement.

A *conventional shared-ownership scheme* was introduced by a
Conservative government in July 1979 to appeal (together with the sale
of council houses) to up to 6 million tenants.

In the 1980s both the main political parties were in favour of introducing
a *cash incentive scheme* to encourage council tenants to buy in the open
market. Labour had proposed that if it regained power in 1983 it would
provide a cash subsidy (equivalent to an RTB discount) as a contribution
towards a down-payment on a house in the owner-occupied sector.
Almost an identical facility was actually made available by the
Conservative government in 1988 – cash incentives of £200 million being
allocated to enable about 50,000 council tenants to put down a deposit on
private sector housing.

To further extend the owner-occupied sector, the Conservative
government in 1992 introduced the *do-it-yourself shared ownership*
(DIYSO) scheme, whereby households on council waiting lists could buy
property on the open market jointly with a housing association. Under
this scheme the association would buy the property outright and the
household would purchase a 25 per cent share of the equity with the help
of a mortgage, and pay rent to the association for the remainder.
Subsequently, households could increase their share of the equity as their
incomes increased.

However, in the late 1990s the Labour government discontinued the
DIYSO scheme, and, starting in 1999, phased in a *home-buy* scheme.
Under Homebuy the minimum equity that had to be bought was as much
as 75 per cent, but the rest would be covered by an interest-free loan from
an RSL. Although evidence suggests that compared to shared ownership,
Homebuy was preferred by relatively well-off purchasers since it gave
them a greater sense of ownership and independence (4,000 loans were
provided in 1999/2000), it was a comparatively costly option for lower-
income households and took too little account of regional variations in
house prices and incomes (Jackson, 2001; Salman and Bar-Hillel, 2000).

According to its Green Paper (DETR/DSS, 2000), Labour acknowledged
that there was scope for further support for people on the threshold of

home-ownership. It therefore proposed a new *starter home initiative* which would:

- involve repayable interest-free loans;
- be targeted at key workers on lower incomes;
- be focused on dwellings in the bottom quartile of house prices in a local housing market area;
- be available in areas where house price affordability was a significant problem and where there was excess demand for housing; and
- be available in areas of new development or renovation, and might also cover market areas across a number of adjacent authorities – provided that it conformed with local housing strategies.

The initiative (announced in September 2000) would thus facilitate the provision of interest-free top-up loans of up to £10,000, plus cash for deposits and help with shared ownership schemes (where a housing association owned half the equity) to enable up to 14,000 nurses, teachers, police officers and other essential workers on average salaries obtain a mortgage to buy a home – worth no more than £150,000 – in high-priced areas such as London and the South East. However, only time will tell whether assistance to key workers will be adequate. The maximum permitted price under the initiative was well below the average price of houses in the capital (£180,000 in 2001), and, although the initiative was intended to be supported by an additional DETR allocation of £250 million (over the period 2001/02–2003/04) under the government's *Spending Review* (Treasury, 2000), departments other than Environment, Transport and the Regions – such as the Home Office and the Department for Education and Employment – failed to contribute any further funds to the initiative.

A final way of facilitating low-cost home-ownership is through *homesteading* arrangements. Whereas the starter-home initiative is applicable to London and the South East, homesteading is a particularly useful way of promoting owner-occupation in run-down areas of low housing demand, such as the inner cities of the North, where the alternative would involve street after street of substandard housing falling into the hands of unscrupulous landlords intent on letting to tenants on housing benefit. Clearly, while such schemes would not be suitable for families with children, they could be instrumental in providing a low-cost supply of starter homes. The Secretary of State 'needs to think about how groups of poorer residents can acquire their own homes and make a go of it in areas spiralling into decline' (Field, 2001).

Protecting home-owners

Recognising the concern among housebuyers about the inherent risks of borrowing, Labour, on regaining power in 1997, began working in partnership with mortgage lenders and insurers to improve the quality and value of mortgage payment protection insurance (MPPI) – the industry aiming to achieve a take-up of 55 per cent of new home-buyers by 2004. The government also encouraged mortgage lenders to offer more flexible mortgages that allow borrowers to vary their repayments, for example through 'payment holidays' when household finances are tight, or to make larger repayments when incomes permit.

To assist would-be housebuyers, and particularly to tackle gazumping, the government, in its Homes Bill of 2001, proposed introducing a range of measures to speed up the home-buying process and to make it easier and more consumer-friendly. These measures would have included a seller's pack (containing the documents and information required to complete a sale such as a home condition report written in plain English, a local authority search, a draft contract and title deeds); and the creation of a central computerised database of all Land Registry searches to reduce the time it takes to complete sales from three months to 28 days. Clearly, it was crucial for estate agents to ensure that sellers complied with the Act, and for sellers to instruct solicitors as soon as they put their property on the market, rather than when they found a buyer. However, critics argued that the scheme would become unwieldy and impose an excessive burden on sellers, particularly in areas of low house prices. With a seller's pack costing £300–£400, and with houses selling for around £10,000 in, for example, parts of the North, this additional cost of conveyancing would undoubtedly have obstructed rather than eased the functioning of the housing market. For this reason, the DETR, in its Bill, proposed exempting sellers in low-priced areas from the duty to provide a seller's pack.

Second homes and holiday lets

It has been argued increasingly that the demand for second homes in the more popular parts of the countryside should be restricted to ensure that housing is available at an affordable price to local people on average incomes, and to minimise the damage to local communities. Because of the seasonal nature of residence, more and more villages are becoming ghost communities for around six months of the year. By 2001 there were probably as many as 200,000 second homes in England and Wales (compared to only 92,000 in 1991) and 300,000 holiday lets. In the more popular villages, the shortage of homes for the local population is acute; for example, in Troutbeck, in the Lake District National Park, 40 out of 105 dwellings at the start of the new millennium were second homes or holiday lets (Hetherington, 2000d).

Aware of the problem, the Cabinet Office (1999) recommended that special zones should be declared in which second homes would be prohibited, and the Rural White Paper (DETR/MAFF, 2000) proposed – in an attempt to deflate demand and prices – that local authorities should be given powers to levy the full council tax on second homes and holiday lets, instead of the 50 per cent rate currently being paid.

Building societies and the economy

The availability of long-term finance is essential to an ever-expanding owner-occupied sector. From the nineteenth century until the 1990s, the building societies were the principal lending institutions for house purchase, ahead of the banking sector and other mortgage lenders such as insurance companies, local authorities and pension funds. However, in recent years the banking sector has substantially gained ground in terms of net advances and balances outstanding (Table 10.1), in part an outcome of the demutualisation of many building societies (such as the Halifax and the Abbey National) and their conversion into banks. To the banks at a time of recession, mortgage loans posed less risk than lending to industry and commerce, while the demand for loans in these latter sectors collapsed in the early 1980s. Under government encouragement the opportunity to seize a share of the relatively stable home loans market was irresistible, and by 1993 banks eventually overtook building societies as the principal providers of home loans, consolidating this role in the late 1990s. To be leading lending institutions, the building societies – and subsequently the banks – had to attract the lion's share of personal savings. Whereas in the past, the building societies invariably offered higher rates of interest to their savers than those offered by banks, in recent years the banks have demonstrated an ability both to compete for funds and to expand their role as mortgage lenders.

Unlike social housing investment, which – historically – has mainly financed housebuilding, most institutional mortgage finance is used to

Table 10.1 *Institutional funds for house purchase*

	Net advances			
	1989		1999	
	£m	%	£m	%
Banks	7,169	21	21,492	56
Building societies	24,002	70	10,638	28
Other specialist mortgage lenders	2,951	9	6,405	17
Other funds	−66	—	−153	—
Total	34,056		38,382	
	Balances outstanding			
	1989		1999	
	£m	%	£m	%
Banks	79,193	31	345,032	70
Building societies	152,542	59	113,469	23
Other specialist mortgage lenders	17,413	7	33,753	7
Other funds	8,855	3	2,184	—
Total	258,003		494,438	

Source: CML, *Housing Finance*

facilitate the resale of 'second-hand' housing, indicating its relative and absolute inefficiency. Also, one of the principal criticisms of housing finance in the past was that funds were being attracted away from manufacturing industry into the home loans market, and were thus contributing to the deindustrialisation of much of the United Kingdom.

The house price cycle

Rising house prices are associated with increasing numbers of fiscal incentives (government measures to extend owner-occupation and mortgage availability), but they are also related to increases in average earnings. Table 10.2 shows that in 1989 (at the end of the house price boom of the late 1980s), average house prices soared by 21 per cent, average earnings increased by 9.1 per cent, and the house price/earnings ratio (HPER) was as high as 4.9. During the subsequent slump, average house prices decreased by as much as 3.8 per cent in 1992, average earnings increased by only 3 per cent, and the HPER fell to 3.4 in 1995/96. In 1999, with the upturn in the house price cycle, average house

Table 10.2 *Average house prices, average earnings and retail prices, 1989–99*

	Average house prices		Average earnings		House price/ earnings ratio (HPER)	Retail price index % increase	Real house prices % increase
	£	% increase	£	% increase			
1989	70,400	21.0	14,268	9.1	4.9	7.8	12.2
1990	69,500	–1.3	15,655	9.7	4.4	9.5	–9.8
1991	68,600	–1.4	16,832	7.5	4.1	5.9	–6.9
1992	66,000	–5.2	17,824	5.9	3.7	3.7	–7.2
1993	64,300	–1.1	18,363	3.0	3.5	1.6	–4.0
1994	66,300	3.1	18,991	3.4	3.5	2.5	0.6
1995	66,700	0.7	19,637	3.4	3.4	3.4	–2.6
1996	69,100	3.6	20,373	3.7	3.4	2.4	1.1
1997	75,600	9.4	21,345	4.8	3.5	3.1	6.0
1998	83,900	10.9	22,298	4.5	3.8	3.4	7.3
1999	93,500	11.5	23,360	4.8	4.0	1.6	9.8
2000	106,900	14.3	—	—	—	—	—

Source: CML, *Housing Finance*

prices rose by 11.5 per cent, average earnings increased by 4.8 per cent, and the HPER reached 4.0. Whereas increases in house prices in cash terms were markedly greater than increases in real house prices during the boom years, house prices in cash terms fell less dramatically than real house prices during the slump.

The extent to which house prices change is also determined by the house price/income ratio (HPIR), advance/income ratio (AIR) and advance/purchase price ratio (APPR). A slump normally occurs – and housebuying becomes more affordable – when the HPIR and AIR decrease and the APPR increases. Conversely, a boom normally takes place – and housebuying becomes less affordable – when the HPIR and AIR rise and the APPR falls. Table 10.3 reveals that during the house

Table 10.3 *House price increases, house to income ratios, advance to income ratios and advance to purchase price ratios, 1995–99*

	House price increase (%)	House price to income ratio (HPIR)	Advance to income ratio (AIR)	Advance to purchase price ratio (APPR)
1995	0.7	2.84	2.09	73.7
1996	3.6	2.85	2.08	72.9
1997	9.4	2.92	2.11	72.3
1998	10.9	3.00	2.13	71.0
1999	11.5	3.10	2.18	70.2

Source: CML, *Housing Finance*

price boom of the late 1990s there was a broad positive correlation between the increase in average house prices and the increase in the HPIR and AIR, but an inverse correlation between house price increases and the APPR.

The house price boom of the late 1980s

In the late 1980s there were as many as 1,580,000 transactions in the owner-occupied market in England and Wales alone, and average house prices in the United Kingdom increased substantially, considerably ahead of increases in average earnings and the retail price index. The causes of the boom were numerous, but essentially derived from an increase in the general level of demand for home-ownership.

First, there was a large increase in the number of people wanting to buy. The high birth rates in the early and mid-1960s resulted in a surge in demand from new households wanting to set up home in the late 1980s.

Second, liberal mortgage lending policies, encouraged by financial deregulation (for example by means of the Building Societies Act of 1986), made it easier for housebuyers to obtain larger and larger mortgages (Bank of England, 1991; JRF, 1991). The number of mortgages increased from 7.7 million to 9.1 million, 1985–89, while net advances increased from £19.1 billion to £40.1 billion. Mortgage debt as a proportion of GDP therefore increased from 32 per cent to 58 per cent, compared to only 20 per cent in France, Germany and Japan, and the HPER soared to 4.9:1 in 1989. In the United Kingdom in the late 1980s, 100 per cent mortgages were common, whereas in France mortgages were no higher than 80 per cent and in Germany and Japan rarely exceeded 60 per cent. Since both realised and unrealised capital gain were greater than mortgage interest payments, the strong speculative motive in house purchase encouraged buyers to maximise their borrowing. With the end of multiple mortgage interest relief from 1 August 1988 (announced in the previous spring's Budget) there was a last-minute hike in demand between March and the end of July which substantially accounted for the 22 per cent increase in average house prices in 1988.

Third, coming towards the end of the boom, there was a big rise in post-tax incomes, in part due to the reduction in the basic rate of tax from 30 to 25 per cent and the top rate from 60 to 45 per cent, enabling borrowers to obtain more substantial mortgage loans.

Fourth, the house price boom can also be attributed to a deficiency of supply. Housebuilding was increasing more slowly than the number of new households, in part as a result of the planning constraints imposed by county councils (in attempts to protect green belts, other areas of high landscape value and the alleged interests of existing communities), and possibly also because of private landowners withholding sites from the market for speculative reasons.

Finally, in the late 1980s it was predicted that the replacement of domestic rates by the community charge (poll tax) in 1989 and 1990 would inflate house prices still further, particularly at the top end of the housing market, since the perceived cost of home-ownership would be reduced. Wealth would consequently be distributed in favour of owner-occupiers (most notably in much of the South East and West Midlands) and against the interest of tenants. The inflationary effect of the poll tax, however, was subsequently masked by the slump in house prices.

Effects of the boom

One attribute of the boom was that it had a very variable spatial effect on house prices. There were very wide regional disparities in the rate of increase in average house prices, 1985–88, and in HPERs (Table 10.4). Whereas in East Anglia and the South East (outside of Greater London) prices increased by 81 and 79 per cent respectively, in Northern Ireland and Scotland the rate of increase was as low as 30 and 17 per cent, and HPERs ranged from 3.6:1 to 2.21:1 in 1988. Differential increases clearly led to a divergence of average house prices in the late 1980s. By 1988, prices were as high as £77,700 and £72,600 in Greater London and the rest of the South East respectively, but as low as £30,200 in the Northern region and £29,900 in Northern Ireland. Regional disparities clearly resulted in a 'mobility trap' (Murphy and Muellbauer, 1990). Households wishing to move from areas of high to low unemployment, for example from the Northern region to the South East, not only would have had to pay substantially more for a house but also would have had to secure more remunerative employment to afford their housing needs, as regional price and income statistics suggest (Table 10.4). Conversely, few households would have been willing to move from high- to low-price regions (assuming jobs were available) since they would have had to forgo opportunities for substantial capital accumulation, and might also have feared that they would be priced out of the house market should

Table 10.4 North–South disparities in house price change and earnings during the boom and slump, 1985–92

	Average house prices			Average earnings of borrowers, 1988 (£)	House price/ earnings ratio 1988 (£)	Average house prices			Average earnings of borrowers, 1992 (£)	House price/ earnings ratio, 1992 (£)
	1985 (£)	1988 (£)	% increase, 1985–88			1992 (£)	% increase, 1988–92			
South										
East Anglia	31,661	57,296	81	15,936	3.60:1	64,610	7	18,920	3.41:1
South East (excl. GL)	40,487	72,561	79	20,180	3.60:1	75,189	8	23,786	3.16:1
South West	32,948	58,457	77	16,672	3.51:1	77,416	4	19,950	3.88:1
Greater London	44,301	77,697	75	24,409	3.18:1	77,446	6	26,597	2.91:1
West Midlands	25,855	41,700	61	15,130	2.76:1	58,405	17	19,601	2.98:1
East Midlands	25,539	40,521	59	14,533	2.79:1	56,527	14	19,818	2.98:1
North										
Yorkshire & Humberside	23,338	32,685	40	13,387	2.44:1	54,699	31	18,213	3.00:1
Wales	25,005	34,244	37	12,886	2.66:1	49,502	15	17,797	2.79:1
North West	25,126	34,074	36	12,257	2.57:1	58,169	38	19,172	3.03:1
Northern	22,786	30,193	33	12,693	2.38:1	48,624	30	17,672	2.75:1
Northern Ireland	23,012	29,875	30	13,490	2.21:1	39,240	30	18,009	2.18:1
Scotland	26,941	31,479	17	13,822	2.28:1	52,274	48	19,950	2.62:1
UK	31,103	49,355	59	16,040	3.08:1	62,265	14	20,819	2.99:1

Source: CML, *Housing Finance*; DoE, *Housing and Construction Statistics*

they wish to return. The mobility trap is far less evident in other countries. Smith (1990) pointed out, for example, that manual workers in the United States were 18 times more likely to move from one region to another than their counterparts in Britain.

A further effect of the boom was that much mortgage finance leaked out of the housing market into the rest of the economy. Even before the boom was fully under way, the Bank of England (1985) estimated that about £2 billion per annum leaked from the house purchase/improvement market through 'premature equity release'. An increasing proportion of mortgages were being used neither for trading up nor for improvement but for the purpose of general consumption. By 1988, at the peak of the house price boom, the mortgage institutions lent a total of £25 billion more in mortgages than was being spent on buying or improving homes – one of the reasons why consumption grew 2 per cent faster than the rate of growth of the gross national product, with subsequent adverse effects on the rate of inflation and the balance of payments.

Building societies and other mortgage lenders, whose activities either caused or were in response to house price inflation, clearly helped to create a 'new class of wealth owners' (Hamnett, 1988b). Whereas the total value of inherited residential property was £465 million in 1968/69, it increased to £7.2 billion in 1988/89. According to the Housing Research Foundation, 27 per cent of this inheritance was reinvested in housing, 24 per cent was spent on consumption and the rest was saved. The Henley Centre of Forecasting estimated that the value of inherited buildings and land (most of this being residential property) would soar to £21.5 billion by the year 2000 and particularly benefit households inheriting estates in the higher-priced regions of the South – widening the 'North–South divide' still further. In 1989, for example, the average value of an estate in Greater London was £87,000 compared to only £45,000 in the North East (Durham, 1990). This disparity was even more serious than these figures suggest since, compared to the South, there was a low level of owner-occupation in some of the northern regions and therefore relatively fewer households had any property to leave their heirs. Social divisions thus not only widened regionally but also between tenures, adding to the problems of the growing underclass (Williams, 1990).

The effects of the house price boom on the macroeconomy were no less important. Since house prices were an important element in the cost of living of housebuyers, Muellbauer (1986) argued that soaring house prices in 1984–86 put pressure on the labour market and added 4 per cent to the level of real wages (and predicted even larger increases in nominal wages in the late 1980s). Because wage agreements are often nationally based, house prices in the South East thus influenced wage levels throughout Britain. The view that house prices during a house price boom in effect determine wage levels (rather than vice versa) could be supported by the notion that people buy the most expensive house they can afford with the largest mortgage they can raise, but then need to earn more to maintain their customary level of consumption and savings.

The house price slump

Average house prices in the United Kingdom decreased from £70,400 in 1989 to £64,300 in 1993, and in England and Wales alone the number of residential transactions plummeted from 1,467,000 to 1,144,000 over the same period (Wilcox, 2000). The slump was particularly severe in the South East, East Anglia, the South West and Greater London, where house prices fell continually between 1989 and 1993.

A substantial decrease in demand, the principal cause of the slump, was caused by many factors. First, in response to an increase in bank base rate, interest rates on new mortgages rose from 9.5 per cent in April 1988 to 15 per cent (or above) in February 1990 and remained above 10 per cent until September 1992. Second, unemployment increased from 5.5 per cent in May 1990 to 10.9 per cent in January 1993, reducing the ability of many potential home-owners to buy, while lowering the purchasing confidence of others (Figure 10.1). Third, other would-be buyers awaited further falls in prices, and by remaining outside of the housing market helped to bring about the price reductions expected. Fourth, in response to falling prices and less security, mortgage lenders were increasingly reluctant to provide 100 per cent mortgages, and advances fell from £40 million to £27 billion, 1988–91. Fifth, the ending of multiple mortgage interest relief in 1988 and the end of the 'stamp duty holiday' in August 1992 both decreased demand (from December 1991, stamp duty on transactions from £30,000 to £250,000 was suspended to encourage housebuying). Sixth, the high level of demand in the late 1980s was very unstable and it was probable that house prices had risen to an unrealistically high level in relation to incomes during the previous boom and were destined to fall (Table 10.2). The HPER reached a peak of 4.9:1 in 1989 compared to an average of 3.5:1 throughout the

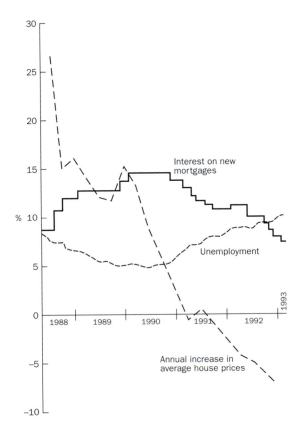

Figure 10.1 *Mortgage interest rate, unemployment and house prices, 1988–92*

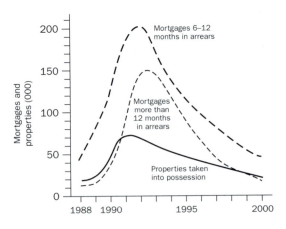

Figure 10.2 *Properties taken into possession and mortgages in arrears, 1985–2000*

Source: CML, *Housing Finance*

1980s and an average of only 3.3:1 throughout the period since the 1950s. In total, demand decreased from 1.9 million to 1 million transactions, 1988–92. Seventh, the replacement of the community charge (poll tax) by the council tax in April 1993 almost certainly helped to keep house prices depressed, since in many areas (particularly in the South) and at the upper end of the market the new tax was up to 50 per cent more than the former charge. Eighth, the reimposition of 1 per cent stamp duty in August 1992 on all house purchases over £30,000 (after a six-month suspension in an unsuccessful attempt to boost demand) almost certainly curbed any chance of a house price recovery for the time being.

The slump was also due to an excess of supply of housing for sale. Many households inflicted with unemployment, failed businesses or bankruptcy were unable to sell their homes at an acceptable price (or at all) and were thus unable to repay their mortgage loans. The number of properties repossessed (and often put on to the market at greatly deflated prices) increased from 18,510 in 1988 to 75,540 in

1991 (Figure 10.2). Although the number of repossessions fell slightly in 1993, there was a marked increase in the number of mortgages in arrears for 12 or more months, with the possibility of a resulting future increase in repossession and sale. There was also an increasing number of unsold newly built houses on the market (230,000 in 1992), which helped to depress prices, while 160,000 inherited properties also contributed to the slump.

Effects of the slump

The regional impact of the slump was the reverse of that resulting from the former boom. There was now a regional convergence (instead of a divergence) in the rate of change in average house price and rates of unemployment (Table 10.4 and Figure 10.3). Whereas Scotland, the North and Northern Ireland continued to enjoy a boom in the period 1989–93, the South West, East Anglia, the South East and Greater London experienced a very severe slump. Despite convergence, however, house prices were still markedly higher in, for example, the South East than in much of the rest of the country (and unemployment in this region was now also high). Thus convergence, as such, did little to encourage any notable increase in household mobility.

The effects of the slump in house prices on owner-occupiers and the macroeconomy were, however, substantial. By 1992, 21 per cent of all households (disproportionately in Greater London, the rest of the South East and East Anglia) were caught in the 'negative equity trap'. Since the value of their property was £6 billion below their mortgage debt (Bank of England, 1992), they were thus generally unable to sell their properties and buy elsewhere – the market, to an extent, ceasing to work. By October 1993 over 25 per cent of all those who bought their houses between 1988 and 1991 were caught in the negative equity trap, with as many as 41 per cent in London and the South East owing more on their mortgages than they could raise by selling their homes (Dorling and Gentle, 1994). There was evidence, moreover, that the negative equity trap did not affect only first-time buyers; as many as one in three households in negative equity were second- or third-time buyers of which a high proportion were in professional occupations with high incomes (Forrest et al., 1994).

The increase in the number of repossessions (referred to above) was clearly not only a cause of the house price slump, but also an effect.

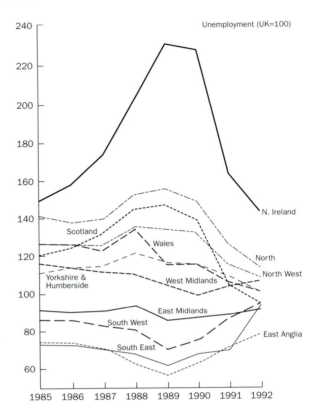

Figure 10.3 *North–South divergence in unemployment and house prices, 1985–92*
Sources: CML, *Housing Finance*; Department for Education and Employment, *Employment Gazette*

Had house prices been buoyant, mortgagors facing repayment difficulty could have sold their properties and traded downward. With the malfunctioning of the market, this was no longer possible and, at worst, dispossessed owner-occupiers found themselves homeless. To a significant extent, the substantial increase in homelessness and the increase in the number of homeless households in temporary accommodation (from 94,000 and 16,000 respectively in 1985 to 148,300 and 62,700 in 1992) was attributable to repossession.

With falling house prices and higher unemployment in the South East, pressures on the labour market of the region eased in the early 1990s with a stabilising effect on national wage settlements. The housebuilding industry, in particular, responded to falling house prices by decreasing output, reducing employment and lowering the rate of increase in wages in the industry and thus helped to lower general consumer demand. Falling house prices in the southern regions of Britain, moreover, deterred housebuyers from taking out second mortgages to purchase

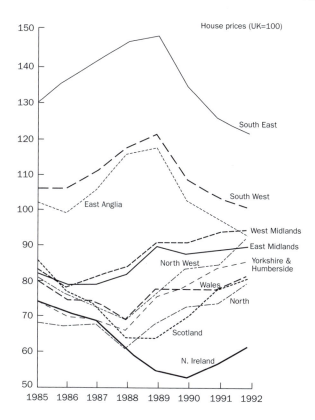

House prices (UK=100)

consumer goods and services, and nationally equity withdrawal was further constrained by the value of inherited properties rising less rapidly in the early 1990s than in the 1980s. With positive equity falling from £25 billion in 1988 to zero in 1993 (Hutton, 1994), consumer demand was consequently reduced, with disinflationary effects on consumer prices.

If the government had failed to recognise the connection between rising house prices and inflation in the late 1980s, it was evident that, in the 1990s, it was beginning to acknowledge the link between the housing slump and the economic recession. In December 1991 the government introduced a £1 billion rescue package intended to help the victims or potential victims of repossession, many of whom had recently become unemployed or had suffered a significant reduction in income for other reasons associated with the recession. Cheap loans became available to housing associations to enable them to acquire repossessed homes in order to rent them back to their former mortgagors; funds were allocated to building societies to facilitate the conversion of existing mortgages into (lower) rental agreements; and income support for unemployed mortgagors (amounting to about £1.25 billion per annum) was paid directly to mortgage lenders to reduce the extent of mortgage arrears and the risk of foreclosure. There was little evidence, however, to suggest that these measures had more than a very marginal effect on the number of dwellings in danger of repossession, and by 1994 it seemed likely that income support for unemployed mortgagors would be phased out.

After the 1992 General Election the Conservative government introduced a number of measures to increase the level of demand for home-ownership as well as to decrease excessive supply. First, building societies were permitted to raise the amount they could lend unsecured from £10,000 to £25,000 per person in an attempt to enable owner-occupiers with negative equity in their homes to sell and subsequently to buy elsewhere – a scheme soon adopted by some of the larger mortgage lenders such as the Abbey National with its Negative Equity Mortgage Scheme. Second, the Budget of March 1993 raised the transaction threshold from £30,000 to £60,000 on which stamp duty (at 1 per cent) would apply. Third, the government, regarding low interest rates as central to monetary policy, withdrew the United Kingdom from the Exchange Rate Mechanism and brought down the base rate from 10 per cent in August 1993 to 6 per cent in January 1993. Interest rates on new mortgages consequently fell from 10.75 per cent in September to 7.99 per cent by February 1993 – the lowest since the 1960s and, other things being equal, this should have stimulated demand. Finally, in an attempt to reduce excess supply, the government allocated £750 million to the Housing Corporation (in the Autumn Statement of 1992) to facilitate housing association acquisition of unsold housing. By March 1993 over 21,000 empty houses were thereby removed from the market, although most were newly built houses rather than repossessions, most of the £750 million was spent in the South East, and the number bought was a negligible proportion of the estimated 200,000 houses on the market.

There was little evidence, however, to suggest that these measures had more than a very marginal effect on the number of dwellings in danger of repossession; and from April 1995 ISMI was changed in a way which could only have been detrimental to households at risk from repossession. Previously, unemployed mortgagors were eligible for ISMI to cover half their mortgage interest payments for the first six months, and thereafter to meet the full interest payments on mortgages up to

£120,000, but from the beginning of the financial year 1995/96 ISMI was available only to existing mortgagors who became unemployed, and only then after two months from the claimant becoming unemployed and in respect of only half the mortgage interest paid on mortgages not in excess of £100,000. Anyone taking out a new mortgage after October 1995 was ineligible for support for the first nine months after becoming unemployed and thereafter for half support. This change in policy was intended both to cut public expenditure in this area, and to encourage mortgagors to take out private MPPI. At the time of the ISMI changes, only 12–13 per cent of all mortgagors were covered in this way, but by 1999 the proportion had increased to roughly 25 per cent (Quilgars, 1999).

By the mid-1990s the risk of repossession had diminished. After the 1992 General Election, the Conservative government had put in place a

---

Box 10.2

Owner-occupation and the welfare system

The housing benefit system undoubtedly discriminates against home-owners. Whereas in 1997 the average housing benefit payment to tenants amounted to £44 per week, low-income owner-occupiers – eligible for income support for mortgage interest (ISMI) – received only £37.16 per week, and then only after the first nine months of their claim. With the abolition of mortgage interest tax relief (MITR) in April 2000, low-income home-owners subsequently received only a limited amount of government support relative to low-income households in other tenures (Burrows and Wilcox, 2000). In aggregate, all forms of assistance to low-income home-owners were cut substantially during the 1990s, and by 1999 'home-owner households received only 8 per cent of total government support provided to assist low-income households with their housing costs' (Burrows and Wilcox, 2000: 47).

Clearly, a more tenure-neutral system of housing support for low-income households would help to ensure the sustainability of the owner-occupied sector. Although a mortgage support scheme broadly equivalent to the current housing benefit scheme for tenants was proposed by Kempson and Ford (1997), the Housing Green Paper (DETR/DSS, 2000) – adhering very largely to the *status quo* – suggested that ISMI should be extended from 9 to 14 months, on the presumption that there would be an effective and affordable mortgage payment protection insurance (MPPI) alternative. However, this extended period of delay 'would inevitably lead to higher levels of mortgage arrears and possessions' (Burrows and Wilcox, 2000: 53), and hence would reduce rather than enhance the ability of low-income home-owners to meet their mortgage and repair costs.

number of measures that had increased the level of demand for home-ownership, thereby eliminating excessive supply.

From slump to further boom

It might have seemed by 1995 that the stage was set for a recovery in the housing market. Owner-occupation had not been so affordable for generations. Apart from comparatively low mortgage interest rates, the HPER – at 3.4 – had fallen to its lowest level since the 1950s. Building societies and banks, moreover, had become more sympathetic to mortgagors with negative equity who wished to move: the Halifax Building Society, for example, was beginning to provide mortgages which covered the remaining debt (of up to £25,000) on the property to be sold and up to 100 per cent of the purchase price of the new property.

The number of residential transactions in England and Wales consequently rose from 1,047,000 in 1995 to 1,220,000 in 1998 (Wilcox, 2000), while average house prices in all regions rose dramatically, particularly in Greater London, the South East and South West, where there were increases of respectively 67, 49 and 40 per cent in the second half of the decade (Table 10.5). The boom was undoubtedly the result of average earnings increasing from around £19,600 in 1995 to £23,400 in 1999, HPERs rising from 3.4 to 4.8, building society average mortgage rates decreasing from 8.00 to 6.49 per cent, and unemployment falling from 8.0 to 4.3 per cent during the same period (Figure 10.4).

By May 2001, following a cut in bank base rates to 5.25 per cent, mortgage interest rates for new borrowers fell to around 6 per cent, the lowest rates since 1962. House prices were thus set to go on rising at a fairly rapid rate, even though, according to Land Registry data for England and

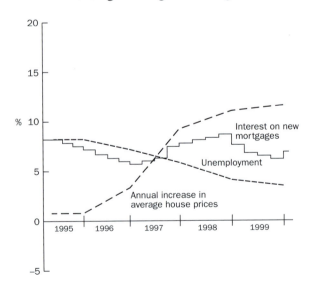

Figure 10.4 *Mortgage interest rate, unemployment and house prices, 1995–99*

Table 10.5 *North–South disparities in house price change during the slump and boom, 1989–99*

	Average house prices			Average house prices		
	1989 (£)	1993 (£)	% increase, 1989–93	1995 (£)	1999 (£)	% increase, 1995–99
North						
Northern Ireland	35,000	42,000	20	48,800	71,500	47
Scotland	46,400	60,700	31	60,500	71,200	18
Wales	52,200	53,600	3	53,600	66,900	25
Northern	42,800	52,000	21	51,100	63,600	24
North West	49,700	56,500	14	57,300	69,300	21
Yorkshire and Humberside	49,700	55,900	12	54,800	65,300	19
South						
East Midlands	59,200	55,000	–7	56,100	69,900	25
West Midlands	62,100	60,500	–3	62,200	78,800	27
East Anglia	75,300	58,700	–22	61,800	81,100	31
Greater London	98,700	82,200	–17	89,100	149,100	67
Rest of South East	98,200	75,900	–23	81,100	120,700	49
South West	78,400	62,300	–21	65,000	90,700	40
UK	70,400	64,300	–9	66,700	93,500	40

Source: CML, *Housing Finance*

Wales, they were currently increasing by 'only' 7.6 per cent per annum, compared to 10.1 per cent in 2000. In some regions, however, the market was overheating. Average house prices were rising by as much as 13.8 per cent in East Anglia and 12.9 per cent in the South East.

Macroeconomic determinants of the demand for home-ownership

In general, the extent to which households can afford to buy a house is mainly determined by the HPER and the mortgage interest rate. Normally, when ratios and/or mortgage interest rates are high, there is a relatively low level of affordability, but when ratios and/or interest rates are low, affordability is relatively high. But ratios and interest rates do not indicate the varying degrees of affordability among different sorts of households. When ratios were high, for example 4.9 in 1989 or 4.0 in 1999 (and in the former year when the mortgage interest rates exceeded 14 per cent), households with above-average incomes were willing to afford a higher level of housing expenditure in order to trade up for speculative reasons, believing that house prices would continue to rise in the foreseeable future. But when ratios were comparatively low, for example 3.4 in 1995 (and when in the same year the rate of mortgage interest had fallen to less than 8 per cent), households with below-average incomes (including many first-time buyers) and those with insecure incomes would have found house prices unaffordable, even though prices might have fallen dramatically from previous peaks in real and even in cash terms. National HPERs and mortgage interest rates also fail to show substantial regional variations in affordability (see Chapter 12).

Research undertaken by Bramley (1990) suggested that during the boom conditions of 1989 affordability among first-time buyers under 30 years old was even less than hitherto acknowledged. He showed that nationally, only 22 per cent of potential new home-owners earned enough to buy a new three-bedroom house, while in London the proportion was as low as 10 per cent (compared with 38 per cent in the North and North West). It was therefore a cause for concern that only one-third of the 100,000 houses needed each year in the subsidised social sector were being built, exacerbating the problem of affordability.

However, during the following decade – and in respect of all first-time buyers – it was clear that as boom turned into slump, both average

Table 10.6 *Average monthly incomes and average mortage repayments, 1990–99*

	Average monthly repayments (£)	Average repayment as % of average income
1990	381.02	26.9
1991	337.93	23.0
1992	284.27	19.1
1993	261.92	17.5
1994	277.12	18.0
1995	288.29	18.5
1996	281.45	17.1
1997	337.94	19.0
1998	360.77	19.0
1999	381.78	18.1

Source: Wilcox (2000)

monthly repayments and average repayments as a percentage of average incomes decreased, and owner-occupation became more affordable. In 1990 average first-time buyers repaid £381 per month, equivalent to about 27 per cent of their income, whereas in 1993 average monthly repayments had fallen to £262, or only 17.5 per cent of their income. But as slump turned into boom at the end of the decade, average monthly repayments and average repayments as a percentage of average incomes increased, and home-ownership became less affordable. By 1999 average monthly repayments amounted to £382, and average repayments as a proportion of average incomes rose to 18 per cent (Table 10.6).

At the beginning of the new millennium it was clear that the problem of affordability was particularly acute in London, the South East and parts of the South West, most notably among key public sector employees. Table 10.7 reveals that in the housing market there were substantial shortfalls in average earnings and the earnings of teachers, nurses and bus drivers.

Notwithstanding the very real problem of affordability in a large part of England, over 1.6 million households became owner-occupiers in Britain for the first time in the 1990s, and even during the boom years of 1995–99 the number of owner-occupiers increased by around 520,000. Even in London, where house prices were generally unaffordable to many key public sector employees at the start of the new millennium, there were areas of affordable housing in boroughs such as Bexley, Lewisham, Newham and Waltham Forest, where the average

Table 10.7 *Unaffordable house prices in London, the South East and parts of the South West, 2000*

	Average house price earnings (£)	Average earnings borrowing (£)	Teachers' borrowing shortfall (£)	Nurses' borrowing shortfall (£)	Bus drivers' borrowing shortfall (£)
London					
Kensington & Chelsea	496,977	−356,464	−365,040	−384,934	−415,411
Westminster	527,878	−342,105	−395,741	−415,635	−446,112
Camden	285,098	−127,816	−153,161	−173,055	−203,532
Richmond	260,072	−127,769	−128,135	−148,029	−178,506
Islington	226,970	−88,610	−95,033	−114,927	−145,404
Wandsworth	215,791	−87,498	−83,854	−103,748	−134,225
Barnet	186,408	−69,811	−54,473	−74,363	−104,843
Merton	176,550	−67,058	−44,613	−64,507	−94,984
Kingston	173,053	−53,838	−41,116	−61,010	−91,487
Harrow	163,634	−43,646	−31,697	−51,591	−82,068
Ealing	165,621	−42,158	−33,684	−53,578	−84,055
Bromley	148,772	−39,232	−16,835	−36,729	−67,206
South East					
Windsor & Maidenhead	222,407	−81,042	−104,123	−123,069	−149,614
Surrey	191,182	−57,885	−72,898	−91,843	−118,389
Wokingham	180,383	−45,092	−62,079	−81,024	−107,570
Oxfordshire	145,718	−32,618	−27,434	−46,379	−72,795
Hertfordshire	140,590	−17,074	−22,383	−40,171	−69,059
Hampshire	126,308	−14,033	−8,024	−29,969	−53,515
East Sussex	104,271	−9,257	+14,013	−4,932	−31,478
West Sussex	121,967	−8,983	−3,683	−22,628	−49,174
South West					
Poole	124,076	−17,197	−4,191	−27,268	−55,974
Wiltshire	116,714	−17,886	−3,170	−19,906	−48,612
Dorset	115,365	−20,801	−4,519	−18,557	−47,263

Source: Labour Research Department, 2000

Note: Average house prices are taken from Halifax statistics and compared to average local earnings taken from the *New Earnings Survey*. The shortfall is the difference between the average house price locally and the maximum mortgage that a two-adult household, working in the jobs specified, could raise (2.5 times joint earnings).

price of property was no more than £100,000, while in Barking and Dagenham house prices averaged £78,157 in December 2000, less than the average price of a property in Glamorgan in south Wales (Williams, 2000).

Conclusion

On regaining office in 1997, Labour was determined that there would be no return to the speculation, instability and negative equity that characterised the housing market in much of the 1980s and 1990s. An attempt was therefore made to keep the demand for home-ownership under control by increasing stamp duty on residential transactions (or imposing it where it had not previously existed) and phasing out what remained of mortgage interest tax relief (MITR). Stamp duty was thus increased in July 1997 from 1 per cent to 1.5 per cent on transactions in excess of £250,000, and to 2 per cent on sales above £500,000. In April 1998 stamp duty was further increased to 2 per cent on transactions above £250,000, and to 3 per cent on sales above £500,000. In April 2000 duty of 1 per cent was imposed on transactions of £60,000–£250,000, on sales of £250,000–£500,000 it was increased to 3 per cent, and on sales above £500,000 it rose to 4 per cent. Meanwhile, MITR was reduced from 15 per cent to 10 per cent in April 1998 as a prelude to its complete abolition in April 2000. Whereas the sector was the recipient of MITR of £1.84 billion in 1997/98, by 2000/01 owner-occupiers had not only lost this subsidy but were liable to stamp duty of £2.82 billion in 2000/01, a £4.6 billion difference (Dwelly, 2000b).

Although these measures came at the right time in the market cycle (i.e. as house prices were rising rapidly), instability was rampant (Dwelly, 2000b). Regional divergences in house prices and unemployment rates reappeared, indicating that – at least in terms of these indicators – the 'North–South divide' was still very apparent (Figure 10.5).

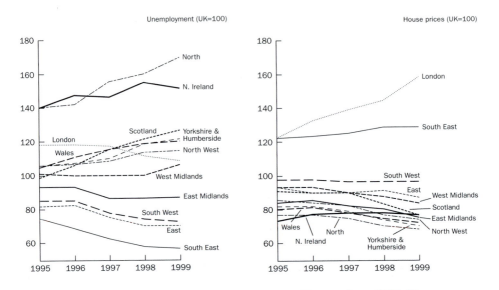

Figure 10.5 *North–South divergence in unemployment and house prices, 1995–99*

Sources: CML, *Housing Finance*; Department for Education and Employment, *Employment Gazette*

Further reading

Ball, M. (1983) *Housing Policy and Economic Power*, London: Methuen. Contains a detailed and comprehensive analysis of the owner-occupied sector, exploring both the supply and the demand determinants of provision and price.

Ball, M. (1986) *Home Ownership: A Suitable Case for Reform*, London: Shelter. A summary of the author's earlier analysis (1983), including proposals for reform.

Boddy, M. (1980) *The Building Societies*, Basingstoke: Macmillan. Although a little dated now, a useful introductory text on the role of building societies in the home-ownership market.

Boddy, M. (1989) 'Financial deregulation and UK housing finance', *Housing Studies*, Vol. 4, No. 2. Partly fills the gap in the literature on the major changes affecting building society activity in the late 1980s.

Boleat, M. (1983) *The Building Society Industry*, London: Allen & Unwin. A useful account of the role of building societies from the perspective of an insider.

Booth, P. and Crook, T. (eds) (1985) *Low Cost Home-Ownership*, Aldershot: Gower. A collection of papers on various aspects of low-income home-ownership – inevitably a little dated now.

Doling, J., Ford, J. and Stafford, B. (1988) *The Property Owning Democracy*, Aldershot: Avebury. A valuable analysis of owner-occupation in the late 1980s, focusing on mortgage arrears.

Ermisch, J. (ed.) (1990) *Housing and the Economy*, Aldershot: Avebury. Contains papers on the relationship between owner-occupied housing and the national economy.

Ford, J. and Kempson, E. (1997) *Bridging the Gap? Safety Nets for Mortgage Borrowers*, York: Centre for Housing Studies, University of York/York Publishing Services. An examination of whether private insurance has bridged the gap left by the withdrawal of state support, and a review of the consequences of the changing relationship between state and private safety net provision.

Forrest, R., Murie, A. and Williams, P. (1990) *Home-Ownership: Differentiation and Fragmentation*, London: Unwin Hyman. A valuable introduction to the complexities of owner-occupation.

Forrest, R., Kennet, T. and Leather, P. (1997) *Home Owners on New Estates in the 1990s*, Bristol: Policy Press. An assessment of how attitudes and circumstances had changed in two newly built estates as a result of depression in the housing market in the early to mid-1990s.

Gough, T. (1982) *The Economics of the Building Societies*, Basingstoke: Macmillan. A detached and critical view of the economic role of building societies in the home-ownership market.

Hamnett, C., Harmer, M. and Williams. P. (1991) *Safe as Houses*, London: Paul Chapman. An exploration of the relationship between home-ownership and the accumulation of wealth.

Jackson, A. (2001) *Evaluation of the 'Homebuy' Scheme in England*, York: Joseph Rowntree Foundation/York Publishing Services. An examination of whether or not the 'Homebuy' scheme had been successful during its first 15 months.

Karn, V., Kemeny, J. and Williams, P. (1985) *Home Ownership in the Inner City: Salvation or Despair?*, Aldershot: Gower. A research report on home ownership in inner Birmingham and Liverpool.

McCrone, G. and Stephens, M. (1995) *Housing Policy in Britain and Europe*, London: UCL Press. Contains a useful examination of owner-occupation in Britain in a European context.

Maclennan, D. (1994) *Housing Policy for a Competitive Economy*, York: Joseph Rowntree Foundation. Explores the linkages between housing and the national economy in the context of boom and bust in the late 1980s and early 1990s.

Maclennan, D. and Gibb, K. (eds) (1993) *Housing Finance and Subsidies in Britain*, Aldershot: Avebury. A wide-ranging collection of papers on housing finance, including a consideration of finance in the owner-occupied sector.

Maclennan, D., Gibb, K. and More, E.A. (1991) *Paying for Britain's Housing*, York: Joseph Rowntree Foundation. Examines the distribution and methodological issues relevant to the housing tax debate.

Maclennan, D., Meen, G., Stephen, M. and Gibb, K. (1997) *Fixed Communities, Uncertain Incomes: Sustainable Home Ownership and the Economy*, York: Joseph Rowntree Foundation/York Publishing Services. Taking into account the volatility in the UK housing market and concerns about the future sustainability of owner-occupation, this study provides an analysis of local surveys and macroeconomic models, and suggests a range of policy options.

Merrett, S. with Gray, F. (1982) *Owner-Occupation in Britain*, London: Routledge & Kegan Paul. A detailed review of the history of home-ownership in Britain and an examination of contemporary issues.

Saunders, P. (1990) *A Nation of Home Owners*, London: Unwin Hyman. A study of the experience of and attitudes towards home-ownership in three English towns.

Stephens, M. (1993) 'Housing finance deregulation: Britain's experience', *Netherlands Journal of Housing and the Built Environment*, Vol. 8, pp. 159–75. A report of two useful surveys comparing European mortgage finance with deregulation in the United Kingdom.

Terry, R. and Joseph, D. (1998) *Effective and Protected Housing Investment*, York: Joseph Rowntree Foundation/York Publishing Services. A research report that contains an analysis of the development of flexible tenure and low-cost home-ownership within a rationalised system of housing subsidies and taxes.

Part III   **Single issues in
housing policy**

11 Affordability

In each of the rented sectors, without the provision of housing benefit there would be serious problems of affordability, and within the owner-occupied sector, where mortgage interest relief was progressively phased out in the 1990s, problems of affordability have, to an extent, worsened in numerous areas of the United Kingdom at the beginning of the new century. To analyse this issue, this chapter:

- examines the relationship between average rents and average incomes in the private unfurnished sector and the social housing sectors;
- explores the effect of house price/earnings ratios and house price/income ratios on affordability;
- examines affordability among first-time buyers and low-income owner-occupiers; and
- concludes by considering some of the ways in which affordability could be increased in both the rented and the owner-occupied sectors.

The rented sectors

In the private rented sector the 1991 Census revealed that as many as 40 per cent of households were non-earning, only one in five contained more than one earner, and as many as one-half of all households in the sector had disposable incomes of less than £8,500 (ONS, 1994), yet under the Housing Act of 1988, fair rent tenants were being increasingly brought into the assured tenancy or assured shorthold system (with minimum lettings of only six months), and tenants suspecting that their rents might be above the market level lost their right to refer the matter to the Rent Assessment Committee (DoE, 1995b). As Table 11.1 illustrates, private sector rents (as a percentage of average incomes) increased substantially throughout the 1980s and 1990s. In some areas of Britain, where the local authority stock had been depleted, the upward pressure on private sector rents might have been even greater had Conservative government proposals to permit local authorities to accommodate a proportion of the homeless in private rented housing (DoE, 1995a) been fully implemented.

Table 11.1 *Average rents and incomes, private unfurnished sector, Great Britain, 1980–98*

	1980	1990	1998
Average rent (£)	10.85	24.00	59.24
Average income (£)	60.00	110.00	231.00
Average rents as % of average incomes	18.1	21.8	25.6

Source: Wilcox (2000)

Note: Average income data are mean averages of gross income of household heads in the sector.

As shown in Table 11.2, local authority housing rents as a proportion of average incomes also rose rapidly in the 1980s and 1990s, largely to reduce the scale of Exchequer subsidies and to eliminate subsidisation from rates. The privatisation of much of the local authority housing stock also inflated rents, since the loss of stock (particularly in older surplus-yielding states) has reduced the extent to which newer high-cost council housing can be cross-subsidised through the process of rent pooling.

Table 11.2 *Average rents and incomes, local authority sector, Great Britain, 1980–98*

	1980	1990	1998
Average rent (£)	7.70	18.82	42.24
Average income (£)	68.00	93.00	130.00
Average rent as % of average income	11.3	20.2	32.5

Source: Wilcox (2000)

Note: Average income data are mean averages of gross income of household heads in the sector.

Housing association tenants similarly experienced an increase in rents as a proportion of incomes (Table 11.3). Since the Housing Act of 1988 required an increasing proportion of capital expenditure in this sector to be funded by private institutions (56 per cent by 1997), rents have risen rapidly to ensure a competitive return on investment.

Table 11.3 *Average rents and incomes, housing associations, Great Britain, 1980–98*

	1980	1990	1998
Average rent (£)	12.52	25.00	54.51
Average income (£)	66.00	94.00	142.00
Average rent as % of average incomes	19.0	26.60	38.40

Source: Wilcox (2000)

Note: Average income data are mean averages of gross income of household heads in the sector.

If it is assumed that an 'affordable rent' should be a rent that does not exceed 25 per cent of the net income of the head of a household – as suggested by both the NFHA (1994) and the SFHA (1993) – then (according to data shown in Tables 11.1–11.3) rents in all sectors clearly became 'unaffordable' by the late 1990s. It is clear from these tables that housing association rents were the least affordable throughout the 1980s and 1990s, but although both local authority and housing association rents were – according to the above definition – affordable in the 1980s, this was no longer the situation by 1998. It must also be taken into account that the tables show only rents as a proportion of average incomes, and therefore affordability would have been significantly less in each sector in respect of households with lower-paid jobs or none.

Box 11.1

Affordability in London

In London the affordability of housing in the owner-occupied sector decreased in the late 1990s because of escalating house prices (see Garrratt, 2001). The house price/income ratio (HPIR) in the capital peaked at 3.46 in 1999 compared to a ratio of only 3.10 in the United Kingdom as a whole (Table 11.4). The advances to income ratio in London also increased to an extent, from 2.16 to 2.37, while nationally it remained relatively constant because of lower rates of house price inflation elsewhere in the United Kingdom, and similarly the deposit to income ratio increased in London at a faster rate than in the United Kingdom as a whole.

However, although access to the owner-occupied sector in terms of entry costs has deteriorated, the costs of servicing mortgage debt in London in relation to income have remained low by historic standards, although not quite as low as in the United Kingdom as a whole – a situation greatly helped by low mortgage interest rates (Table 11.4), an outcome of a strong economy.

continued

Table 11.4 Indicators of affordability in the owner-occupied sector, London, 1995–99

| | House price to income ratio | | Advance income ratio | | Deposit to income ratio | | Mortgage payments to income (%) | | Mortgage interest rates (%) | Average house price (£) | Increase in average house price (%) |
	London	UK	London	UK	London	UK	London	UK	UK	London	London
1995	2.95	2.84	2.16	2.21	0.33	0.28	15.0	14.1	7.98	89,100	1
1996	2.99	2.85	2.18	2.22	0.26	0.25	13.3	12.4	7.00	91,600	3
1997	3.13	2.92	2.23	2.23	0.32	0.28	15.9	14.7	8.16	105,300	15
1998	3.25	3.00	2.30	2.24	0.56	0.46	16.1	14.6	7.75	120,800	15
1999	3.46	3.10	2.37	2.27	0.72	0.57	14.9	13.5	6.88	149,100	23

Sources: CML, *Housing Finance* (various)

The owner-occupied sector

In owner-occupation, it is possible to ascertain the degree of affordability by a number of indicators: specifically the house price/earnings ratio (HPER), the house price to income ratio (HPIR), the advance to income ratio (AIR) and the mortgage cost to income ratio.

Figure 11.1 shows that the HPER reached peaks of 4.9 and 4.4 in respectively 1989 and 1990, but fell to 3.4 in 1995 and 1996. Peaks in the HPER were associated with house prices rising at a more rapid rate than earnings, whereas troughs broadly coincided with price increases rising more slowly and more in line with increases in earnings. Using the HPIR, it is clear that owner-occupied housing was more affordable for first-time buyers in the mid-1990s than in 1989 and 1999, although existing owner-occupiers (whose properties had soared in value in 1989 and 1999) would have found the cost of 'replacement' housing reasonably affordable in their particular location or in a cheaper price area.

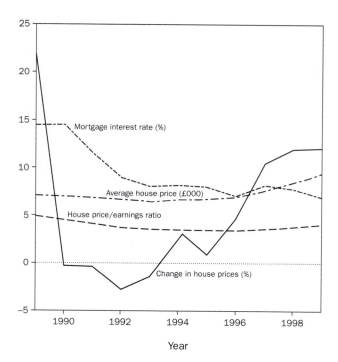

Figure 11.1 *House price/earnings ratios, mortgage interest rates and house prices, 1989–99*
Source: CML (2001)

In comparing the HPER and the HPIR, it was evident that the former indicator rose more quickly than the latter during the boom years of 1989 and 1999. In London in particular, the total income of first-time buyers was growing more quickly than the average earnings of the working population (partly attributable to the increase in the number of working women), and consequently house prices grew more in line with the income of mortgagors than with average earnings (Garrett, 2000).

In contrast to both the HPER and the HPIR, the AIR has remained comparatively stable over time – necessitating first-time buyers having to make larger initial deposits and thereby adversely affecting affordability among some potential first-time buyers (Garrett, 2000).

Historically low interest rates have had a major impact on affordability in recent years. Not only have mortgage interest rates fallen in recent years from 14.48 per cent in 1990 to 6.88 per cent in 1999 (Figure 11.1), but as a percentage of income, mortgage costs (largely consisting of interest charges) roughly halved, enhancing affordability among all mortgagors (except those on fixed-interest loans).

Taking the four indicators together, Garratt (2000) suggested that affordability improved slightly towards the end of the 1990s since the impact of lower mortgage costs outweighed the substantial growth in house prices. Since mortgage costs are largely dependent on interest rates, these in turn are clearly dependent on the strength of the economy, and as long as the economy remains strong, affordability should be maintained.

Affordability among first-time buyers

Although research undertaken by Bramley (1990, 1991) had revealed that affordability among young first-time buyers was low not only during the boom conditions of 1989 but also during the subsequent slump, 400,000 households nevertheless became owner-occupiers for the first time in 1991, and over 350,000 in 1992 (Boleat, 1992). The Conservative government was undoubtedly achieving its aim of increasing property ownership. By 1993 the HPER and debt-servicing costs had fallen to their lowest levels since the 1950s and 1960s and thus houses were more affordable than at any time over the past three or four decades. Towards the end of 1993, AIRs had fallen to a low level. The *TSB Affordability Index* (Trustee Savings Bank, 1993) indicated that typical first-time buyers needed to spend only 26 per cent of their take-home pay on mortgage payments, as against 67 per cent in the first quarter of 1990, and even in the most expensive region, the South East, the proportion was barely higher at 30 per cent. Only unemployment – or the fear of unemployment – and a general lack of confidence in the market prevented a noticeable increase in housebuying and an upturn in house prices in 1993 and 1994.

By the end of 1996 it was evident, however, that house prices were rising by at least 7.5 per cent per annum nationwide – the prelude to further increases in the late 1990s. However, although affordability was still constrained by the tightening up of building society and bank lending, by many first-time buyers still having negative equity (a mortgage debt greater than the value of the property), by lower-income households failing to gain from economic growth in the 1980s and 1990s, and, increasingly, by single-person households making up a higher proportion of total households, in 1998 and 1999 it was evident that affordability had increased significantly – as in the case of mortgagors in general – largely as a consequence of lower mortgage costs.

Affordability and low-income owner-occupiers

Mortgage costs to income ratios for low-income buyers are particularly high – about 28 per cent in the case of the lowest income quintile of homebuying households in England or, proportionately, twice the average level for all mortgagors (Burrows and Wilcox, 2000). If repair costs are also taken into account, and the total cost of housebuying is compared with the net rather than the gross income of the mortgagors, then the poorest 20 per cent of buyers will spend as much as 42 per cent of their disposable income on housebuying – a considerably greater outgoing than tenants would expect to pay (see Tables 11.1–11.3).

Conclusion

There are clearly indications that rents became increasingly unaffordable in all rented sectors in the 1990s, while in the owner-occupied sector there is evidence that among low-income housebuyers there were low levels of affordability. However, while affordability increased among first-time buyers across most of the United Kingdom in the late 1990s, largely because of lower mortgage interest rates, in London it decreased dramatically, where house prices were not only the highest in the country but increasing at a rapid rate.

While the most obvious way of increasing affordability would be for the government to cap or decrease both rents and house prices, these approaches are incompatible with the Blair government's propagation of an increasingly neo-liberal welfare regime. It is possible, nevertheless, that affordability in the rented sectors could be increased in the short term by a more effective system of housing benefit (see Chapter 5), and in the owner-occupied sector by the introduction of a new form of housing allowance targeted at low-income home-owners, while in London assistance – already being given to key workers to help facilitate house purchase – could be significantly increased (see Chapter 12). However, since object subsidies such as these have the tendency to underpin or inflate rents or house prices (see Chapters 6, 9 and 10), an emphasis on increasing the supply of social and owner-occupied housing might prove to be a more effective way of ensuring an adequate future provision of affordable housing, particularly if aided by 'bricks and mortar' subsidies and/or by a means of ensuring an adequate supply of low-cost land.

To this end, it is imperative that an average of over 82,000 affordable rented dwellings and 143,500 owner-occupier and other rented units are built each year over the period 1996–2016 (see Holmans and Brownie, 2001). An increase in the supply of housing of this magnitude would reduce pressures on all sectors of the housing market and help decelerate the rate at which rents and house prices rise. However, it is very probable that at the current level of public investment in affordable housing, insufficient dwellings will be constructed to have much of an impact on rents and prices in the near future, and at the time of writing there were no plans to reduce the cost of land acquisition in areas of low affordability.

Further reading

Bramley, G. (1990) *Bridging the Affordability Gap in 1990*, London: Association of District Councils and House Builders Federation. A useful report on affordability at the end of a house price boom.

Bramley, G. (1991) *Bridging the Affordability Gap*, London: Association of District Councils and House Builders Federation. An report on affordability at the beginning of a house price slump.

Burrows, R. and Wilcox, S. (2001) 'Half the poor? The growth of low income home-ownership', *Housing Finance*, No. 47. A valuable review of the development of affordable owner-occupation.

Garratt, D. (2000) 'Affordability', *Housing Finance*, No. 47. Presents a useful analysis of the affordability in the owner-occupied market.

Garratt, D. (2001) 'London's housing market', *Housing Finance*, No. 49. A thorough exploration of the relationship between house prices and affordability in an inflated housing market.

Maclennan, D., Gibb, K. and More, A. (1990) *Paying for Britain's Housing*, York: Joseph Rowntree Foundation. Contains a description of patterns of affordability across case study areas, and provides an overview of a number of analytical difficulties that have to be confronted in identifying 'expenditure', particularly in the owner-occupied sector.

Whitehead, C. (1991) 'From need to affordability: an analysis of UK housing objectives', *Urban Studies*, Vol. 20, No. 6.

12 Regional disparities and problems

For much of the time since the Depression years of the 1920s and 1930s there has been a prominent 'North–South divide' in the prosperity of England, notwithstanding pockets of prosperity in the North (for example in parts of Cheshire or North Yorkshire currently) and localised areas of poverty in the South (such as in parts of inner London for generations). There has also been a disparity between the prosperity of Scotland, Wales and Northern Ireland on the one hand, and England on the other. Since it is broadly agreed that demand rather than supply is the principal determinant of property values, the disparity in living standards across the United Kingdom is inevitably a fundamental cause of housing surpluses in some regions and housing shortages in others, but also a major cause of regional variations in house prices and rents.

To examine these connections, this chapter:

- identifies regional disparities in living standards;
- explores regional disparities in housing demand;
- discusses housing market outcomes;
- analyses low demand in the North and Midlands;
- examines housing shortages in London;
- considers such aspects of disparity in Scotland, Wales and Northern Ireland; and
- concludes by commenting on the impact of disequilibrium in housing markets on the wider economy.

Regional disparities in living standards

In the English regions, as elsewhere, GDP per capita is broadly synonymous with living standards. However, as Figure 12.1 illustrates,

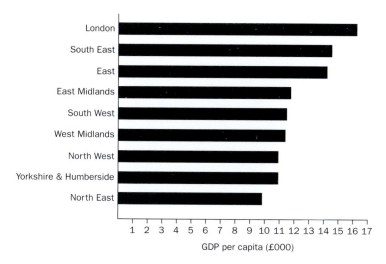

Figure 12.1 *GDP per capita, English regions, 1998*
Source: ONS, *Regional Trends*

there is a wide inter-regional disparity in GDPs per capita and hence major variations in the standard of living across England, ranging in 1998 from around £9,800 in the North East to £16,200 in London. Taking the Severn–Wash line as its boundary, a 'North–South' divide is clearly apparent. The five northern regions all have GDPs per capita below the national average, while three of the four southern regions have per capita GDPs above this level – influencing the magnitude of demand and supply accordingly.

Among local authorities, regional disparities in 1998 were even more marked – with a major impact on local housing markets. Indices of GDP per capita ranged from as low as 63 in Sefton and Wirral (in Merseyside) to 153 in Swindon and 418 in parts of London (Table 12.1). A total of 24 northern local authorities had per capita GDPs below 90 per cent of the national level, while in the South 14 authorities had per capita GDPs of more than 10 per cent above this level. Half of the northern authorities experienced a comparative decrease in their per capita GDPs between 1994 and 1998, with a deflating effect on demand, whereas in the South, nine authorities enjoyed a comparative increase, thereby inflating demand.

Clearly, regional disparities in GDP per capita are interrelated with regional disparities in housing demand and housing market outcomes, and are particularly associated with a surplus of housing in much of the North and Midlands, and dire shortages of affordable housing in London.

Table 12.1 *GDP per capita of selected local authority areas, 1994 and 1998 (UK = 100)*

	1994	1998
The North		
Sefton	62	63
Wirral	63	63
South Nottingham	68	67
Barnsley, Doncaster & Rotherham	65	68
Blackpool	72	69
East Derbyshire	73	70
Greater Manchester North	72	72
East Merseyside	74	72
Durham County Council	75	74
North Nottinghamshire	80	78
Shropshire County Council	79	78
Sunderland	80	79
Northumberland	79	81
Staffordshire County Council	77	81
Walsall & Wolverhampton	89	82
Calderdale, Kirklees & Wakefield	85	83
East Riding of Yorkshire	81	83
Dudley & Sandwell	87	84
Bradford	85	85
Sheffield	91	87
Liverpool	93	87
South & West Derbyshire	89	87
Blackburn & Darwen	90	88
Darlington	94	88
The South		
Oxfordshire	113	110
Buckinghamshire	109	113
Cambridgeshire County Council	112	114
Inner London – East	120	115
Surrey	112	120
Bristol, City of	121	123
Thurrock	122	125
Southampton	123	126
Portsmouth	123	126
Peterborough	123	128
Milton Keynes	136	136
Berkshire	138	138
Swindon	151	153
Inner London – West	423	418
England	102	102

Source: ONS, *Regional Trends*

Regional disparities in housing demand in England

Whereas in aggregate, the demand for housing is determined by GDP per capita, in a more disaggregated sense demand is influenced by the rate of population growth, migration, the level of unemployment, activity rates and incomes. Table 12.2 reveals that whereas in the North East the population decreased by 1.3 per cent 1981–91 and by 0.5 per cent 1991–98, lowering the level of demand, in London the population grew by 1.2 per cent and 4.3 per cent in the same respective periods, raising the level of demand. Except in the East Midlands, the rate of population increase in the North was slower than in England as a whole (or was negative), whereas in the South the population of all regions (except London between 1981 and 1991) grew faster than the national aggregate.

Table 12.2 *Population change, 1981–98*

	Population (000)			Population growth (%)	
	1981	*1991*	*2000*	*1981–91*	*1991–98*
North East	2,636.2	2,602.5	2,589.6	−1.3	−0.5
Yorkshire & Humberside	4,918.4	4,982.8	5,042.9	1.3	1.2
North West	6,940.3	6,885.4	6,890.8	−0.8	0.1
East Midlands	3,852.8	4,035.4	4,169.3	4.7	3.3
West Midlands	5,186.6	5,265.5	5,332.5	1.5	1.3
East	4,854.1	5,149.8	5,377.0	6.1	4.4
London	6,805.6	6,889.9	7,187.2	1.2	4.3
South East	7,245.4	7,678.9	8,003.8	6.0	4.2
South West	4,381.4	4,717.8	4,901.3	7.7	3.9
England	46,820.8	48,208.1	49,494.6	3.0	2.7

Source: ONS, *Regional Trends*

North–South disparities were also apparent in respect of migration (Table 12.3). Net out-migration occurred in four out of the five northern regions between 1991 and 1998, lowering the level of demand, whereas net in-migration affected three out of four southern regions in the same period, raising the level of demand. At a local level, migration (rather than variations in birth and death rates) has been the principal determinant of an increase or decrease in total population. Table 12.4 shows that whereas in the North, the population decreased by as much as 11.2 per cent in

Table 12.3 *Migration within England, 1981 and 1998*

	1981			1998		
	Inflow (000)	Outflow (000)	Net migration (000)	Inflow (000)	Outflow (000)	Net migration (000)
North East	31	39	−8	39	44	−5
Yorkshire & Humberside	68	73	−5	93	98	−5
North West	102	122	−20	104	116	−12
East Midlands	77	72	+5	108	97	+11
West Midlands	67	79	−12	93	101	−8
East	121	104	+7	143	124	+19
London	155	187	−32	171	218	−47
South East	202	166	+36	226	207	+19
South West	108	88	+20	139	111	+28
England	94	93	−1	111	111	0

Source: ONS, *Regional Trends*

Knowsley and 10.7 per cent in Liverpool, in the South it increased by as much as 35.1 per cent in East Cambridgeshire and 61.3 per cent in Milton Keynes. With 25 northern local authorities witnessing a decrease in their population of at least 3.0 per cent, 1981–98, and 25 southern local authorities experiencing an increase in their population of 20 per cent or more in the same period, a clear North–South divide was evident. With regard to average weekly earnings, regional disparities were again evident. As Figure 12.2 shows, the earnings of both men and women (and the labour force as a whole) were below the national level in all northern regions, whereas in the South, earnings were substantially above the national level in London and the South East, but – as in the North – below the national level in the East and South West. In regions with disproportionately low earnings, the demand for housing would have been at a comparatively low level, whereas in regions with disproportionately high earnings, demand would have been comparatively high.

Regional variations in unemployment are often cited as the principal attribute of the North–South divide. However, although the rate of unemployment ranged from as much as 10.1 per cent in the North East to as little as 3.6 per cent in the South East in 1999/2000, with corresponding effects upon demand, in the East Midlands the rate of

Table 12.4 *Population change in selected regions and local authority areas, 1981–98*

	Population change, 1981–98 (%)
The North	
Knowsley	−11.2
Liverpool	−10.7
Salford	−8.6
Redcar & Cleveland UA	−8.3
Easington	−7.9
Manchester	−7.1
Gateshead	−6.7
Sefton	−6.2
Sandwell	−6.2
Wolverhampton	−5.8
Coventry	−4.7
South Tyneside	−4.6
Sedgefield	−4.5
City of Kingston upon Hull UA	−4.4
St Helens	−4.2
Barrow in Furness	−4.2
Wirral	−4.0
Copeland	−3.9
Middlesbrough UA	−3.6
Pendle	−3.6
The South	
Eastleigh	21.1
Kensington & Chelsea	21.3
Halesham	23.3
East Hampshire	21.5
Breckland	21.5
South Cambridgeshire	21.7
Fenland	22.3
Cherwell	23.3
Hart	23.3
North Dorset	23.4
Wokingham UA	24.2
North Cornwall	24.5
Teignbridge	24.6
Tower Hamlets	24.8
Huntingtonshire	24.9
Suffolk Coastal	25.1
Bracknell Forest UA	30.7
Forest Heath	31.4
East Cambridgeshire	35.1
Milton Keynes UA	61.3
England	5.7

Source: *ONS, Regional Trends*

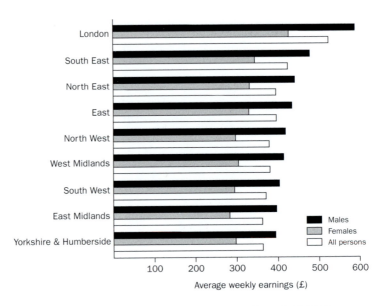

Figure 12.2 *Average weekly earnings, English regions, April 1999*
Source: ONS, *Regional Trends*

unemployment (5.2 per cent) was lower than the national rate but in London (at 7.6 per cent) it was higher (see Figure 12.3). There were also clear North–South disparities in long-term unemployment (Figure 12.3).

Figure 12.4 similarly shows that there are clear regional differences in activity rates among men and women (and among the population as a whole), with the lowest rates occurring generally in the North and the highest rates in the South. However, as with unemployment, the East Midlands and London are exceptions to this pattern, with the activity rate in the former region being disproportionately high and its counterpart in the capital being disproportionately low. Clearly, high activity rates are associated with a high demand for housing, while low activity rates depress demand.

At a local authority level, North–South disparities in unemployment, activity rates and – to an extent – average gross weekly earnings are very much in evidence. Table 12.5 reveals that unemployment ranged from as much as 14.5 per cent in Middlesbrough to as little as 2.8 per cent in Surrey in 1998/99, with 16 northern authority areas experiencing unemployment of more than 6 per cent in 1998/99 and 13 southern authority areas witnessing unemployment of 5 per cent or less – with corresponding effects upon demand.

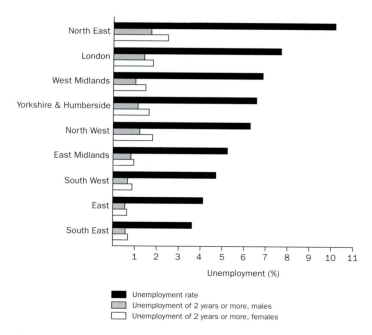

Figure 12.3 *Unemployment rates, English regions, 1999/2000*
Source: ONS, *Regional Trends*

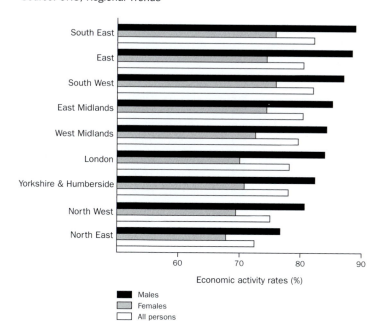

Figure 12.4 *Economic activity rates, English regions, 1999*
Source: ONS, *Regional Trends*

Table 12.5 *Unemployment, activity rates and average gross weekly full-time earnings, selected regions and local authority areas, 1998/99*

	ILO unemployment rate 1998/99	Activity rate 1998/99	Average gross weekly full-time earnings, 1999
	(%)	(%)	(£)
The North			
Middlesbrough	14.5	60.9	362.3
City of Kingston upon Hull UA	13.8	64.5	344.3
Halton	11.7	64.0	379.5
Merseyside (Met. County)	11.0	62.2	371.8
North East Lincolnshire UA	9.0	66.3	359.1
Tyne & Wear	9.5	65.2	352.6
South Yorkshire (Met. County)	9.1	68.8	350.6
West Midlands (Met. County)	8.6	70.3	387.9
Leicester	8.6	67.2	351.7
Durham County	8.1	67.1	343.4
Stockton-on-Tees	8.2	69.9	361.4
Nottingham	7.9	65.7	361.1
Derby	6.7	72.3	414.8
West Yorkshire (Met. County)	6.2	74.2	368.2
Greater Manchester (Met. County)	6.2	70.4	377.0
Cumbria	6.1	72.6	361.1
The South			
Brighton & Hove	5.7	75.5	352.4
East Sussex	5.5	76.6	355.0
Kent County	5.5	76.6	381.4
Suffolk	5.5	77.5	348.9
Bedfordshire County	5.2	79.5	401.9
Buckinghamshire County	4.5	80.6	466.8
Essex County	4.3	76.7	396.8
Devon County	4.3	80.3	328.0
Hampshire County	3.6	80.4	417.5
West Sussex	3.4	80.9	410.5
Oxfordshire	3.3	80.5	410.1
Hertfordshire	2.8	81.5	443.8
Surrey	2.8	82.5	476.1
England	6.0	74.2	405.4

Source: ONS, *Regional Trends*

Table 12.5 also indicates that activity rates ranged from as 60.9 per cent in Middlesbrough to as much as 82.5 per cent in Surrey in 1998/99, with rates in all northern areas except one being lower than the national rate, and in all southern areas being higher – respectively depressing and inflating demand.

The same table further shows that average weekly earnings ranged from as little as £343.4 in Durham County to as much as £476.1 in Surrey in 1999, with corresponding effects upon demand. However, although earnings (and demand) in as many as 15 northern areas were lower than the national level, in the South earnings (and demand) were higher than the national level in only seven areas.

Housing market outcomes

The interaction of demand and supply determines the extent to which there is a surplus or deficit of housing (as well as the pattern of tenure, the level of rents and housing benefits, and house prices).

Table 12.6 clearly shows that in the English housing market, supply exceeds demand in the North, whereas in the South, demand is generally

Table 12.6 *Number of dwellings and households, England, 1981 and 1998*

	Number of dwellings (millions)		Number of households (millions)		Balance (millions)	
	1981	1998	1981	1998	1981	1998
North East	1.02	1.12	0.98	1.09	+0.04	+0.03
North West	2.66	2.94	2.55	2.84	+0.11	+0.10
Yorkshire & Humberside	1.90	2.13	1.83	2.10	+0.07	+0.03
East Midlands	1.48	1.76	1.41	1.72	+0.07	+0.04
West Midlands	1.94	2.19	1.86	2.16	+0.08	+0.03
East	1.86	2.18	1.76	2.22	+0.10	−0.04
London	2.68	3.04	2.62	3.06	+0.06	−0.02
South East	2.75	3.30	2.64	3.30	+0.11	0
South West	1.73	2.10	1.64	2.05	+0.09	+0.05
England	18.03	20.84	17.31	20.54	+0.72	+0.30

Source: ONS, *Regional Trends*

greater than supply. Specifically, the number of dwellings in the North was greater than the number of households in both 1981 and 1998, although crude surpluses diminished over the intervening period. Nevertheless, in the latter year, surpluses were still quite substantial, ranging from 30,000 to 100,000, but these would have been less had the number of concealed households, dwellings in a poor condition and second homes been deducted from the total. In contrast, in the South only one of the four regions showed a surplus in 1998. In London there was a crude deficit of 20,000 dwellings and in the East a shortage of 40,000. Net deficits in these regions would of course have been even greater.

Disparities in the pattern of housing tenure are also the outcome of the interaction of demand and supply. Whereas over 70 per cent of dwellings in the South East (excluding Greater London), the South West and the East Midlands were owner-occupied in 1998, in the North, Greater London, and Yorkshire and Humberside, owner-occupation accounted for fewer than 65 per cent of dwellings (Table 12.7). Local authority housing, on the other hand, was a disproportionately important tenure in the North and North West, where respectively 26 and 20 per cent of all dwellings were council owned, in contrast to the South East (outside of Greater London) and the South West, where less than 12 per cent of dwellings

Table 12.7 *Housing tenure, England, 1997/98*

	Owner-occupied	Local authority rented	Housing association rented	Private rented
	(%)	(%)	(%)	(%)
North	61.6	25.6	4.1	8.7
Yorkshire & Humberside	64.3	17.3	3.2	12.1
North West	67.8	19.9	4.7	7.6
East Midlands	71.1	16.5	2.9	9.6
West Midlands	68.3	18.3	5.2	8.2
East Anglia	69.1	13.3	4.9	12.7
Greater London	56.1	19.6	7.2	17.1
Rest of South East	73.2	11.2	5.8	11.2
South West	72.8	10.2	4.2	10.2
England	68.0	15.9	5.0	11.1

Source: DETR, *Housing and Construction Statistics*

fell into this sector. Private and housing association renting was particularly 'over-represented' in Greater London.

With regard to how much people pay for housing in each of the English regions, Tables 12.8–12.11 and Figure 12.5 reveal that average weekly rents, average house prices, mortgage cost to income ratios, average mortgage payments and the proportion of households in receipt of

Table 12.8 *Average weekly rents by region, by tenure, 1998/99*

	Private sector	*Local authority*	*Registered social landlords*
	(£)	*(£)*	*(£)*
North East	56	36.80	43.40
Yorkshire & Humberside	59	35.10	46.10
North West	74	40.60	43.50
East Midlands	68	38.10	48.20
West Midlands	71	39.80	47.70
East	81	45.60	52.20
London	142	58.00	59.30
South East	95	50.30	58.00
South West	75	43.70	50.50
England	91	43.80	51.70

Source: ONS, *Regional Trends*

Table 12.9 *Average regional house prices, 1980–2000*

	1980	*1990*	*2000*	*Increase %,*
	(£)	*(£)*	*(£)*	*1980–2000*
North	17,710	43,655	66,300	274
Yorkshire & Humberside	17,689	47,231	68,500	287
North West	20,092	50,005	74,300	270
East Midlands	18,928	52,620	77,600	310
West Midlands	21,663	54,694	88,900	310
East Anglia	22,808	61,427	96,900	325
Greater London	30,968	83,821	175,900	468
Rest of South East	29,832	80,525	144,100	383
South West	25,293	65,378	105,600	318
United Kingdom	23,596	59,785	106,900	353

Source: CML, *Housing Finance*

Table 12.10 *Mortgage cost to income ratios for first-time buyers, 1990 and 1999*

	1990	1999
North	22.2	15.5
Yorkshire & Humberside	23.6	16.9
North West	23.6	16.6
East Midlands	27.1	16.5
West Midlands	25.8	17.1
East Anglia	28.9	17.2
Greater London	30.9	20.2
Rest of South East	31.3	19.2
South West	30.0	18.6
United Kingdom	25.5	18.1

Source: Wilcox (2000)

Table 12.11 *Average regional mortgage repayments, 1980–98*

	1980 (£ per week)	1990 (£ per week)	1998 (£ per week)
North	16.30	40.14	58.64
Yorkshire & Humberside	14.14	42.97	61.91
North West	15.48	45.05	58.55
East Midlands	15.89	50.30	61.56
West Midlands	18.82	46.92	65.57
East Anglia	19.63	61.83	66.56
Greater London	23.00	89.56	97.70
Rest of South East	23.70	81.76	88.89
South West	18.58	67.80	64.77
United Kingdom	19.50	60.39	70.63

Source: Wilcox (2000)

housing benefits are all generally lower in the northern regions than in the South – the result of North–South disparities in wage levels, rates of unemployment, activity rates, and housing surpluses and deficits.

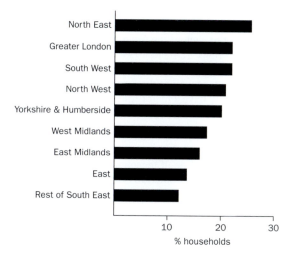

Figure 12.5 *Households in receipt of housing benefits, Great Britain, 1997/98*
Source: DSS, *Family Resources Survey*

Box 12.1

Housing and regional policy in England

Throughout much of the 1980s and 1990s, the government – through the medium of urban development corporations, city grants, City Challenge and the Single Regeneration Budget (SRB) – targeted a significant proportion of its regeneration budget at the clearance and servicing of sites for housing development However, in 1994, with the establishment of government offices for the regions (GoRs) – regeneration took on a new regional dimension. Integrating many of the responsibilities of the DETR and DTI at a regional level, GoRs became not only directly responsible for housing through their role in funding local authority housing investment programmes (HIPs), but also indirectly responsible for housing through their control over the SRB Challenge Fund and English Partnerships (Cameron *et al.*, 1998).

The SRB Challenge Fund, set up in 1994, facilitated a 'rolling programme of bidding rounds for local partnerships to secure resources for up to seven years for a mix of economic, social and physical regeneration schemes' (DETR, 1999: 137), while English Partnerships, formed in 1993, was a national quango which had subsumed a range of regeneration grants and agencies and had grown rapidly to a budget of over £400 million.

continued

In 1998 the government established regional development agencies (RDAs) 'to promote the long-term economic development of each of the [nine] English regions through the application of integrated funding solutions' (DETR, 1999: 135). Although not directly responsible for housing, the RDAs – in inheriting the SRB Challenge Fund and most of English Partnerships' Investment Fund (as well as the financial resources of the Rural Development Commission) – inevitably have a major role to play in the application of housing policy (Cameron et al., 1998).

The Housing Corporation also became organised on a regional basis in the 1990s, and through its regional offices it funds and regulates registered social landlords (RSLs) and establishes regional priorities for the allocation of the annual approved development programmes (ADPs). There is also some co-ordination of HIPs and ADPs within the framework of local authority housing strategies (Cameron et al., 1998). However, there remains the need for local government and the RDAs to work closely together and for much greater flexibility in tackling matters of concern.

Two housing issues are particularly relevant to decision-making at regional policy level. The first concerns the future spatial distribution of population and the location of new housing provision – a matter of considerable importance since recent forecasts predict the need to accommodate 4.4 million extra households by 2016 (TCPA, 1996); while the second is related to social exclusion and the supply of housing land, and involves a consideration of whether greenfield land or urban brownfield sites should be used to satisfy the needs.

Although the institutions of the European Union (EU) have no direct responsibility for housing in its member states, the Regional Development Fund provides support for economic development and infrastructure which indirectly affects housing development, while the URBAN Community Initiative is potentially a source of funding for social housing in disadvantaged neighbourhoods.

In the future, regional policy in England might address, for example, the problems of low demand for housing in the North (such as high vacancy rates, dereliction and social exclusion) and the availability of affordable housing in the South East – including the issue of greenfield versus brownfield development.

Low demand in the North and Midlands

Some cities, owing to deindustrialisation, lost a significant proportion of their population during the last quarter of the twentieth century, while greenfield housing development proceeded apace. The population of both Manchester and Newcastle fell by 20 per cent between 1975 and 2000, while – because of out-migration – the population density of Manchester, Liverpool, Newcastle and Birmingham is now only half that of London (Power, 1999). In the inner areas of Manchester and Newcastle, among the population that remained, nearly 50 per cent of households were classified as poor at the beginning of the new millennium, compared to 20 per cent nationally, and over one-third of children were in lone-parent families (Power, 2000).

However, even in the inner cities, there is a polarisation between prosperity and poverty. Research undertaken by Robson (2001) showed that although recent development (such as penthouses, restaurants and upmarket stores) has brought about the regeneration of the central areas of Leeds, Liverpool and Manchester, only a short distance away – in the inner suburbs – there are extensive areas of impoverished terraced housing and half-empty council estates where unemployment soars to unacceptable levels. Whereas, for example, new canal-side loft apartments in central Manchester sold for up to £1 million at the beginning of the new millennium, in parts of eastern Manchester, houses would have sold – if at all – for as little as £10,000. Blighting, however, is not just confined to the inner cities. Nevin et al. (2000) estimated that along the M62 corridor (connecting Liverpool and the Manchester conurbation) around 280,000 dwellings were blighted, of which 100,000 were privately owned.

It was not altogether surprising that research undertaken by Heriot-Watt University and commissioned by the DETR (2000b) revealed that although, nationally, 12 per cent of all local authority dwellings, 8 per cent of all housing association dwellings and 2.6 per cent of RSL dwellings were affected by weak or inadequate demand, the proportions were considerably greater in the northern regions and in the Midlands (see Table 12.12). Cole et al. (1999) revealed that the scale of low demand was even greater than that estimated by the Heriot-Watt survey. Cole and his researchers found that almost half a million local authority and housing association dwellings in England (about 10 per cent of the social stock), together with more than 400,000 private dwellings, were blighted by low demand. In the social sector, according to Cole et al., the excess supply of housing – in relation to demand – was attributable both to the failure of Conservative governments to obtain appropriate information before injecting money into sink estates, and to some local authority housing managers failing to reveal the true level of demand (see Hetherington, 2000b).

Table 12.12 *Housing affected by weak or inadequate demand, 1999*

	Tenure					
	Local authority		Registered social landlords		Private	
	No.	%	No.	%	No.	%
North East	64,000	32	7,900	16	20,200	2.6
Yorkshire & Humberside	58,500	13	10,500	14	55,300	3.8
North West	113,000	13	32,600	19	173,700	8.6
East Midlands	43,000	11	6,100	8	59,800	3.9
West Midlands	31,500	11	6,700	13	33,500	2.5
East	12,000	4	5,900	9	8,300	0.9
London	32,500	6	7,000	3	7,300	0.4
South East	10,500	4	9,200	4	6,000	0.2
South West	9,000	4	2,600	3	11,400	0.7
England	377,000	12	92,100	8	375,000	0.4

Source: Heriot-Watt postal survey of local authorities and registered social landlords

---

Box 12.2

Housing and regional policy in Scotland, Wales and Northern Ireland

Established in 1885, the Scottish Office subsequently gained an increasing number of country-wide administrative responsibilities ranging – until its demise in 2000 – from economic development to housing, local government, community care, health, education, urban policy, new towns, town and country planning, economic development, forestry and fisheries, nature conservation and the countryside, and environmental protection.

In recent years, however, the Scottish Office devolved some of its responsibility for regional development to the Highland and Islands Development Board (HIDB), created in 1965; the Scottish Development Agency (SDA), set up in 1975; and Highlands and Islands Enterprises (HIE), which superseded the HIDB in 1991. In the same year the SDA merged with the Training Agency in Scotland and was renamed Scottish Enterprise. Although regional development was often synonymous with industrial development, it often had an indirect effect on the demand for, and supply of, housing by enhancing job opportunities and incomes, and facilitating infrastructure improvement. Within a deprived inner city area, however, the SDA was instrumental in improving housing conditions through the medium of the Glasgow East Area Renewal (GEAR) project (see Chapter 13).

Although the Scottish Office had had the role of implementing policy in Scotland that emanated from Westminster, in 2000 the responsibility for primary housing legislation (and the other legislative responsibilities of the Scottish Office) passed from London to the Scottish Parliament in Edinburgh, where it might be expected that over the course of time, a greater co-ordination of housing policy and regional development could be achieved.

In Wales a broadly similar system of government emerged. Although a Welsh Office was established comparatively late, in 1964, it had almost the same responsibilities for Wales as the Scottish Office did for Scotland, including a responsibility for housing. There were also similar development quangos, notably the Welsh Development Agency (WDA) and Land Authority for Wales (LAW), both established in 1975, and a Development Board for Rural Wales (DBRW), set up in 1977.

However, after devolution in 2000, the newly constituted Welsh Assembly – unlike the Scottish Parliament – only had powers to introduce secondary legislation which fleshes out the detail of primary legislation produced in London, including legislation concerning housing. As part of the process of devolution, it is intended that the role of the WDA be strengthened by enabling it to take over the powers of the LAW and the DBRW. It can only be hoped that this will stimulate economic development more effectively than hitherto, and indirectly have a beneficial effect on housing in much of the Principality.

Although in general the administration of Northern Ireland has been in a state of flux since the 1970s (direct rule superseded devolved government in 1972, and this in turn was replaced unpredictably by the Northern Ireland Assembly in 1999), housing has remained a responsibility of the Northern Ireland Housing Executive (NIHE) throughout most of this time, which by definition assumed provincial rather than purely local responsibilities. Working together with both the Department of the Environment Northern Ireland and the Department of Economic Development, the NIHE has the potential to tackle problems of housing need across the province.

Housing shortages: London

Whereas in much of the North and the Midlands there is low demand for housing and consequently low house prices, in Greater London there is a serious shortage of affordable housing. Standard house prices in 1999–2000 ranged from as little as £55,500 in Yorkshire and Humberside to as much as £145,000 in the capital, and North–South disparities were widening markedly. In 1999–2001, when standard house prices decreased by 2.3 per cent in Scotland or rose by only 1.4 per cent in the North, in Greater London and the South East prices soared by 19.5 and 22.7 per cent (Table 12.13).

Table 12.13 *House price changes, United Kingdom, 1999–2000*

	Standard price (£)	Annual increase or decrease, 1999–2000[a] (%)
South East	129,453	22.7
Greater London	145,104	19.5
East Anglia	83,751	18.9
South West	95,677	19.4
Northern Ireland	67,021	10.4
West Midlands	78,561	10.1
East Midlands	68,407	10.0
Wales	64,034	8.4
North West	62,040	6.9
Yorkshire & Humberside	55,494	1.8
North	56,228	1.4
Scotland	60,689	–2.3
United Kingdom	84,422	11.3

Source: Halifax

Note: [a] June 1999–June 2000

Within Greater London – the highest-priced region – there are, however, very wide disparities in property values. Table 12.14 reveals that in Kensington and Chelsea, and Westminster, average house prices exceeded a quarter of a million pounds in 2000 – requiring would-be buyers to have an income of at least £85,000 to secure a purchase, whereas in Newham, and Barking and Dagenham, prices were below £80,000, necessitating an income of around £27,000 to buy. However, even at £80,000, house prices in the cheapest areas of London are higher than those in the Midlands and the North.

The inadequate supply of affordable housing in London is of increasing concern since the number of households in the capital is projected to rise by 564,000 by 2016 (McCarthy, 2000). Not only are there relatively few low-price houses in the owner-occupied sector, but the supply of low-cost rented housing is severely limited. Shortages in the social housing sector are largely attributable, first, to public funds in the 1980s and 1990s being directed at the repair and maintenance of the council stock rather than at new housebuilding; and second, the fact that the Labour government – post-1997 – has the worst record of any government since 1945 in facilitating the construction of social housing (Livingstone, 2000). During the 1990s the number of low-rent houses fell by more than 50,000 – largely because of the Right to Buy policy.

Table 12.14 *House prices and earnings necessary to buy, London, 2000*

London borough	Average house price (£)	Earnings necessary (£)
Kensington & Chelsea	318,722	106,241
Westminster	263,281	87,760
Hammersmith & Fulham	239,273	79,758
Camden	229,533	76,508
Richmond	227,699	75,900
Islington	212,102	70,701
City of London	187,614	62,538
Wandsworth	187,462	62,487
Barnet	161,278	53,759
Kingston upon Thames	157,154	52,385
Tower Hamlets	155,055	51,685
Ealing	152,820	50,940
Harrow	146,947	48,982
Hounslow	146,466	48,823
Merton	146,371	48,790
Southwark	142,153	47,384
Lambeth	141,813	47,271
Haringey	140,521	46,840
Bromley	136,533	45,511
Brent	135,449	45,150
Hackney	129,891	42,297
Hillingdon	119,644	39,881
Sutton	113,231	37,744
Redbridge	109,379	36,460
Enfield	108,971	36,324
Greenwich	107,709	35,903
Croydon	106,060	35,354
Havering	101,957	33,986
Bexley	94,826	31,609
Lewisham	94,782	31,594
Waltham Forest	86,906	28,969
Newham	78,288	26,096
Barking & Dagenham	66,999	22,333

Source: LHF

An inadequate supply of affordable housing is – to a large extent – responsible for homelessness in the capital. Table 12.15 shows that London has the greatest concentration of homeless households in England, and although the number of households 'accepted' as homeless in London decreased between 1992 and 1999 (from around 37,500 to 27,840), in

Table 12.15 *Homeless households, England, 1992 and 1999*

	Homeless acceptances		Number of homeless households in temporary accommodation	
	1992	*1999*	*1992*	*1999*
North East	7,570	4,830	470	1,110
Yorkshire & Humberside	14,430	8,210	1,890	1,740
North West	16,900	10,860	2,230	1,830
East Midlands	10,450	7,300	1,560	1,920
West Midlands	17,070	13,360	1,660	1,570
Eastern	9,300	8,550	4,200	4,250
London	37,550	27,840	39,580	36,330
Rest of South East	13,030	12,380	8,110	8,220
South West	8,990	9,490	3,020	4,870
England	138,740	104,770	63,070	62,170

Source: DETR, *Homelessness Statistics*

1999 there were still about 36,000 homeless households placed in temporary accommodation by the London boroughs. By February 2000 the number of homeless families placed in bed and breakfast accommodation exceeded 6,000 for the first time, with the shortage of places so severe that many families were being accommodated outside the capital (McCarthy, 2000). It is remarkable that in recent years, governments have been willing to finance the provision of temporary accommodation for the homeless, rather than invest – more cost-effectively – in permanent social housing.

Conclusion

Clearly, the surplus of dwellings in parts of the North and the shortage of housing in London are among the starkest manifestations of inter-regional imbalance. North of the Severn–Wash line, not only is there an abandonment and consequential waste of social and private assets, but also there are under-occupied neighbourhoods and boarded up dwellings that not only blight the physical environment of an area but result in the further out-migration of households, and an increase in vandalism, insecurity and crime. Ironically, steps to deal with vandalism and crime, and make properties more secure, often worsen the image of an area. In many cities, households have moved out of local authority housing estates and into private rented housing – imposing an extra strain on the housing benefit system.

Some social landlords recognise that it might be necessary to invest more money into their estates to prevent weak demand degenerating into complete abandonment, while Kirklees (Huddersfield) is attempting to attract homeless households from London with the offer of good-quality housing built in the 1950s and 1960s – which locally is unable to compete with newer housing association dwellings (Hughes, 1999).

There is a view, however, that many housing estates, particularly in the North, are beyond redemption and should be demolished. Both Lord Rogers's Urban Task Force report (DETR, 1999) and the Transport and Regional Affairs Select Committee of the House of Commons recognise that many urban sink estates should be demolished and replaced by middle-income developments to attract the middle classes back into the cities, and – through the creation of mixed communities – lay the basis for an urban renaissance.

In London, instead of there being a surplus of housing as in the North and Midlands, the housing market is overheated, with at least as serious – though different – knock-on effects on the local economy. Since there are substantial shortages of affordable housing in the capital, key public sector workers and highly skilled private sector employees are driven out in search of low-price or low-rent housing elsewhere . In focusing on the need to reduce housing shortages, the newly established Housing Commission of the Greater London Authority (GLA) reported that 31,800 new houses would be required each year until 2010, plus a further 11,200 houses for the homeless or those living in poor or overcrowded conditions (GLA, 2000). It urged that 65 per cent of all new houses (28,000) should be built for people on low or moderate incomes in the form of shared ownership, discounted home ownership and discounted rented housing.

However, it is far from certain whether private developers will be willing to meet this demand. If public policy is amended to require developers to hand over more than 25 per cent of their sites for housing association provision – the current requirement – and if, as an outcome of the Urban White Paper (DETR, 2000a), developers are expected to build to higher levels of quality than hitherto, they may no longer have any incentive to build in London. However, if the Corporation of the City of London promotes the development of skyscraper office development in the near future, it could ensure that adjacent areas benefit from planning gain.

Developers might be willing, for example, to pay for the provision of affordable housing in the deprived east London boroughs as a quid pro quo for planning permission in the square mile.

With regard to enhancing affordability in London, the government – in its *Spending Review* (Treasury, 2000) – announced that a £250 million fund would be set up to help key workers acquire housing in the capital over the period 2001/2–2003/4. Assistance would involve a combination of interest-free loans of up to £50,000 (which would be used as a top-up for the purchase of a house of less than £125,000 in value) and low-cost ownership schemes that would enable local authority or housing association tenants to convert rents into mortgages.

Further reading

Armstrong, H. and Taylor, J. (2000) *Regional Economics and Policy*, 3rd edn, New York: Harvester Wheatsheaf. Provides a very useful account of the development of policy and its application.

Cameron, S., Baker, M., Bevan, M., Hull, A. and Williams, R. (1998) *Regionalisation, Devolution and Social Housing*, Coventry: National Housing Forum. Draws together some key issues concerning the relationship between devolution and social housing.

Damesick, P.J. and Wood, P.A. (1987) *Regional Problems, Problem Regions and Public Policy in the United Kingdom*, Oxford: Clarendon. Explores the complex relationship between regional problems and regional policy.

Hall, P. (1982) *Urban and Regional Planning*, 3rd edn, Harmondsworth: Penguin. A valuable introduction to the history of urban and regional planning, a classic in its field.

Moore, B., Rhodes, J. and Tyler, P. (1986) *The Effects of Government Regional Economic Policy*, London: HMSO. An evaluation of regional policy in Britain, comparing outcomes under Labour and Conservative administrations.

Niner, P. (1999) 'Insights into low demand for housing', *Foundations*, Joseph Rowntree Foundation, July. Draws on the findings of a number of research projects to provide a useful insight into the phenomenon of 'low demand' for housing, particularly in parts of the urban North.

Parsons, D.W. (1988) *The Political Economy of British Regional Policy*, London: Routledge . Provides a useful overview of regional policy from a political as well as an economic perspective.

Power, A. and Mumford, K. (1999) *The Slow Death of Great Cities?
Urban Abandonment or Urban Renaissance*, York: Joseph Rowntree
Foundation/York Publishing Services. A detailed study of low housing
demand in inner-city areas and its relationship to poverty, joblessness and the
quality of housing.

Townroe, P. and Martin, R. (1992) *Regional Development in the 1990s: The
British Isles in Transition*, London: Jessica Kingsley . A useful examination
of regional development within the context of cyclical economic change.

Wannop, U. (1995) *The Regional Imperitive: Regional Planning and
Governance in Britain, Europe and the United States*, London: Jessica
Kingsley. A valuable text exploring the relationship of regional policy to
other regional initiatives.

13 Urban regeneration

Since the Second World War much of the population of the inner cities of the United Kingdom has been disadvantaged by a plethora of economic, social and physical factors which have adversely affected its standard of living and life-chances. For at least half a century it has had to endure poor housing or, worse still, homelessness, while employment has been increasingly relocated to greenfield sites. Abandoned inner city land and buildings have become derelict and former industrial sites are often contaminated. Although professional, managerial and skilled manual workers and their families have migrated to outer suburban and rural areas, semi-skilled and unskilled groups have remained trapped in the inner city, where rates of unemployment have been consistently higher than elsewhere. Incomes in the inner city are normally below the national average; educational attainment up to secondary school level has been poor; there are high rates of crime; incidences of morbidity and mortality are higher than in more affluent areas; the physical environment is unsatisfactory; and there is a substantial need for infrastructural improvement.

Not only is urban regeneration deemed necessary in order to reconstruct the economy and physical fabric of the inner city, but it is also required to reduce the extent to which many of the people of the inner city are excluded from opportunities to raise their living standards, particularly in terms of employment and housing. Although the term 'social exclusion' is of recent origin (and is considered specifically in Chapter 14), its relevance to the urban economy in modern times can be traced back to at least the 1940s, when the first of a range of 'top-down' regeneration strategies was introduced. In reviewing the many attempts by successive governments to promote the regeneration of the United Kingdom's cities since the the Second World War, this chapter:

- traces the development of the many strategies that were applied during the second half of the twentieth century – from the slum clearance and New Town policies in the immediate post-war years to the regeneration programmes funded by the Single Regeneration Budget in the 1990s;
- explores ways in which public policy could ensure that most new development occurs on brownfield sites by around 2010; and
- concludes by examining the deficiencies of the Urban White Paper (DETR, 2000a).

Development of urban regeneration strategies, 1946–96

From the mid-1940s to the early 1970s, a *filtration* strategy was applied as the principal means of promoting urban regeneration. A total of 28 new towns were developed under the New Towns Acts of 1946 and 1952, and 70 expanded towns were built under the Town Development Act of 1952, mainly in greenfield locations, to absorb a proportion of the population and economic activity of the inner cities. Simultaneously, slum clearance and redevelopment was undertaken in the decongested cities – often involving the construction of high-rise housing at comparatively low densities – to accommodate households remaining behind or to house new immigrant populations.

From 1969 to the present, the bulldozer has very largely given way to renovation on grounds of cost-effectiveness. A *rehabilitation* strategy has been implemented through the medium of the Housing Acts of 1969, 1974 and 1980, the Local Government and Housing Act of 1989 and the Housing (Grants, Reconstruction and Regeneration) Act of 1996 (see Chapter 4). Grants have been used – with varying degrees of success – to renovate private sector housing, not least within the inner cities. The strategy to an extent was based on area improvement and, latterly, also on means-testing (see Chapter 4).

While the former strategies were concerned very largely or wholly with housing, the *replacement* strategy – introduced under the Inner Urban Areas Act of 1978 – was intended to promote employment and industrial development in the inner cities. Through the UP, and within a framework of a tripartite partnership between central and local government and the private sector, public resources were targeted at 57 urban priority areas (UPAs) in an attempt indirectly to induce private investment into regeneration schemes.

It was in the 1980s, that governments employed an *enterprise* or 'property-led' strategy in an attempt to regenerate large swathes of the inner city. The strategy was applied in two parts. First, in order to encourage home-ownership or to attract private capital into rented housing, local authority housing was increasingly privatised. RTB schemes were introduced – initially under the Housing Act of 1980 – to enable tenants to buy their own homes at a discount, and – following legislation in 1986 – local authority estates were sold off to housing associations, mainly through the process of LSVTs. Under the Housing Act of 1988, estates in poor condition were – to an extent – transferred to HATs for the purpose of renovation, prior to being sold off (subject to tenant consent) to alternative landlords (see Chapter 8). Secondly, commercial, industrial and – to some extent – residential development was promoted through the medium of UDCs, derelict land grants, urban development grants, urban regeneration grants and city grants, while enterprise zones were employed to attract mainly non-residential development. This approach was highly dependent on the principal of 'leverage', as pioneered in regeneration schemes in the United States. Essentially, the government was willing to invest a specific sum in regeneration on condition that the private sector would, in response, invest a multiple of that sum – leverage ratios of 1:4 being the norm. While the enterprise strategy undoubtedly led to an increase in the stock of private property and a substantial amount of development, arguably it did little to reduce the extent of social exclusion; for example, waiting lists for social housing remained lengthy in many inner city areas, and comparatively few jobs were created specifically for the working population of the inner city, so that unemployment remained high.

By the early 1990s it was evident that the regeneration policies of the 1980s lacked coherence, 'especially in the financing of the varied and varying policies, and . . . [the] wide variation in outcomes' (Robson, 1999). It was consequently thought that policy might be more successful in regenerating the deprived urban areas if there were a shift of emphasis from a 'top-down' to a 'bottom-up' approach whereby the form of regeneration would be selected and managed by the local authority. The City Challenge initiative was therefore introduced in 1992 to encourage all 57 UP authorities to bid for regeneration funds (including funds for housing renovation), but although 31 authorities were successful in their bids and were consequently able to undertake imaginative and well-managed five-year regeneration programmes, the 26 unsuccessful authorities were deprived of funds, to the disadvantage of those in housing need and the unemployed. Nevertheless, from 1992 to 1998 (when the initiative came to an end), recipient authorities levered £1 billion of resources from the private sector, and – as well as reclaiming derelict land and developing commercial and industrial property – built or improved 40,000 houses (Robson, 1999).

The problem of coherence was tackled in 1994 when the Single Regeneration Budget (SRB) was established to bring together budgetary segments of the Department of the Environment, the Department of Trade and Industry, the Department of Employment, the Home Office and the Department of Education to fund approved regeneration programmes within the UPAs. While a large proportion of SRB funds were allocated to government initiatives (such as the UDCs, HATS, English Partnerships and City Challenge), resources for up to seven years were also allocated – through a Challenge Fund – to a rolling programme of bidding rounds for local partnerships to facilitate a mix of economic, social and physical regeneration. The administration of the SRB was devolved to 10 integrated government offices for the regions (GoRs), which were created to ensure that public expenditure on urban regeneration would be substantially rationalised at a regional level.

Whereas the enterprise strategy of the 1980s largely marginalised local authorities, both 'City Challenge and, to an even greater extent, the SRB gave bidders real incentives to establish effective partnerships across the domains of the public sector, private sector and voluntary/community sector' (Robson, 1999: 175–76). Hence bottom-up initiatives were placed – at the very least – on a par with top-down strategies, but much still needed to be done to regenerate the deprived urban areas in social as well as in economic and physical terms.

An urban renaissance?

After coming to power in May 1997, and aware of the very mixed effects of urban policy under the previous administration, the Labour government commissioned Lord Rogers to chair an Urban Task Force to examine ways in which public policy could ensure that 60 per cent of all new development would occur on brownfield land within 10 years. In its report, *Towards an Urban Renaissance* (DETR, 1999), the Task Force called for a new approach to urban regeneration, and – in the light of studies undertaken in Barcelona (where high-density housing development was central to regeneration) and in Germany (where environmental impact taxes had been introduced) – proposed that:

Box 13.1

Urban regeneration and housing in Manchester

The population of Greater Manchester – the oldest industrial conurbation in the world and the centre of the largest industrial region in Europe – decreased by 1.6 per cent, 1981–99. In the cities of Manchester and Salford – at the core – the population decreased by as much as 7.1 and 8.6 per cent respectively, more than in any other district within the conurbation (Table 13.1).

Table 13.1 *Population change, Greater Manchester, 1981–98 (%)*

Bolton	2.0
Bury	3.5
Manchester	−7.1
Oldham	−0.1
Rochdale	0.0
Salford	−8.6
Stockport	0.8
Tameside	0.9
Trafford	0.6
Wigan	0.1
Greater Manchester (total)	−1.6

Source: ONS (2000)

The conurbation's economy remained stagnant throughout much of the 1980s and 1990s, and there was an outward migration of population. The North West region, with Greater Manchester at its heart, had a GDP per capita of only £11,000 in 1998 (well below the national figure of £12,500), and the average price of housing units in the region was as low as £65,500, less than half the average price of housing in London.

Because of its tightly drawn boundaries within the conurbation of Greater Manchester, the City of Manchester (with a population of around 430,000) 'has become one large, impoverished inner city struggling to revive its core within a ring of decayed, disinvested and evacuated inner neighbourhoods' (Rogers and Power, 2000: 9). On the basis of the DETR's Index of Urban Deprivation the city ranked as the second most deprived district in the north of England after Liverpool in the 1990s (Table 13.2), and more deprived than any inner London borough except Newham (see Box 13.2). Unlike most other cities in the North, Manchester did not experience any reduction in its degree of deprivation between 1991 and 1996.

Although Manchester is rebuilding its centre, both commercially and residentially, architects, planners and developers face 'a more difficult and, many think, a less

Table 13.2 *Northern city rankings from the 1991 and 1996 Index of Deprivation*[a]

District	1991 index	1996 index	1991 rank	1996 rank
Liverpool	399	401	1	1
Manchester	375	363	2	2
Knowsley	359	337	5	9
Salford	210	266	10	23
Oldham	306	248	11	33
Middlesbrough	303	264	12	24
Bradford	300	259	14	28
Sunderland	293	269	17	21
Hull	292	261	18	26
Newcastle upon Tyne	289	280	21	19

Source: Robson (1999)

Note: [a] The index measured 12 socio-economic indicators of deprivation to produce a composite value of the degree of deprivation in each district. The higher the score, the greater the deprivation.

attractive proposition' (Rogers and Power, 2000: 206) when attempting to recycle the 'marginalised' city of the poor. In east Manchester in particular (where the population declined from 86,000 at its peak to less than 20,000 by 1999), there are many increasingly abandoned 'neighbourhoods that are so poor and run down that no one with any choice is willing to move in or invest' (Rogers and Power, 2000: 16), rendering pump-priming government grants or EU assistance largely irrelevant. Clearly, solutions need to be found to the problems associated with thousands of hectares of Manchester's inner neighbourhoods that are losing people and property to the detriment of the metropolitan economy.

In east Manchester there are hectares of boarded up, unlettable houses, vast numbers of empty industrial buildings in the process of demolition, and an increasingly derelict urban landscape. Yet even here it is possible that the district's ten devastated and half-abandoned neighbourhoods could grow again, given better transport, integrated mixed-use development and a greener environment, and favourable knock-on effects from the development and future use of the 2002 Commonwealth Games facilities.

However, in Manchester and Salford as a whole there were around 30,000 vacant housing units in 1998, 17,500 outstanding planning applications for new housing in 1998/99, an annual flow of 2,500 new dwellings allocated in development plans, and an equivalent increase in the forecast number of households. It is not surprising, therefore, that house prices in much of Manchester and Salford are the lowest in the country. Communities are being destroyed in many parts of inner Manchester and in Salford by the lure of new housing in the surrounding small towns and countryside beyond the city's boundaries. Without doubt, all the new homes that might be required for the next two generations could be developed within the existing urban framework of Manchester itself.

A number of regeneration initiatives, however, have been undertaken in recent years. For example, close to the core, an extensive area of nineteenth-century high-density

continued

terraced housing in Moss Side was cleared and replaced by the Hulme estate during the 1960s. Consisting of deck-access blocks designed for 12,000 people (a fraction of those who lived in the area in the 1930s), the estate was badly managed and failed socially, structurally and environmentally. Because of these failures it was in turn demolished in the 1980s and replaced by traditional terraced development of the sort residents preferred, with the aid of government funding of £37.5 million under its City Challenge facility.

1 Value added tax should be harmonised on residential conversions and new housebuilding to eliminate the cost advantage of predominantly greenfield development (in 1999 value added tax was 17.5 per cent on the conversion of old houses but zero on new housebuilding).

2 Private capital should be levered into urban regeneration schemes, in much the same way as in the former UDCs during the 1980s.

3 Regional development agencies (RDAs), like UDCs, should have the freedom to promote regeneration through a variety of processes, without day-to-day consultation with the government.

4 There should be 100 per cent brownfield regeneration by 2030, and at comparatively high density.

5 Tax incentives should be used to encourage the setting up of 'urban regeneration companies'.

6 Local authorities should be permitted to keep a share of tax revenues derived from regeneration.

7 Local authorities should be allowed to initiate a system of uniform environmental impact fees.

8 Like the former UDCs, local authorities should be enabled to compulsorily purchase land without having to specify an actual development and its viability.

9 Financial instruments should be introduced to attract institutional investment into the private rented sector.

10 Full council tax liability should be extended to owners of homes that remain empty and derelict for year after year.

11 The General Development Order should be amended to prevent owners from using empty brownfield sites for car parks and other low-density purposes.

12 Utilities, such as Railtrack (until 2001 the company responsible for railway track and stations), should be encouraged to release their extensive urban landholdings to facilitate large-scale regeneration.

13 Vacant urban land should be subject to taxation as an inducement to development.

Box 13.2

Urban regeneration and housing in London

While Greater London had a GDP per capita of around £16,000 in 1998, well above that for the United Kingdom as a whole (£12,500), and the average price of housing units in the capital was as high as £150,0000 (compared to £94,000 in England and Wales as a whole), ten boroughs of Inner London have for many years been amongst the most deprived in the United Kingdom, yet it is in these areas that population growth – particularly among ethnic minority communities – has been substantial in recent years (Table 13.3).

Table 13.3 *Population change, Greater and Inner London, 1981–98 (%)*

Inner London	
Camden	5.3
City of London	−3.6
Hackney	5.2
Hammersmith & Fulham	4.1
Islington	7.8
Kensington & Chelsea	21.3
Lambeth	6.6
Lewisham	3.1
Newham	8.7
Southwark	6.2
Tower Hamlets	6.2
Wandsworth	1.4
Westminster	17.3
Greater London	5.6

Source: ONS (2000a)

Although parts of Inner London – for example in Docklands – are blessed by a number of major regeneration schemes enhancing the role of the core as a hugely wealthy 'international' city, the deprived boroughs of east London (home to hundreds of thousands of poor people trapped in municipal housing estates) grapple

> with extraordinarily high levels of unemployment resulting from the highest rate of job losses in the country during the 1980s. As a consequence, there has been an exodus of better-off people, racial polarisation and the concentration of people with few skills.

> (Rogers and Power, 2000: 49)

continued

Table 13.4 *London rankings from the 1991 and 1996 Index of Deprivation*[a]

Borough	1991 index	1996 index	1991 rank	1996 rank
Newham	378	386	2	2
Hackney	363	352	4	4
Tower Hamlets	303	343	13	6
Southwark	300	337	15	8
Islington	312	322	9	10
Greenwich	270	316	25	11
Lambeth	290	316	20	12
Haringey	318	315	8	13
Lewisham	265	294	29	14
Camden	286	282	22	17

Source: Robson *et al.* (1998)

Note: [a] The index measured 12 socio-economic indicators of deprivation to produce a composite value of the degree of deprivation in each district. The higher the score, the greater the deprivation.

Despite London's enormous wealth, half of the 20 most deprived local authority areas in 1996 in the country were in the capital (Table 13.4), and it was evident that over the preceding five years deprivation had got worse rather than better in several of London's inner city boroughs. Clearly, urban generation policy was having little effect on the overall well-being of these areas.

Manifestly, a substantial amount of 'market-led' regeneration took place in the 1980s and early 1990s through the medium of enterprise zones, derelict land grants, urban regeneration grants and urban development grants, but little consideration was given to the satisfaction of the employment and housing needs of the local population. However, on a small scale a number of potentially successful regeneration schemes were put in place in inner east London in an attempt to create competitive local economies. In the Holly Street neighbourhood of Hackney, for an example, an area of high-rise housing, dogged by unemployment, crime and vandalism, was replaced – under a Comprehensive Estates Initiative (CEI) in 1992 – by 1,000 units of new mixed-tenure housing and the development of employment and training opportunities (DETR, 1999). For its success the CEI relied heavily on a series of partnerships between Hackney Borough Council, the residents of Holly Street and the private sector rather than a partnership between the public and private sector alone of the sort that all too often ignore community interests.

Conclusion

Arguably, the Urban White Paper (DETR, 2000a) failed, 'by a demonstrable margin, to meet the recommendations of the Urban Task Force' (Hall, 2001: 2). Although some of the ideas of the Task Force were, to an extent, incorporated in the White Paper (for example, value added tax was reduced from 17.5 per cent to 5 per cent on house conversions; £100 million was allocated to lever private capital into regeneration and over £1 billion was set aside over five years in the form of various incentives to developers and investors; and RDAs began to enjoy greater spending flexibility as a result of combining different RDA budgets into a Single Programme Budget), there were several instances where the White Paper only marginally reflected the thinking of the Task Force or subjected proposals to further review.

Thus, instead of attempting to achieve 100 per cent regeneration by 2030, the government aimed to reclaim only 17 per cent of brownfield land by 2010 – about a half of what was called for; and rather than encouraging the establishment of regeneration companies (the White Paper suggested that 12 should be set up by 2003), the government – at the time of writing – might have discouraged their early formation by failing to announce how they would be helped; while a number of Task Force recommendations were subject to further review; for example, proposals relating to local authorities retaining tax revenues derived from regeneration, initiating a system of environmental impact fees, and permitting local authorities to exercise compulsory purchase powers more easily.

A number of Task Force recommendations failed completely to be included in the White Paper, notably the proposals to attract institutional investment into private renting, extend council tax liability to empty and derelict properties, modify the General Development Order, compel utilities to release their vast landholdings, and tax vacant urban land.

Both the Task Force report and the White Paper failed to address the issue of out-migration. Both people and employment are still drifting away from the cities, and until the government – in policy terms – focuses on job creation in the inner cities, it is not adequately promoting sustainable regeneration (Breheny, 2000).

Further reading

Atkinson, R. and Moon, G. (1994) *Urban Policy in Britain: The City, the State and the Market*, Basingstoke: Macmillan. Presents a useful analysis of the interrelated roles of public intervention and the free market in the application of urban policy.

Blackman, T. (1995) *Urban Policy in Practice*, London: Routledge. Provides a clear review of urban policy, focusing particularly on the 1980s and early 1990s.

Deakin, N. and Edwards, J. (1993) *The Enterprise Culture and the Inner Cities*, Oxford: Oxford University Press. A valuable insight into the development of the free market within an inner-city environment.

Department of the Environment, Transport and the Regions (1999) *Towards an Urban Renaissance*, Final Report of the Urban Task Force, chaired by Lord Rogers of Riverside, London: E. & F.N. Spon. Presents a detailed identification of the causes of urban decline in England and suggests a wide range of practical solutions to attract people back into urban areas.

Hall, S. and Mawson, J. (1999) *Challenge Funding, Contracts and Area Regeneration: A Decade of Innovation in Policy Management and Coordination*, Bristol: Policy Press. A study that brings together the findings of research on the evolution of area regeneration policy in the 1990s.

Imrie, R. and Thomas, H. (eds) (1993) *British Urban Policy and the Urban Development Corporations*, London: Paul Chapman. Presents a useful analysis of the role and performance of UDCs in inner-city regeneration.

Jacobs, J. (1961) *The Death and Life of Great American Cities*, Harmondsworth: Penguin Books. A classic polemic on city planning in the United States, and one which – despite the book's age – is very apposite to the current debate on urban regeneration in the UK.

Lawless, P. (1989) *Britain's Inner Cities*, London: Paul Chapman. One of the best general introductions to the problems of the inner cities and policy responses, although it may now be a little dated.

Public Sector Management Research Unit (1988) *An Evaluation of Urban Development Grant Programme*, London: HMSO. Examines the renovation of privatised council estates and the development of private housing in inner-city locations.

Robson, B.T. (1988) *Those Inner Cities: Reconciling the Social and Economic Aims of Urban Policy*, Oxford: Clarendon Press. A valuable analysis of the relationship between the social and economic aims of inner-city regeneration.

Rogers, R. and Power, A. (2000) *Cities for a Small Country*, London: Faber & Faber. A sequel to the Urban Task Force report (DETR, 1999). It sets out the problems of cities from architectural and social perspectives and proposes radical solutions.

Thornley, A. (1993) *Urban Planning under Thatcherism*, London: Routledge. Provides a detailed appraisal of urban planning during the 1980s.

14 Social exclusion

Clearly, people's experience of housing and homelessness affects virtually every aspect of their social well-being. The recent discussions on social exclusion have become increasingly important to housing policy and practice. It is clear that both market forces and public policy have failed to satisfy the essential needs of many households, not least the need for decent and affordable housing. This chapter:

- examines some of the various definitions of social exclusion;
- discusses the production and distribution of housing;
- explores the concept of poverty and its relation to social exclusion;
- analyses public policy and practice in relation to social exclusion; and
- concludes by examining neighbourhood management and self-help.

Definitions of social exclusion

The original concept of 'social exclusion' is attributable to René Lenoir, a member of Jacques Chirac's government in France in the 1970s. He estimated that around 10 per cent of the population in France were 'excluded'. The excluded group consisted of people such as lone parents and people with disabilities who were not covered under the French social insurance system at that time (Silver, 1995). In the United Kingdom attention was instead being paid to the narrower definition of 'poverty', Townsend (1979) defining the term as 'a level of income below which people are unable to participate in the normal life of society' (Levitas, 1996: 7). This definition refers to people who are excluded from society solely due to their level or lack of income.

In the 1990s, however, there were many debates about the definition of the term 'social exclusion' and whether it differs from the definition of poverty. Room (1995: 105) stated that whereas poverty is

> primarily focused on distributional issues, the lack of resource at the disposal of an individual or a household . . . notions such as social exclusion focus primarily on relational issues in other words inadequate social participation and social integration and lack of power.

The emphasis on social participation can be related to citizenship in terms of social, political and civil rights. It is therefore necessary to consider the impact of factors such as housing, education and employment in order to ascertain the processes that can compound and trap certain people in disadvantaged situations (Marsh and Mullins, 1998). Somerville (1998: 762) also commented that the definition of social exclusion can relate to that of social citizenship and the denial of this right to certain groups due to 'the processes of stigmatisation and restrictive or oppressive legislation and law enforcement and forms of institutional discrimination'.

According to Levitas (1996) and Somerville (1998), debates on social exclusion in Europe have focused on a concern with lack of income and unemployment. Thus, policies adopted in the United Kingdom such as Welfare to Work were based on the notion of removing social exclusion. As Tony Blair (1997) stated, 'our task is to reconnect that working class to bring jobs, skills, opportunities and ambition to all those people who have been left behind by the Conservative years'.

Ratcliffe (1998) similarly argues that social exclusion (evident in a number of different forms) either excludes certain groups from the rest of society or denies them fundamental citizenship rights. First, social exclusion could be confined to a fixed social location whereby there is a group which is separate from the rest of society, and in this definition the term is seen as spatial as well as social. Second, social exclusion can be seen as a desire by an individual to be separate. Ratcliffe (1998) explains that this refers to those sections of society who appear to opt out of the values which the rest of society follows. This group will tend to have similar attitudes of welfare dependency and antisocial behaviour.

A problem with this definition is that a state of self-exclusion may occur with regard to black and Asian communities, who may have a different lifestyle or values and beliefs from the rest of society. Where black and Asian minority groups have moved out of the neighbourhoods where they have lived historically and were safe they may encounter racial harrassment. The issues around racial harassment can be linked back to social exclusion that is based within a system of social processes.

Parkinson (1998) makes the point that while poverty is usually related to low income and material want, social exclusion refers to a wider concept which is based on the lack of access for some groups – such as black and Asian households – to social, economic and political activities in mainstream society. Page (2000) similarly feels that social exclusion is about more than disadvantaged people and neighbourhoods. It is also about the lack of employment and the effect it has on people's ability to participate in society. He also points out that the migration of more mobile people away from disadvantaged areas results in the more disadvantaged and socially excluded remaining behind – not least in areas of poor housing. In the view of Ratcliffe (1998), however, social exclusion is more often seen in terms of the denial of rights such as employment, education and housing due to factors such as gender, race and age.

Social exclusion and the production and distribution of housing

Social exclusion through housing can happen through the production and distribution of housing (Somerville, 1998). With regard to the production of housing, council properties were provided in order to encourage the inclusion of groups of people who found that owing to shortages in housing supply, many people could not afford to buy or rent adequate and appropriate housing. At one time, council housing provided tenants with good-quality and appropriate accommodation at affordable rents. However, there is evidence that since the 1970s many households have been excluded from obtaining the good-quality and affordable housing they need.

In terms of housing distribution, social exclusion could be related to housing tenure. For example, council housing is seen to be less exclusionary as rents are not determined by the market and access to the sector is determined by need and the ability to wait (Harloe, 1995). However, there are still groups who are excluded from the sector such as young single people. Owner-occupation is exclusionary as the market determines prices, and income and wealth determine access to the tenure.

The exclusion of the poorest households from the owner-occupied sector is a reflection of the inequalities that currently exist in British society (Somerville, 1998). Somerville argues that exclusion from entry into this sector is not necessarily social exclusion as some households will be able to obtain rent in the private rented sector and not be excluded socially. In addition, some tenants may still be socially excluded if they are unable to empower themselves or if their rent is too high for the household and they fall into rent arrears. Moreover, institutional discrimination can lead to the exclusion of some groups of people on the basis of factors such as race or gender. For example, single women may find it more difficult to obtain a mortgage than single men.

A number of writers have argued that residential space is an important cause of social exclusion. The incidence of factors such as antisocial behaviour, drug abuse and a high proportion of people who are unemployed can have an effect on social exclusion. This is the case where interaction may be limited to a particular neighbourhood, and also where some families, who may see themselves as respectable, ensure that they do not mix with the families who are seen as being of a lower status.

Social exclusion, race and housing

Harrison (1998: 798) suggests that a discussion regarding race and social exclusion cannot safely ignore:

- patterned and widespread racist practices (reinforced through stereotyping, labelling and violence) in institutions such as housing organisations; and
- controlled processes through practices such as housing allocation and management based on maintaining social order, deterrence and risk minimisation.

Poverty and social exclusion

The prospects for sustainable housing renewal within the deprived areas of Britain are severely constrained by poverty and social exclusion. Rahman *et al.* (2000), in producing their third annual report of indicators of poverty and social exclusion for the New Policy Institute, showed that just over 14 million households in the United Kingdom are living below half the average income after taking into account housing costs. Just over 8 million households had incomes that are 40 per cent less than the average wage. Given that the working families tax credit and the minimum income guarantee for pensioners were in place at the time of this study, the government's attempt to reduce social exclusion must be seen as a long-term project. The study also identified that nearly a half of working-age long-term claimants have disabilities or suffer from long-term sickness. Moreover, Rahman *et al.* (2000) found that there were 4.5 million children in households with below half the average income after accounting for housing costs. In addition, around 2 million children live in households where there is no economically active adult.

Although the number of school exclusions has fallen, African and Caribbean pupils are still over four times more likely to be excluded than white pupils are. Furthermore, 150,000 pupils still fail to obtain any GCSE qualifications above grade D each year.

Approximately half a million young adults aged 16 to 24 were economically inactive, which as a proportion is more than double that for the population as a whole, and were paid at less than half of the male median hourly earnings. Around 60,000 young adults aged 18–20 years old had a criminal record in 1999 and the suicide rate among the 15–24 age group stood at 10 per 100,000. Around 4 million adults were economically inactive, and almost half of all lone parents were not employed, compared to one in twenty couple households with children. Black and Bangladeshi people were found to be twice as likely to be excluded from employment as the white population.

With regard to social exclusion and health inequalities, Rahman *et al.* (2000) found that women from the manual class are 1.5 times more likely to be obese than women from the white-collar class. In addition, unskilled manual workers are 1.5 times more likely to have a long-term illness or disability compared to the professional class. It was also found that the poorest two-fifths of the population are more likely to be at risk of developing a mental illness than the richest two-fifths.

The difference between poverty and social exclusion is unclear when related to older people; 1.5 million pensioners are totally reliant on the state pension and benefits.

Overcrowding in the social rented sector is now at three times the level for those with a mortgage. Moreover, the number of households in temporary accommodation continues to rise sharply. One in six of the poorest households were found not to have any type of bank or building society account, compared with one in twenty households on average incomes.

Johnston *et al.* (2000) conducted a study into the reality of social exclusion and the alleged emergence of a welfare-dependent underclass. The research focused on young adults living in a socially excluded neighbourhood. The research found that the frame of reference for these young adults was the local neighbourhood, and that the opportunity to work or obtain training was linked by its proximity to their neighbourhood. However, they did suffer from negative labelling of the area. Johnston and co-workers found that while objectively the neighbourhood may be perceived to have the indicators of social exclusion, the subjective experiences of the young adults were nonetheless of social 'inclusion'. Many of them were economically active outside the formal employment system. The researchers found that contrary to popular images of disadvantaged neighbourhoods, the young adults were keen to gain employment and thought that government training schemes were a good opportunity to gain qualifications. However, there was a high level of drug use among the young adults. It often began around the age of 13 with glue-sniffing and progressed in some cases to heroin. There was also evidence of relatively high levels of criminal behaviour such as shoplifting, progressing in some instances to domestic and commercial burglary. Johnston *et al.* (2000) concludes that in order to break the cycle of social exclusion, it is important to look at the lives of young people and remove the non-participation and underachievement that they often develop.

Page (2000), in another study of social exclusion, found that those who live in disadvantaged areas and vulnerable to social exclusion were happy with where they lived and preferred to stay in close proximity to their neighbourhood. He too found that the residents did not see themselves as disadvantaged or socially excluded. An 'estate culture' was evident in terms of activities such as crime, drugs, antisocial behaviour, low personal and educational achievement, and estate norms that were different from those of the rest of society, with pressure from their peers to conform. Page found that the front-line workers in the areas of the study thought that social exclusion was a part of the culture of the neighbourhood and that residents were destined to become disadvantaged as a direct result of their surroundings. Staff felt that the lack of resources added to the feeling of disadvantage on estates, and this made many of the front-line workers feel demoralised. They did not feel that they made the life of residents better and were often preoccupied with the problems within their own organisations such as repeated staff restructuring and the prospect of redundancies. Page felt that the lack of involvement on the part of residents was due to their lack of trust of the professionals. The residents felt that the professionals had their own agendas and that the professionals had not been interested in listening to them in the past. Page (2000) concluded that social exclusion was a structural problem, and hence much wider than dealing with disadvantaged individuals. The community and neighbourhood were a very important factor, but resources are required to halt the decline in such areas. Many of the residents had grown up in dysfunctional families and had experienced disadvantaged circumstances such as low educational attainment. Often they had a parent who had a succession of partners. This pattern is likely to be repeated in later life by the young people who live and experience these circumstances.

Financial services and social exclusion

The lack of access to financial services – not least in relation to housing – is a major attribute of social exclusion. Kempson and Whyley (1999) stated that around 1.5 million (7 per cent) of households in Britain do not use any financial services, while 4.4 million (20 per cent) of households used just one or two services. Financial exclusion is an important issue as the ability to budget the household finances is restricted and can add to the disadvantage of the household. The financial services from which some households are excluded include a current account, insurance, long-term investments and a pension. Kempson and Whyley (1999) found that being in receipt of income-related benefits was a strong predictor of financial exclusion. Low household income and the length of time for which the head of the household had been economically inactive followed this. Other factors seen to be more likely to lead to financial exclusion included renting a property, being single, having a Pakistani or Bangladeshi background, or having left school at the age of 16. The region in which the individual resides also has an effect: those who live in Scotland, Wales, Greater London or one of the 50 most deprived local authorities in England and Wales were more likely to be financially excluded.

Financial exclusion is a dynamic process: many households move into and out of financial exclusion. Those people who had only one or two financial products (Kempson and Whyley, 1999) were most likely to have a current account or savings account with a building society or bank. Among those who had never used financial services, the main characteristics included factors such as households whose members had never had a secure job; elderly people who were accustomed to dealing with cash; young households who may use the services in the future; young single mothers; and people in Pakistani or Bangladeshi households who did not use them owing to language barriers, religious beliefs or lack of knowledge. In addition, the research found that among the reasons why people stopped using financial services were a fall in household income and, for many women, the loss of a partner through separation or death, especially when the partner held all the financial products in their name.

When a household experienced a drop in income, there was a tendency to withdraw from financial services in order to maintain tight control over the family finances. Some people were found to withdraw from financial services once they fell into financial difficulties and never attempt to reapply for replacements, or apply but are turned down or do not bother to apply as they fear that they will be refused. The research found that once these people became financially secure in the future, they would reapply for financial services.

Kempson and Whyley (1999) found that the barriers to obtaining financial services include the following:

- Products such as house contents insurance may be too expensive as premiums may be increased in areas classified as deprived or more likely to experience crime.
- Conditions attached to financial products offered may make them unsuitable, such as the offer of a current account without a cheque-book, cheque guarantee card or cash withdrawal card.
- Many financial institutions are not keen to attract low-income customers.

People without access to financial products stated that there were two important areas where financial services were required: for day-to-day money management and for long-term financial security. Kempson and Whyley (1999) found that people did not consider medium-term financial products such as insurance, a need for savings and consumer credit to be important. They found that households without a current account had to deal exclusively in cash transactions, which resulted in problems with paying some bills as organisations would request an extra charge for dealing with cash payments. Not having a current account also makes it hard to cash or issue a cheque. The researchers found strong resistance to consumer credit, even though there was acceptance that some expenditure, such as major repairs, could not be met without it. The lack of access to short-term loans makes money management difficult and so can lead to arrears or resorting to expensive moneylenders.

Kempson and Whyley (1999) found that the barriers to financial services could be reduced by the provision of a current account for day-to-day money management. There should be a buffer zone for flexibility and to allow for the paying of bills. Access to this type of account should not be dependent on credit scoring. There should be products intended to allow for long-term financial security; these should not be easily accessible to users, who might otherwise be tempted to cash them in when they require short-term cash. In addition, the contributions should be flexible in order to enable the user to maintain regular and automatic payments even when in financial difficulties.

The same researchers (Kempson and Whyley, 1999) felt that medium-term financial products would be used by households if insurance premiums could be spread throughout the year and if they offered second-hand replacement value rather than new for old. They also felt that low-income households should have access to short-term loans with fixed and automatic repayments.

The low levels of ownership of telephones and computers in many households ruled out the possibility for them to gain access to facilities such as online banking and telebanking. Kempson and Whyley therefore considered it important for intermediary services to be provided, which are often cheaper. For example, house contents insurance is sometimes available through local authorities and is paid as part of the rent.

Social inclusion: policy and practice

The concept of social exclusion gradually became a cause of concern within the European Community during the 1970s and 1980s. However, initiatives such as the first and second European anti-poverty programmes (1975–80 and 1986–89) were marred by the fact that different members had different definitions of poverty, and by the United Kingdom and West Germany denying that poverty was a problem within their countries since they had efficient labour markets and welfare systems. However, the third European anti-poverty programme (1990–94), focusing on the 'least privileged', explicitly addressed social exclusion. This was found to be an acceptable compromise to countries such as the United Kingdom and Germany as the vagueness of the term allowed member states of the European Union (EU; formerly the European Community) to be more prepared to participate in the combating of exclusion. Article 137 of the Treaty of Amsterdam, which took effect in 1999, provides a legitimate basis for the European Union to aid 'initiatives aimed at improving knowledge, developing exchanges of information and best practices, promoting innovative approaches and evaluating experiences in order to combat social exclusion' (European Federation of National Organisations Working with the Homeless, 1998: 11). This provides the EU with the opportunity to work legally towards the social inclusion of the most disadvantaged people in society.

In Great Britain, as the 1990s progressed it was increasingly recognised that urban regeneration should not be confined solely to the renewal of the built environment. It was evident that past attempts to apply a 'bricks and mortar' approach to inner city problems had in a large number of cases manifestly failed. Power (1997) argued that the construction of mass housing in the 1960s and 1970s often made conditions worse. It was a Big Brother solution to slum problems that – together with unemployment, cuts in local authority budgets and the break-up of the traditional working-class family – led to deepening crises in the early 1980s, provoking inner city riots in, for example, Brixton (south London), Toxteth (Liverpool), Moss Side (Manchester) and St Paul's (Bristol). By the late 1990s, research had revealed that the country had as many as 2,000 estates on the verge of becoming ghettos with high incidencies of unemployment, benefit dependency and crime (Power, 1997).

In a holistic attempt to improve the standard of living on 1,400 of the most deprived estates in England (including 879 in London), a Social Exclusion Unit (SEU) was launched in December 1997. Its original remit included looking at areas of concern such as problem estates, truancy, exclusions from schools, and rough sleeping. As a 12-strong unit located in the Cabinet Office and drawn from the Civil Service, business, the voluntary sector and the police, the SEU was designed to encourage the development of 'bottom-up' programmes aimed at facilitating social inclusion, but it was not initially charged with the task of housing renovation. However, critics argued that the unit 'will have to tackle the physical fabric of estates as well as the people who live on them' (Wintour, 1997) since it was clear that there is a relationship between poor housing, motivation and the willingness to work.

In October 1998 the DETR introduced the New Deal for the Communities (NDC) programme in which 17 'pathfinder' authorities were selected to undertake pilot regeneration schemes (Table 14.1). The DETR also established 18 cross-cutting 'policy action teams' (PATs) to report on problems such as job and skill shortages, business revival, housing abandonment and antisocial behaviour. PATs subsequently reported that as many as 11 per cent of all council houses in England (377,000 units) and 8 per cent of housing association dwellings (90,000 units) were unpopular or blighted, while DETR research revealed that 1,370 estates could be classified as 'run-down', their inhabitants suffering from bad housing, high unemployment, low educational standards and ill health. According to the SEU, a high proportion of these estates are located within the 44 local authority districts with the highest concentration of deprivation in England, with unemployment rates two-thirds higher than the national average; under-age pregnancy and lone parenting 50 per cent above the national average; mortality rates 30 per cent higher than average; and three times the level of poor housing (Hetherington, 2000e).

Table 14.1 *New deal for the communities: pathfinder authorities*

Birmingham	Bradford
Brighton and Hove	Bristol
Hackney	Kingston-on-Hull
Leicester	Liverpool
Manchester	Middlesbrough
Newcastle upon Tyne	Newham
Norwich	Nottingham
Sandwell	Southwark
Tower Hamlets	

Through the medium of a *neighbourhood renewal fund* (co-ordinated by a special unit in Whitehall), £800 million was set aside by the government in October 2000 – for the period 2001–04 – to improve housing, raise school standards and create jobs in 88 local authority districts, specifically with the aim of addressing the North–South divide. As many as 60 of these districts were located in the North and Midlands – the North East alone (with only 5.3 per cent of England's population) receiving as much as 13.25 per cent of the allocation. Areas of deprivation in London and elsewhere received a less than proportionate allocation.

Power and Rogers (2000) suggested that as a prelude to reversing the fortunes of the socially excluded, new experiments should be undertaken such as:

- the recycling of empty, under-used buildings;
- the development of housing and other uses at higher densities to allow more services to be supplied;
- the provision of more mixed uses at street level;
- the establishment of independent housing companies to perform an important role in residential development;
- the establishment of neighbourhood management;
- the introduction of neighbourhood warden schemes; and
- the provision of cheap and more efficient public transport services.

Clearly, these experiments will succeed 'only if we make cities more physically attractive, more carefully designed, more compact and more cohesive' (Power and Rogers, 2000: 14).

In redesigning inner cities to make them more attractive both socially and economically, it is vital (in the view of Power and Rogers, 2000) to base a plan of action on the voice of existing communities, particularly with regard to safer, cleaner streets, better schools and jobs, and the wider interest of the city. There is also the need to create a secure, cared-for environment which will offer a suitable location for schools, shops, health centres, bus routes and other facilities that make an area come alive. Finally, it is important to create a denser texture in inner city neighbourhoods to compensate for earlier demolitions and population out-migration, to attract middle-income commuters back there, and to link the neighbourhoods to revitalised city centres. This approach was first advocated by Jacobs (1961) in respect of urban renewal in American cities but has rarely been applied to cities in the United Kingdom.

Box 14.1

Social exclusion

Despite major investment initiatives over many years, a local authority, the London Borough of Brent, was still having major problems with addressing social, environmental and economic problems on one of its estates, South Kilburn. The area was in long-term decline, stigmatised, with majority high-rise public housing dating from the 1960s, but was conveniently located. The 2000 Index of Multiple Deprivation showed that the estate was one of the UK's most deprived wards because of poor housing and local environment, children underachieving in school, poor health and high levels of crime. Around a fifth of the residents were unemployed, particularly those aged under 20 years, and two-thirds had an annual income of less than £10,000. The estate comprised a skewed population by both age (particularly older and young people) and race in relation to the rest of the district. The area was defined by its exclusion, and it was clear that there needed to be a re-engineering of active citizenship in housing. The question was how.

The local authority already had existing partnership schemes, through the Single Regeneration Budget and Urban Programme, to address some of the underlying causes of economic and social deprivation and exclusion. Such initiatives had helped establish the residents' resource centre, which has helped more than 1,000 people find work or vocational training through a local labour register, as well as links with local employment agencies and through running training programmes, building training capacity and confidence. The resident participation team supported people on the estate and other initiatives such as community organisations, young people producing their own magazine, funding for health workers and drug advice centres as well as initiatives to cut crime. The private sector has also responded through supporting community events such as the Healthy Living Day and Community Week.

Initial community survey work for the New Deal for the Communities showed that residents' priorities included addressing wholesale housing renewal, unemployment, crime, the environment, and the accessibility and standard of community facilities. The NDC funding is being combined with the millions already spent locally to promote a long-term change in the way of life. For example, it operates in partnership with training and enterprise funding from the Department for Education and Employment to provide computer training on the estate.

The authority felt that keeping tenants involved and engaged would fundamentally involve them, and that they would see real change through housing improvements, which they had identified as a key issue. Tenants were involved throughout in determining the nature of renovation works, such as by being pivotal in option appraisal issues including design, employment, tenure mix, income generation and private finance, security and space standards, increasing energy efficiency, and so on, as well as having the opportunity to learn new building skills, enabling them to gain access to employment.

Residents also felt that estate-based facilities and amenities were important, to regain a sense of 'community'. Shops, social facilities and recreation areas were upgraded and had a major impact in helping to ensure a sense of belonging, ownership and commitment, and making the estate feel more humane to live in. Fundamental organisational and management changes, such as introducing a tenant management board into decision-making, helped regain a sense of estate control. Locally based management with enhanced caretaking responsibilities and new relationships with front-line staff helped to address issues in the early stages and forged new, more personal relationships with residents. The local authority was able to glean good-practice lessons from elsewhere about tackling crime and health issues from the NDC unit within the DETR, finding out both what had worked and what had not.

The authority is currently in the early stages of a ten-year programme, but progress is already notable as tenants are beginning to take charge of their lives.

Neighbourhood management

Neighbourhood management is one of a number of concepts that the Labour government began using in 1998 to tackle social exclusion (Taylor, 2000). The Social Exclusion Unit has identified that the notion of neighbourhood management should include the following:

- a designated local manager for the neighbourhood;
- maximum participation from the local communities and organisations;
- public services at the centre of renewal strategies; and
- targeted assistance from government.

According to Taylor (2000), government sees neighbourhood management as part of a wider means of developing joined-up strategies at local authority level and on the part of central government that are aimed at making things better, not worse. There are still too few neighbourhood management schemes based on the approach that the government is proposing, but her research does show that some current models are service led or top-down.

The government has seen planning agreements as an opportunity to achieve sustainable developments and remove social exclusion. Sustainable developments can be defined as attempts to ensure that the quality of life for communities is improved and maintained for the future. These developments are based on the integration of economic, social and environmental factors, implemented in a balanced and considered way. Macfarlane (2000a) conducted a study into the use of planning agreements to provide employment and/or training for groups of people who are economically inactive either during or after construction or improvements in deprived areas. Planning agreements offer three practical ways to achieve employment and training:

- They involve action to get unemployed people back into the workforce.
- They reduce the risks to existing businesses that can arise from granting new business planning permissions.
- They encourage local recruitment, thereby contributing to environmental sustainability.

Thus, the ability to assimilate people who are economically inactive back into employment is an important part of sustainability. Macfarlane (2000a) suggests three basic effects that favour this concept: social justice; reducing the effect of unemployment and other associated deprivation on the economy; and preventing the skills shortages and wage inflation which occur when the economy begins to expand.

Planning agreements can include a range of initiatives in an attempt to combat social exclusion. These include the following:

- the targeting of employment and training opportunities at those people who are economically inactive;
- a process that ensures that this commitment is inherent in agreements and contracts with agencies;
- on-site recruitment and training facilities; and
- funding to support local training and recruitment initiatives.

Macfarlane (2000a) argues that planning agreements can also be used to reduce the risks to existing local businesses. This is possible by ensuring that new local business planning permissions include as a requirement the training of local unemployed people. Local businesses need to be required to recruit some of their staff from this group of people and so reduce the competition for the small pool of existing skilled workers available in the local neighbourhood. This is in line with the concept of sustainability, as local employers should be responsible for training their workforce.

However, Macfarlane (2000a) found that only 13 per cent of local authorities had used or attempted to use planning agreements for employment matters and that a maximum of 1 to 2 per cent of the planning agreements which are created each year have a local employment element contained within them (see Table 14.2). Table 14.2 indicates that the number of local authorities using planning agreements ranged from 13 in London to 9 in the English districts and 1 each in the English counties, Scotland and Wales. Macfarlane (2000a) believes that policy and legal issues can explain the low usage of planning agreements to reduce social exclusion (see Table 14.3). Many local authorities are apparently unaware that it is not unlawful for a planning agreement to include matters that are in excess of what is necessary, relevant and 'reasonable', as stated in Circular 1/97 (DETR, 1997), 12/96 (Scottish Office, 1996) and 13/97 (Welsh Office, 1997). Court cases have helped to clarify that a lawful planning agreement need only:

- be for a planning purpose;
- be connected to the development site; and
- be logical in practice.

Table 14.2 Use of planning agreements to reduce social exclusion

	English counties		London boroughs		Metropolitan authorities		England		Scotland		Wales		Total
	No.	%	No.	%	No.	%	No.	%	No.	%	No.	%	
Yes	1	5	13	52	3	12	9	6	1	5	1	6	28
Tried	1	5	1	4	2	8	1	0	1	5	0	0	6
Not used	19	90	11	44	21	81	136	93	17	89	15	94	219
Base	21		25		26		146		19		16		253

Source: Macfarlane (2000a)

Table 14.3 *Reasons stated for not using planning agreements for training and employment*

	% or respondents[a]
Never really thought about it	44
It would be beyond our powers	33
It could not be enforced	29
No recent relevant planning agreements	21
It is wrong to use planning agreements in this way	20
It might discourage investment	8
Employment is not a priority for this area	6
It is not our policy/local plan	3

Source: Macfarlane (2000a)

Note: [a] Respondents could select several reasons.

Thus, planning agreements related to employment matters make a measurable contribution towards reducing social exclusion and achieving sustainable development.

Local Labour in Construction schemes have emerged as a way of linking urban regeneration schemes and local residents not in active employment (Macfarlane, 2000b). The concept is to ensure that regeneration programmes generate new investment in terms of dealing with the physical deterioration in both commercial properties and housing. It is also important, however, to ensure that social exclusion within the neighbourhood is tackled with the use of training, employment opportunities and skill shortages.

Self-help in housing management

Social exclusion within the context of housing is often associated with households' inability or unwillingness to undertake comparatively simple repair and maintenance tasks. Williams and Windebank (1999) conducted research into three types of self-help activity:

- household work and DIY – where the work was completed by members of the household, or on an unpaid basis where households helped each other;
- mutual aid – where the work was completed by someone outside the household such as a neighbour or voluntary group; and

- informal exchange – where goods and services are exchanged for money or gifts which are outside the formal structure of tax, social security or labour laws.

Their research found that 47 per cent of households were unable to complete what they saw as basic tasks such as outdoor and/or indoor painting and gardening. Fifty-six per cent had not had the outside of their property decorated in the past five years even though 90 per cent of households wanted to have this work done. Of the tasks that were completed, 85 per cent were conducted by self-help: 76 per cent by DIY, 4 per cent by mutual aid and 5 per cent by informal exchange.

The households' ability to complete tasks was found to be related to the job status of the household. Households with a member of the family with a job were more likely to complete as many tasks as households without employment. In addition, Williams and Windebank (1999) found that there was also a strong North–South divide whereby households in the South completed smaller proportions of tasks than did households in the North.

Six key factors were identified as barriers to household completing self-help activities (Stenhouse and Henderson, 2001):

- economic capital: lack of money and equipment to complete the tasks;
- social network capital: lack of people to approach for help;
- human capital: lack of skills, confidence or the physical abilities to complete tasks;
- institutional barriers: fear (especially by unemployed people) that if they engage in mutual help with others, this may get back to the authorities and consequently they could lose their benefit entitlement;
- environmental barriers: owing to the negative perception of the neighbourhood, some households may decide not to communicate with their neighbours; and
- time capital: lack of time on the part of employed people, but also among the unemployed, who feel pressured to use their time to find employment.

Conclusion

Social exclusion and social inclusion are concepts that are difficult to quantify and define. As indicated above, both terms can have different interpretations. This is further compounded by the fact that the individuals who may be identified by academics and policy-makers as socially excluded may not define themselves in that manner. However, it is clear that social exclusion is wider than just poverty and is concerned with processes as well as outcomes. Therefore, policy-makers must continue to attempt to remove the factors which can result in people being excluded from mainstream society.

Further reading

Anderson, I. and Sim, D. (2000) *Social Exclusion and Housing: Context and Challenges*, London: Chartered Institute of Housing. An excellent introduction to the key discussions surrounding social exclusion/inclusion and the impact on housing policy and practice.

Clapham, D., Evans, A. *et al.* (1998) *From Exclusion to Inclusion: Helping to Create Successful Tenancies and Communities*, London: Hastoe Housing Association. Research commissioned by seven housing associations; provides useful information about registered social landlords and their experiences of social exclusion in practice.

Madanipour, A., Cars, G. and Allen, J. (1999) *Social Exclusion in European Cities: Processes, Experiences and Responses*, London: The Stationery Office.
A useful introduction to the processes which result in social exclusion within Europe, and possible solutions.

Parkinson, M. (1998) *Combating Social Exclusion: Lessons from Area-Based Programmes in Europe*, Bristol: Policy Press. An informative report that explores the experience of social exclusion and successful strategies that have been used in Europe.

15 **Housing and community support**

In relation to housing, community support is high on the social and political agenda in the United Kingdom. In recent years it has been argued that the growth in the use of private and independent agencies by government could result in the gradual decline of the provision of community support within the public sector. Within this context, this chapter:

- defines the term 'community care';
- discusses developments in its application;
- considers housing for special needs;
- examines community support in practice; and
- analyses the Supporting People programme – the most recent policy initiative concerned with community support.

Definitions of community care

Until recently, community support was known as community care. There are a number of definitions of the term 'community', which has evolved from meaning 'local' in the nineteenth century to a definition based on a more middle-class notion of providing a service to the community. The term is never used in a disparaging sense or viewed unfavourably (Williams, 1983).

Despite the number of meanings there are two elements which remain important in the defining of community from within a social care context: first, the focus on local social relations within a geographical area; and second, the concept of belonging. However, being a part of a geographical neighbourhood does not mean that one is necessarily a part of the wider community, because of the factors of inclusion and exclusion (Bulmer, 1987). Cowan (1999) points out that the physical boundaries of what constitutes a community have been increasingly replaced by symbolic boundaries that distinguish the 'insiders' from the 'outsiders', for example in terms of ethnicity.

There is a requirement to define 'care' in relation to the concept of 'need' (Cowan, 1999). Doyal and Gough (1991) identify the basic universal needs as having good physical health and having personal autonomy. They regard these basic universal needs as constituting the preconditions for being able to participate in social life. Bradshaw (1977) defined the different types of needs as follows:

- normative need – defined by experts;
- felt need – equating to want;
- expressed need – the effective demand for a service; and
- comparative need – according to the characteristics of those receiving a service.

It is argued that the notion of 'care' has no inherent connection with the term 'community'. This is due to the fact that women tend to provide much of the care in the community in an informal nature and often in private. Care comprises physical tending, material and psychological support, and a more general concern for others' welfare (Bulmer, 1987).

Cowan (1999) sees community care policies as being based on the concept of familialism whereby it is assumed that caring is a 'natural' role for women but an 'unnatural' role for men. State policies are based on the concept of not interfering with traditional family patterns of caring. Thus, those in need, such as older people, become dependent on family members, especially female family members, when they are no longer capable of looking after themselves.

Development of modern community care

Community care became an official aim of government in the 1960s. Two government health and community care plans proposed growth both in the number of personnel such as health visitors and district nurses and in the number of sheltered housing schemes. The community care plans were aimed at groups such as people with mental health problems, people with learning difficulties, people with a physical disability and older people. In addition, pressures were placed on health service institutions to reduce the numbers of users, while local authorities were encouraged to expand facilities such as sheltered housing schemes (Jones, 1994).

People with mental health problems

The first use of the term 'community care' was in the Royal Commission on Mental Illness and Mental Deficiency report in 1959. The ensuing Mental Health Act 1959 saw the beginning of the closure of the large mental hospitals, while procedures for non-compulsory admissions, compulsory treatment and detention were made clearer (Barnes, 1997). The number of mental hospital beds was reduced by 50 per cent in 1961.

By 1974 there were 60,000 fewer residents in large mental hospitals than in 1954. However, there were insufficient services in the community, and so many ex-patients simply disappeared from the official statistics since no one followed up their progress (Murphy, 1991). The result was a disaster with long-term repercussions. The Department of Health and Social Security published a report in 1975 which admitted that the original 1960s plans for community care had proved to be unsuccessful, with a rise in admission rates, the continuing use of large mental hospitals and poor staffing levels.

People with learning difficulties

The Mental Health Act 1959 was seen as the basis for community care for people with learning difficulties. Local government health departments were expected to provide facilities such as residential, day and family support services. Progress in delivering a service to people with learning difficulties was slow and admissions remained high owing to a number of factors, which included the breakdown of family and community support, the lack of good information for parents and carers, and the low priority given to people with learning difficulties (Cowan, 1999).

People with a physical disability

The Chronically Sick and Disabled Persons Act 1970 with the Local Authority Act 1970 legally required that local authorities should gather information on the scale of problems for people with disabilities. While this did not result in major changes for people with disabilities, it did highlight a change in local authority and voluntary agency perspectives and in public attitudes. However, the Act was under-funded and so many local authorities simply developed sample surveys, despite the need to

redress the neglect which people with disabilities have endured
(Midwinter, 1994).

Older people

The development of new social services departments as a result of the
reorganisation of local authorities after 1971 resulted in older people
being given a higher priority by social workers. This period saw high
spending on social services in terms of resources and staff. Local
authorities were required to offer a greater variety of domiciliary care,
and by 1975 there were 17 home helps, and 15,250 meals per annum for
every 1,000 people over the age of 75 (Cowan, 1999).

Housing for special needs

Because of its physical design, the absence of organisational skills in
running a household, and the need for specialist care, mainstream
housing is often unsuitable for people with mental health problems,
people with learning difficulties, people with physical disabilities and
older people. On the other hand, there has been much dissatisfaction with
many of the traditional larger institutions which have provided
accommodation and care in the past for persons with special needs.
Because of these deficiencies, community care has evolved as a means
of enabling people with special needs 'to enjoy as near "normal" a life as
possible, recognising that this is what most people want. A related benefit
lies in people having contact with, and supplied by, the community at
large' (Gibb *et al.*, 1999: 223).

However, in the 1980s the funding system was not only tailored towards
the provision of residential care (even for those for whom it would have
been inappropriate), but also exceedingly expensive (Gibb *et al.*, 1999).
In reviewing community care policy, the Griffiths Report (1988)
recommended ways in which the organisation and funding of community
care could be improved, while the Wagner Report (Secretaries of State,
1989) focused on the mechanisms that could be employed to enhance
quality and choice in community care.

The Wagner Report, published as the 1989 White Paper *Caring for
People: Community Care in the Next Decade and Beyond* (Secretaries of
State, 1989) identified six key objectives for social care as follows:

1 provision of services for people at home;
2 provision of services for carers;
3 the assessment of need and the provision of individual care packages;
4 a mixed economy of public and independent sector care and to encourage social services to become enablers;
5 a clarification of responsibilities of agencies; and
6 ensurance of better value for money.

Cowan (1999) states that the White Paper expected health authorities and family practitioner committees to collaborate with social services for assessment and preparation of community care plans. GPs and community nursing staff were to be fully involved in community care development. The White Paper proposed that local authority social services departments set up complaint procedures and institute a system of inspection and registration for residential homes in all sectors. The National Health Service and Community Care Act 1990 'emphasised the key role of social services in the co-ordination of community care, involving needs assessment and planning, and the promotion of joint work and networking among agencies, particularly health, social services and housing' (Gibb *et al.*, 1999: 223–24).

The need for joint strategic planning was emphasised further in Circular 10/92 (DSS, 1992), which stated, 'housing authorities and social services authorities are asked to co-operate fully in the planning and assessment procedures'. The circular also identified the need for training to increase knowledge of services provided by the other partners in community care. The circular stated that 'housing authorities should seek opportunities to discuss with social services authorities and those working in the voluntary sector the availability of joint training to build mutual understanding and confidence and to provide a basis for working together'. Clearly, the ability to work together will be improved if staff is made aware of the problems, roles and working methods of the other partners delivering community support. The government was also committed to close most long-stay institutions and replace them with 'more individually tailored living arrangement within the community' (Gibb *et al.*, 1999: 224). Arguably, appropriate housing and support packages would enable even people with severe disabilities to have more independent and satisfying lives.

Unfortunately, because of funding problems it has proved difficult to adopt this approach. Since there is a very wide range of needs, it has not been easy to design a unified system of care and a procedure for rationing and prioritising the use of resources. The varying extent to which families, neighbours and others exercise care on a voluntary basis has to be taken into account, and the degree to which housing managers enable vulnerable people to remain independent for as long as is desirable should also be considered. Housing agencies might also help in this respect by providing specially designed housing dedicated to the needs of specific care groups, but the provision of this form of facility varies substantially in relation to individual needs and location.

To help identify housing and care needs and – in order of priority – the services required, the community care planning system has become central to the task of providing accommodation and supporting care users. The ability of the system (involving agencies such as health, social services and housing) to function effectively is of course dependent upon capital and revenue resources – the former being derived from the budgets of local authority social service departments and housing association HAG/SHG funding, and the latter being covered by housing benefit, special needs management allowances (in England and Wales) and special needs allowance packages (in Scotland). If these forms of assistance are insufficient to match rents (which can often be very high since they have to cover both running and care costs), recipients will be expected to make a contribution themselves, seek assistance from social services for some elements of care, or attract help from charitable organisations.

Since this system is both complex and fragile for housing providers managing an existing development or contemplating undertaking a new development – in terms of both capital and revenue considerations – there are often 'long periods of uncertainty when the future financial security of projects cannot be assured' (Gibb et al., 1999: 229). Financial uncertainty has been exacerbated since the government is no longer willing for housing benefit to cross-subsidise the cost of care in respect of sheltered housing. By the late 1990s the principle was 'established that housing benefit should cover only the cost of providing accommodation, including the cost of managing it, and not the cost of care' (Gibb et al., 1999: 230).

Community support in practice

Community support reforms have tended to concentrate on the choice, dignity and independence of users and carers. However, the emphasis is placed on those who have been assessed as having the greatest need and a nationwide strategy has not been developed. Although people with a high level of need must be given a high level of priority, it is important to ensure that people with lower levels of needs are not neglected in order to prevent the potential consequences if those needs are not met. The majority of residents in supported housing live in homes owned by registered social landlords or local authorities. Support is normally provided directly by the landlord or statutory partner organisations.

Cameron *et al.* (2001), in their study of the Crossing the Housing and Care Divide Programme, which was launched in 1995 by the Housing Corporation and Anchor Trust, looked at the interface between housing and personal care systems with particular reference to older people. The general aim of the programme was to:

- enable housing to become a more integral part of community support;
- provide a high-quality and cost-effective service delivery;
- develop more inter-agency working; and
- increase the involvement of service users in the planning, delivery and monitoring of services.

Cameron *et al.* (2001) found that agencies that had a history of working together had developed strategies that were effective, whereas organisations that did not have a record of working together found it more difficult to develop effective strategies. Difficulties were particularly noticeable with agencies, particularly local authorities, working in partnership with the health service. The study found that a lack of understanding about the roles of different professions and agencies often prevented such partnerships working well together. There tended to be no clear definition of carers and users, and a lack of clarity about whether the programme was aimed at the carer or user. Furthermore, the success of some projects in achieving user involvement could be linked to the personal attributes and skills of the appointed staff such as local knowledge, the respecting of the users' views, persistence and commitment. It was found that often, involvement was mainly tokenistic (Cameron *et al.*, 2001).

The problems associated with the shift from long-stay institutional care to community settings in general, and ordinary housing in particular, are best illustrated below with regard to people with learning difficulties. (Similar problems relating to the elderly are examined in Chapter 16.)

Community support for people with learning difficulties

Harker and King (1999) found in their research of housing and support for people with learning difficulties that commissioners and service managers were very much aware of the limitations of the available housing and support. However, there was concern that there was insufficient opportunity for choice during a time in which the number of people being assessed as having learning difficulties is increasing, coupled with growing expectations about independent living.

Until recently the main options for people with learning difficulties were residential care or hospital. In 1999 the government announced that it was phasing out remaining long-stay hospitals by 2004. Currently, owner-occupation is the tenure of about 68 per cent of all households, and renting (in both the private and social sectors) accommodates approximately 32 per cent of all households, bur Figure 15.1 shows that only 7 per cent of people with learning difficulties were owner-occupiers or tenants, while 53 per cent (approximately 26,000 adults) were still living with their families. There is evidence that some people with learning difficulties had problems finding options, especially those with complex needs and those showing challenging behaviour. It was also evident that there are some people in residential care who could be living more independently if the correct form of support were provided.

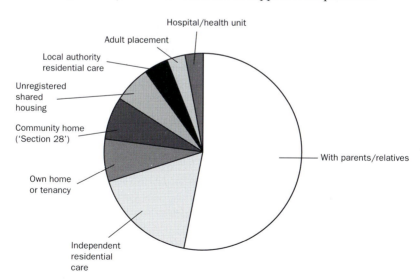

Figure 15.1 *Where people with learning difficulties are supported and housed*
Source: Harker and King (1999)

Harker and King (1999) found that some authorities did not have a strategy in place in order to identify the needs of users and carers, the quality of service delivery and the areas of concern. There was also a lack of benchmarking of service delivery, providers and performance to seek out value for money. The study also found a number of older carers who were coping under stress and in difficulty. It was found that without a structured plan for future caring needs there was a lack of confidence and a high level of anxiety. There was a lack of information for the general public regarding the range of housing or residential accommodation with support.

Concerns with regard to the use of ordinary lettings without adequate supervision that could lead to isolation for the tenant and the risk of harassment were identified as a general problem. In addition, the difficulties in forming effective partnerships with health authorities and trusts were seen as a problem as there was a need for better definition of roles within the partnership. It was found that partnerships with social services and housing departments failed owing to operational weaknesses, and because systems failed to integrate but ran parallel, with old views about the roles of each department tending to persist. Community support was found to be a marginal issue in many housing departments.

A major effect of the National Health Service and Community Care Act 1990 has been the rapid growth in the number of vulnerable people who are living in the community, which can lead to isolation and/or harassment. This can be due to a number of factors such as language and/or cultural barriers, lack of transport, or fear of crime. Problems can also arise when individuals who have spent some time in institutions find it difficult to develop social skills, while bereavement is another cause of individuals becoming isolated.

Dean and Goodlad (1998) studied befriending services that have been developed by voluntary organisations as a way to address the feeling of isolation by offering opportunities for social interaction and a sense of being a part of a community. The study found that 62 per cent of voluntary organisations received funding from social services departments and 22 per cent received funding from health authorities. Only 4 per cent charged the users of their befriending services directly.

Research found that there was a risk of social exclusion based on the need to select and reject some potential users of the services. Selection was based on rejecting people who were thought to be unsuitable or

Box 15.1

Community support: low-intensity support services

Low-intensity support includes services such as emotional and practical support for people with learning disabilities and mental health problems, and support with housework and shopping for frail older people. The typical arrangement is that a support worker provides a few hours of support each week in the client's own home. Non-priority community care groups such as women escaping domestic violence and homeless people may not qualify for full social services assistance but can benefit from low-intensity support.

The support is delivered by using a mixture of registered social landlords and voluntary organisations and is aimed at attempting to help people remain in their homes. It can prevent a crisis that could result in vulnerable people being re-admitted into care institutions and can often help to improve people's quality of life and develop social inclusion in the local community.

KeyRing is an example of an agency that provides low-intensity support services and is one of the pioneers of mutual support. It is a voluntary organisation with 22 living support networks for 170 people with learning disabilities who have basic self-care skills, but also require continuing or irregular support. The network usually contains approximately nine people who know each other's addresses and telephone numbers. The networks are co-ordinated by community living workers, who spend 10–12 hours a week with members and live in the same area. Telephone support is available during the evenings and weekends.

KeyRing also assists tenants with developing links with the local community. Before a network is established, the agency will audit a community. It helps tenants to develop defensive strategies and supports those who have experienced harassment and wish to take further action against the perpetrators. Users of the KeyRing service reported that there was an improvement in self-esteem, confidence and health.

Home-Link is a scheme developed on the KeyRing model to assist people with mental health problems in general needs housing. It provides permanent housing and low-intensity practical support that includes mutual support networks. The aims of the scheme include:

- the integration of people into the community;
- the promotion of independent living and reduction of isolation;
- the provision of increased options in the range of supported accommodation;
- the prevention of deterioration in people's mental health condition; and
- the delivery of flexible support to people in permanent good-quality accommodation.

Among its objectives are:

- to foster the development of social links and contact with outside organisations through a network;

continued

- to allocate or assist in securing good-quality ordinary housing;
- to develop good working partnerships between health, housing, social services and voluntary organisations by using existing networks to deliver integrated housing and support services;
- to provide continuous support; and
- to provide advice and assistance to facilitate access to and take-up of other resources and services.

The housing department allocates secure accommodation units, usually near to each other. Many properties are connected to a 24-hour lifeline service. The users of the service are allowed to contact their workers at any time between visits for advice and/or reassurance. The workers spend 5–6 hours a month with each user and provide help with matters such as budgeting and paying bills. The scheme does not use workers who are mental health professionals in order to ensure that the focus is not on the users' medical condition.

Users of the Home-Link service reported a reduction in anxiety, an improvement in their mental health and that they had become and were able to remain independent.

'difficult'. However, there were some voluntary organisations that were successful in providing support or befriending relationships for challenging users (see Box 15.1). All the befriending organisations were found to use volunteers, as they were cheap and generally have the skills and attributes required. Many organisations had fewer men than women volunteers, and in two out of five schemes more than three-quarters of befrienders were women. In a number of cases some users of the services eventually became volunteers for other users. It was found that the befriending services helped participation in the community, especially in relation to creating social links, encouragement to use local services and facilities, and the development of wider social networks (Dean and Goodlad, 1998).

Children with disabilities

Oldman and Beresford (1998) found that the housing needs of children with disabilities were wider than the simple access issues associated with adults. Many parents were concerned that their children were not getting the exercise or therapy needed as there was often insufficient space within the home for wheelchairs, or walking and/or standing frames, while a quarter of families stated that they had a lack of space for storing equipment. Access was a common problem for a quarter of families as stairs were often difficult for parents and the child with a disability, as often the parent had to carry the child up and down the stairs several times during the day. The majority of homes lacked a downstairs toilet and the bathroom was seen as the most inconvenient room to use. Table 15.1 shows that only 19 per cent of families that had undertaken adaptations to their properties had eliminated all unsuitable factors, while 21 per cent reported that despite the adaptations to their homes they had high levels of housing unsuitability. Many homes lacked adequate gardens and a third of families were affected by limited play space. Safety issues with regard to kitchens, stairs, bathrooms and areas outside the home were also of concern to parents. The study found that families with children with a disability were more likely to live in rented accommodation and to be on low incomes than other families. Some families were found to be living in overcrowded accommodation. Just under a fifth of families were found to be living in cold, damp homes that were in poor repair and/or situated on unfriendly estates (Oldman and Beresford, 1998).

Table 15.1 *Housing adaptations for disabled children and their families*

Have adapted	Number of reported factors making home unsuitable			
	None	*1–3 factors*	*4–6 factors*	*7 or more factors*
Yes (%)	19	46	14	21
In process	11	33	44	11
No (%)	27	36	20	17

Source: Oldman and Beresford (1998)

The Supporting People programme

The government's Supporting People programme was scheduled to be implemented in 2003 and aimed to bring together the previously disjointed funding and policy frameworks for all housing-related support services for vulnerable people. Supported housing services have developed over a number of years and currently provide a range of services which have been developed in response to an assorted range of needs. However, the provision of supported housing in the past has been:

- separate from supported provisions to other local priorities and strategies in probation, social services, health and housing;
- driven by the availability of capital and revenue funds and not by the analysis of local needs; and

- linked to outdated models of 'hostel'-type provision which no longer meet clients' needs.

Despite the lack of strategic nationwide provision there has been the development of a very effective and valuable pool of supported housing which provides much-needed assistance to vulnerable people. The aim of the Supporting People programme is to put the funding and development of the sector on a more secure and co-ordinated basis. The programme is therefore intended to improve the quality and effectiveness of these support services in a range of ways. These methods include focusing provision on local need through the introduction of a more systematic and strategic process to assess the needs within and across local authority areas and supply relevant support services. A further aim of the programme is to improve the range and quality of services by promoting the development of a wider range of support services which are focused on the needs of the people receiving support and based on informed good practice. In particular, the link that existed in the past of support services to tenure will be broken so that more floating support may be introduced where appropriate. This new single budget will be administered by local authorities and will bring together all the disparate funding streams, including housing benefit's support element, and provide a coherent framework of support, irrespective of type of accommodation or tenure. Those who will be affected by the Supporting People programme include older people, people with mental health problems and young care leavers.

The government has clearly defined what is meant by 'support' service as provided within the programme. The new support services are intended to enable people to remain independent, or establish themselves independently in the community, whether in a tenancy, their own home, or specialised supported housing. The aim is to assist individuals to attain and retain a stable place to live in order to enable many vulnerable people to rebuild their lives. Enabling older people to remain in their homes for as long as they wish is a key element of the government's housing strategy for older people.

The government also intends the new support services to be part of the range of preventive strategies being developed by local authorities aimed at giving early help to avoid the need for acute or crisis care. The programme is intended to supply additional resources to community care packages or to subsidise local authorities' current community care budgets, although it can operate in parallel with them. Support within this programme is primarily delivered to people whose needs do not require intensive personal care. It is intended to be an integral element of the emerging government strategies such as combating social exclusion, and for the development of housing services in line with the Housing Green Paper (DETR/DSS, 2000). The provision of housing support can play an important part in the delivery of each of these programmes and each local authority will be expected to identify how best to ensure that the provision of support and supported housing under the Supporting People programme can complement them. Supported housing or support services may form an element of packaged care and support provided to a vulnerable person, with, for example, a short programme of intensive counselling or medical support being provided by a more specialist agency in parallel with longer-term provision by the Supporting People funds.

It is intended that the full ranges of vulnerable groups receive support services. Older people in sheltered accommodation form a large part of the programme and are a key priority for continued support. In addition, the government intends to ensure that current patterns of provision, particularly support for more marginal groups such as homeless people, people who misuse drugs or alcohol, and people under probation service supervision, continue to be an important part of the programme. Identifying and addressing this range of needs will be an essential part of the administrative framework at a national and local level.

The programme will seek to encourage more co-ordination with NHS bodies in commissioning and funding services, recognising that everyday support services can play a vital role in aspects such as the reduction of emergency admission of people with mental health problems, contributing to the health and well-being of older people, and speeding up discharge from hospital to home.

The government also intends that the Supporting People programme should aim to integrate support with wider local strategies, particularly within health, social services, housing, neighbourhood renewal and community safety. In addition, the programme will be monitored and inspected in a more structured way, including integration with the Best Value regime in order to ensure its quality and effectiveness. The programme will also include changes in the arrangements for funding and managing the sector which will lead to effective decision-making and cost-effective administration.

The Supporting People programme will provide generic support services by skilled staff, who in the main will provide very different services from staff trained to provide personal and medical care.

Conclusion

While Cowan (1999) suggests that the users of community support are more likely to be those who were economically and socially excluded in the past, it is also the case that recipients of care services within the community often become subject to social exclusion. It is important therefore that the new Supporting People programme is funded sufficiently and monitored closely to ensure that the users of community support are receiving the best service possible. Also, strategic planning across the country will be necessary in order to prevent the current situation, whereby the level of support the user receives is dependent on where they happen to reside.

Further reading

Bornat, J., Johnson, J., Pereira, C., Pilgrim, D. and Williams, F. (1993) *Community Care: A Reader*, Basingstoke: Macmillan/Open University Press. An excellent book that examines a wide range of issues and perspectives.

Cowan, H. (1999) *Community Care, Ideology and Social Policy*, Englewood Cliffs, NJ: Prentice Hall. A useful introduction to community care strategies and the underlying political assumptions that help to define community care policies.

Daley, G. (1996) *Ideologies of Caring: Rethinking Community and Collectivism*, 2nd edn, Basingstoke: Macmillan. A classic feminist critique of the community care model.

Morris, J. (1996) *Encounters with Strangers: Feminism and Disability*, London: Women's Press. An analysis of the disabled people's movement from a disabled feminist perspective.

16 Housing and the elderly

According to the OPCS (1992a), there were 9.8 million pensioners in Great Britain in 1991 out of a total population of 56.2 million. Of this total, 11.7 per cent were aged from 60 or 65 (in the case of women and men respectively) to 74, 5.5 per cent were aged from 75 to 84 and 1.5 per cent were aged 85 and over. In the group aged 75 and over, 4.6 per cent were women and 2.4 per cent were men. Overall, the percentage of households containing pensioners had increased from 6.8 million to 7 million, and the total number of pensioners had increased from 9.1 million to 9.4 million (1981–91). Of particular interest to all those concerned with the elderly was the increase in the number of pensioners living alone, the proportion increasing from 14.2 per cent in 1981 to 15.1 per cent in 1991. Moreover, another 9.7 per cent of households in Great Britain in 1991 consisted of two or more pensioners and no other persons. Undoubtedly, the 2001 Census will show that the total number of pensioners in Great Britain has risen substantially in the period 1991–2001, and particularly the proportion aged 75 and over.

While a substantial proportion of the growing number of pensioners will continue to satisfy their housing needs in the conventional housing market and will not require or seek any form of care, undoubtedly there will be many hundreds of thousands of elderly people who will require a varying degree of care either within the community or in residential homes. In focusing on these needs and associated policy, this chapter:

- examines household formation and the problem of disability among the elderly;
- considers household mobility;
- comments on the funding of residential care;
- explores the 'staying put' option;
- analyses future housing need; and
- concludes by discussing the costs of residential care.

Household formation and the problem of disability

Since 1945 there has been a growing trend for older people to live either with a spouse or alone, and not with other relatives or friends (Grundy, 1992). Because of continuing widowhood and divorce, significantly more older women than older men have lived alone, but the trend is that more men are already living alone, and more are likely to live alone, into old age (CSO, 1994). Men have tended to remarry if widowed, so that about 77 per cent of older men are living with a spouse at the age of 65–69, and more than 20 per cent are living with a spouse (maybe a different one) at the age of 90. By comparison, less than about 55 per cent of women are living with a spouse at the age of 65–69, while virtually none still have a spouse at the age of 90 (Grundy, 1992). At the age of 65–74 about 19 per cent of men, 37 per cent of women and 29 per cent of all in that age group were living alone in 1992; by the age of 75, 30 per cent of men, 59 per cent of women and 48 per cent of all that age group were living alone (OPCS, 1992a). This means that 29 per cent of the homes needed by those in the still active age group of 65–74 should be one-person dwellings. This trend continues into the age group 75+, with 48 per cent living alone. Two reasons for this trend are the increase in the rate of divorce and the reduction in the remarriage rate of divorced persons (Ermisch, 1990). As a cohort ages, so more of its members may be thought to need an institutional home or to want to live with friends and relatives. However, such moves seem not to be borne out by the facts in the majority of cases. Projections show an increase in women under pensionable age living alone and a very significant increase in the percentage of households occupied by men under pensionable age living alone (CSO, 1994). This tendency has been growing since 1981 and may mean that by the year 2017 there is a considerably larger percentage of single male elderly households. A significant message to housebuilders in view of this demographic change is that more one-person homes will be required, not least, presumably, because more old men have become able to cope on their own, learning housekeeping skills that in a previous generation were considered to be exclusively women's work.

In 1987, 61 per cent of households with heads in the Third Age (from 50–74) were home-owners, with 80.5 per cent living in houses (Victor *et al.*, 1992); and it can be estimated that those without severe medical problems will continue to live as owner-occupiers of houses into their Fourth Age (75+), until death or frailty intervenes. Government policy, as set out in *Caring for People: Community Care in the Next Decade and*

Beyond (Secretaries of State, 1989), encourages people to remain in their own homes as long as possible. There is unlikely to be a relatively large proportion of older people living with their kin. Grundy (1993) showed that in 1980 only 16 per cent of elderly people lived with their children; while in 1985 45 per cent lived with a spouse and 19 per cent with others. The evidence is that elderly people prefer to live independently, rather than with relatives (Grundy 1993), but their health may not allow such independence. Yet only 1 per cent of 65- to 69-year-olds were living in institutions, and 9 per cent of such men and 14 per cent of such women lived with relatives or friends; but for those aged 85 and over, 20 per cent were living in institutions and 25 per cent of males and 20 per cent of females lived with relatives (other than a spouse) or friends (Grundy, 1993). In the period 1992–2007 there will be a 50 per cent increase in the number of those aged 85 years or more (Grundy, 1993).

An indication of the kind of housing that may be in demand, or may be needed, can be deduced from the fact that while the life expectancy rate for men and women at the age of 65 in the United Kingdom in 1988 was 13.7 and 17.6 respectively, the disability-free life expectancy for such people was 7.6 and 8.8 respectively (Robine and Blanchet, 1992). Women live longer than men, but older women generally have a higher rate of morbidity than older men (Victor, 1991; Grundy, 1987a). The disability level increases with age: while 1.8 per cent of women aged 60 to 74 suffer the most severe level of disability in Great Britain, from the age of 75, 10.2 per cent of women and 6.4 per cent of men suffer at this level; and for women aged 60–74 the general disability level is 26.4 per cent compared with 63.1 per cent at the age of 75 and thereafter (Martin *et al.*, 1988). For men aged 75 and over, the general disability level is 53.3 per cent. Such statistics can indicate the number of homes in any one area that may need to be adapted for the use of infirm older people. However, statistics need to be used with care because experience in the United States has shown a lack of consensus in the measurement of the activities of daily living in surveys (Wiener, 1990). It is functional disability in the activities of daily living that is of most concern to specialist builders, such as the inability to wash, dress, move around and up and down stairs and to prepare food. Victor (1991) showed that in 1985 while 9 per cent of those aged 65 had difficulty in bathing or showering, and 8 per cent were unable to cook a main meal, only 2 per cent could not get around the house and to the toilet. Victor estimated that by 2025 there would be 226,000 elderly people in Great Britain unable to get to the toilet and 1 million who had problems with bathing or showering and getting up or down stairs (Victor, 1991). Even out of a total population of 55 to 60 million, 1 million is a significant number of older people requiring special accommodation.

Household mobility

The ability to move house depends first and foremost on income and realisable assets. In aggregate, pensioners' incomes in 1991 were 35 per cent higher in real terms than in 1981, their investment income had doubled in value and the income from occupational pension schemes had nearly doubled (CSO, 1993). But relatively, pensioners were generally income-poor. Whereas average adult weekly earnings amounted to £343 (gross) in 1993 (*Employment Gazette*, 1993), the basic weekly pension in 1993/94 was only £57.60 for a single person and £92.10 for a couple. Because of these comparatively low sums (and taking into account other income and savings), more than 1.6 million people aged 60 or over received income support in 1983 – support ranging from £63.95 per week for a single pensioner aged 60–74 to £99.25 for a married couple in the same age group (with higher payments for persons aged 80 and over) (DSS, 1993). Nearly 60 per cent of pensioner households were thus dependent on state pensions and benefits for at least three-quarters of their income by the late 1980s. It was not surprising, therefore, that where the head of household is aged 65 or over, a disproportionate amount of income is spent each week on housing, fuel and food – 62 per cent in the case of pensioners living alone compared to 39 per cent by all households (CSO, 1993), while, according to an NOP survey, 21 per cent of retired people who owned their own homes had incomes so low they often had difficulty in paying for essential items (British Gas, 1991).

Although income-poor, many pensioners are asset-rich. In 1998/99 about 76 per cent of those aged 55–64, 74 per cent of those aged 65–74 and 61 per cent of those aged 75 or over were owner-occupiers, and most owned outright (Table 16.1). In 1999, the average house price in the United Kingdom was £93,500 (£9,400 more than in 1998), yet at the same time more than a quarter of elderly couples were living on less than £135 per week (Jones, 2000). Equity release, however, would have provided the means by which their incomes or disposable wealth could have been enhanced. Although there are many schemes that facilitate the release of equity (see Box 16.1), these tend to involve home-owners *either* taking out a mortgage which provides the borrower with a cash lump sum (or a

Table 16.1 *Housing tenure of the elderly, United Kingdom, 1998/99*

Tenure	Age group		
	55–64	*65–74*	*75 and over*
Owner-occupied, owned outright	44	65	57
Owner-occupied, with mortgage	32	9	4
Rented from local authority	15	18	25
Rented from housing association	4	4	8
Rented privately, unfurnished	1	—	—
Rented privately, furnished	4	4	6

Sources: ONS, General Housing Survey, 2000; NISRA, *Continuous Household Survey*, 2000

fixed monthly income), *or* selling part or all of the property to generate cash.

While there is little empirical evidence of the total number of elderly wishing to move, the relatively high level of demand for owner-occupied housing in the retirement areas of Britain, the substantial increase in market-led private sector sheltered housing development in the 1980s and very long local authority and housing association waiting lists for sheltered housing all give an indication of the magnitude of potential mobility. In terms of housing economics, the option to move undoubtedly has its merits. It releases large (formerly under-occupied) dwellings for family use, it unlocks assets and liberates many old people from perceived poverty, and the provision of an adequate supply of small dwellings or sheltered housing gives elderly people an alternative to sharing accommodation with their children, other relatives or friends, or moving into very costly residential care. There is clearly a wide range of options available to the elderly wishing to move.

The most active and independent elderly might wish to move within the normal housing markets. While these elderly people may not need any special accommodation, they face the same constraints on mobility as are experienced by other households. With regard to owner-occupation, insufficient income and/or financial and physical assets could impede mobility. Nevertheless, a sizeable proportion of elderly owner-occupiers, and particularly those without an outstanding mortgage on their existing homes, should find it easier in financial terms to move than younger households – not least because most would probably wish to trade down and at the same time to release equity on their former homes. Even elderly owner-occupiers wishing to trade up would find mobility feasible if they were able to take advantage of a shared ownership (part-buy/part-rent) scheme such as that facilitated by the Joseph Rowntree Housing Trust. Elderly council tenants, however, are relatively immobile – normally being little helped by national mobility or local transfer schemes – while mobility among private tenants is impeded by the introduction of assured tenancy and the reintroduction of market rents under the Housing Act 1988.

Box 16.1

Equity release

Equity release is a means by which owner-occupiers who own their homes outright can gain access to a cash lump sum or a flow of cash to 'compensate' them for having much of their wealth locked up in property. At the beginning of the new millennium, mortgage lenders and insurance companies such as Northern Rock and Norwich Union offered a number of schemes designed for the elderly that facilitated the release of equity; for example:

1   A *standard home equity release mortgage* provided a home-owner with a lump sum (usually equivalent to 20–50 per cent of the value of the property) repayable at a fixed rate of interest for the period of the loan.
2   A *capital access plan* also enabled an owner to release a percentage of the value of their property in the form of a lump sum, but there would be no monthly repayments. The owner could continue living in their house until they died or moved into long-term care, when the loan and interest would be repaid from the sale of the property.
3   A *home income plan* would enable an owner to use an equity-based loan to buy an annuity which would provide a guaranteed income for life (net of fixed interest payments on the loan). The loan would be repayable on death from the sale of the property.
4   A *home reversion scheme* would provide an owner with a lifetime income, or a discounted lump sum and lifetime income, in return for the sale of part or all of the property.

The demand for housing varies not only by income and by household structure, but also by locality. Elderly people move from one home or area to another. They are more dominant as a proportion of the whole population in some local authorities than in others. For example, virtually all the districts along the southern coast of England, the western coast of Wales and most of the coast of East Anglia had in 1981 the highest quintile by percentage of the population of pensionable age (Coleman and Salt, 1992). These were mostly people who had retired to such areas.

For 1991 OPCS data show that Christchurch, in Dorset, was the district with the highest percentage of residents of pensionable age (at 34.6 per cent), with Tamworth, Staffordshire, the lowest (at 11.6 per cent). The New Towns, with the immigrants of the 1960s, tend to have the lowest percentages of older people at present, with retirement areas on the coast having the highest percentages. For example, Milton Keynes had 11.8 per cent people of pensionable age, compared with 31.9 per cent for Eastbourne (East Sussex) and 30.5 per cent for Tendring (Suffolk), which includes Clacton, Frinton-on-Sea and Walton. And there is no evidence of a change in this retirement pattern. Thus housebuilders may be wise to continue to construct small homes for the elderly in most coastal districts, within walking distance of the sea, if that is possible and if town planning permission can be obtained. Other favoured locations are towns in the Home Counties beyond Greater London. By contrast, Greater London has been 'the biggest source of retirement migration' (Coleman and Salt, 1992), except that the inner cities generally have many old age pensioners, but usually these are not of a social class with an income to enable them to get into the private housing market. There are regional variations with, for example, the highest proportion of solitary women aged 75 or over residing in Yorkshire and Humberside while Wales has the lowest proportion (10 per cent lower) (Grundy, 1987b). The statistics for Wales are in several respects different from those for England, with, in 1991, 15.7 per cent of pensioners living alone (and 17.1 per cent of all persons having a limiting long-term illness) compared with 13.1 per cent for Great Britain as a whole (OPCS, 1992b).

Overall, the South West and Wales contained the highest proportion of population aged 60/65–79 in 1997, whereas the South West, South East (outside London) and Wales contained the highest proportion of people aged 80 and over (Table 16.2). By contrast, the lowest proportion of the population aged 60/65–79 lived in London and Northern Ireland, while Northern Ireland contained the lowest proportion of people aged 80 and over.

Research undertaken by the JRF (1990) indicates that a year or two after the age of 65 between about 15 and 20 per cent of people migrate, and that about 10 to 12 per cent migrate each year in their seventies. Relevant questions are how many move to a new building, to a converted or an extended one, or to a home vacated by others.

The facts about the tendency of a comparatively small proportion of older people to migrate help to throw light on the wider, geographical and

Table 16.2 *Regional distribution of the elderly population, United Kingdom, 1997*

	Males 65–79 and females 60–79	80 and over
South West	18.4	5.1
South East (outside of London)	16.3	4.4
Wales	18.1	4.3
East	16.7	4.1
Yorkshire & Humberside	16.6	4.0
East Midlands	16.7	3.9
North West	16.6	3.9
West Midlands	16.7	3.7
North East	17.4	3.6
London	13.5	3.6
Scotland	16.7	3.5
Northern Ireland	14.1	3.1
United Kingdom	16.4	4.0

Source: ONS, *Regional Trends 34*, 1999

detailed nature of the demand for accommodation for elderly people. However, it is necessary to distinguish between 'micro' moves between dwellings, for example within one apartment building or on one estate, and 'macro' moves within or between counties, regions or countries. Different research data can be cited for each variety of move. There is also a relationship between type of tenure and frequency of moving. Owner-occupiers were the most likely to have moved between counties; however, the scale of movement has not been great: for example, among men aged 60–69, who were owner-occupiers, 5.6 per cent moved in 1970–71 and 3.4 per cent moved in 1980–81 (Grundy, 1987b). It is noteworthy that '32 per cent of all between-county migrants and 35 per cent of all within-county migrants among those aged 75 and over in 1971 were in institutions ten years later' (Grundy). Although this may not be a guide to the future need for institutions, it is a guide to the need for homes adapted to accommodate disabled old people and their caring.

Another geographical variation can be seen in housing tenure: for example people in the Third Age are more likely to be home-owners in the South of England than in the North (Warnes, 1992). However, these patterns may be changing. The later 1980s were marked by a movement out of London and the rest of the South East, so that in 1988–89 the North of England had a net gain of 7,000 migrants (JRF, 1991). Generally, although in the late 1980s over 10 per cent of people in Britain moved house each year, much of this was movement accounted for by people aged between 16 and 24, and migration rates tended to decline with age, with 'subsidiary peaks around retirement (60–65) and in late age (75 and over)' (JRF). In the 1980s rural areas gained older migrants, and the South Coast resorts may be becoming less popular. Generally there was a 'growth in movements by the "old" elderly away from the most popular retirement areas and into middle-class metropolitan suburbs' (JRF).

In 1992 about 70 per cent of those aged 60–64, about 64 per cent of those aged 65–69, 57 per cent of those aged 70–79 and 49 per cent of those aged 80 and over were already owner-occupiers. In the younger age groups, 30–59, an even larger percentage were owner-occupiers, although many more had a mortgage. Assuming not too many defaults on mortgages, within the next 25 years, an even larger percentage of the elderly will be home-owners. Taking the 78 per cent who were owner-occupiers in the age group 45–59 in 1992, who will be 65–79 in 2012, it is reasonable to assume that about 80 per cent of the 29 per cent (315,000) of the 65- to 74-year-olds who will require one-person dwellings will be able to afford to buy them. This indicates a possible demand in Great Britain for an additional 252,000 one-person homes by the year 2017, with an accelerating demand rate each year. However, a problem with such a projection is that it assumes that only about 20 per cent of elderly people – the lowest quintile – will be dependent on social or private rented housing, whereas income levels suggest that the proportion could be considerably greater than this and also demonstrate why some older people cannot afford expensive house repairs.

Since most pensioners who migrate tended to 'trade down' in order to release assets for day-to-day expenditure, a large proportion of moves involved the purchase of small (one-bedroom) houses or flats. In recent years, the supply of new properties of this sort has greatly decreased. In the private sector the number of completions in England fell from 22,521 in 1989 (at the height of the property boom) to only 6,525 in 1997, but local authority completions plummeted from 7,349 in 1987 to only 20 in 1997, while housing association completions fell from 6,288 in 1987 to 3,887 in 1997. Social rented housing in aggregate was clearly a victim of public expenditure cuts. In total the supply of one-bedroom flats and houses fell by 68 per cent, 1987–97 (Figure 16.1). Because this form of housing normally attracts the elderly (as well as young first-time buyers), its reduced supply undoubtedly constrains mobility among the over-sixties.

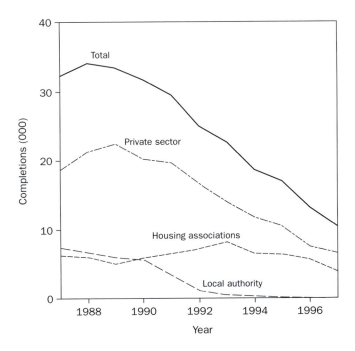

Figure 16.1 *Completions of one-bedroom houses and flats, England, 1987–97*
Source: DETR, *Housing and Construction Statistics*

By 1991, the number of dwellings in Great Britain was 21.2 million, compared with 19.5 million households (CSO, 1994). Yet homelessness grew throughout the 1980s. Many people are unable to compete for housing in the market, and are dependent on social housing or affordable housing to buy or to rent, or on some version of shared ownership. But the government has reduced funding to the Housing Corporation, which finances housing associations, and government policies have virtually eliminated housing starts by local authorities. The Audit Commission was reported as having estimated 'an annual shortfall of social housing in the region of 20,000 units per annum at current building rates' (*Planning Week*, 1994).

Yet the government wants owner-occupation to increase to 80 per cent of the total stock of housing in Britain. The building of new social housing was reduced from 168,000 each year (average) 1961–69 to 33,000 in the years 1990–92 (CSO, 1993). Many of those who could not get tenancy of social housing, or private rented housing (the income-poor but not asset-rich), could not afford to buy a home, in particular in London and in

South East England. Another problem is that in housing occupied by the elderly, there was considerable under-occupation (Murphy, 1989). The metropolitan districts have been losing 'third agers', but Outer London suburbs and Home County towns have attracted people aged 70 plus (Warnes, 1992). There is evidence that many older people want to live within reasonable distance of their children (Warnes *et al.*, 1985), but it may be difficult to find a suitable small home in an acceptable price range because the supply does not meet the demand owing to town planning policies aimed at restricting growth and promoting sustainable development.

Elderly households, instead of buying or renting in the normal housing markets, might seek the acquisition of retirement bungalows or apartments or a move to a retirement village – where there would be the provision of shops, restaurants and social and medical facilities in addition to housing. Retirement villages, however, although common in several countries, are comparatively rare in Britain, owing to their high cost of development (notably, high land costs) and planning resistance, with public policy being generally opposed to a large number of elderly people being concentrated in one place (Bookbinder, 1987).

Only since the 1970s different types of housing specifically designed for the elderly been developed in Britain. Until then, institutional care was the only special provision for the elderly. Whereas the ordinary private housing market and the social sector could be relied upon to satisfy the need for *zero/minimal* and *low dependency* housing (for example, well maintained and heated ground floor flats and bungalows near shops and services), there was a need to develop specific forms of housing more suited to the requirements of the less independent segment of the elderly population. *Medium dependency* housing, particularly in the form of sheltered housing, *high dependency* housing, mainly in the form of 'very sheltered' or extra care housing, and *residential and nursing care* thus expanded rapidly in the 1970s and 1980s in response to changing market and social demand.

Sheltered housing is often the most attractive option open to the elderly wishing to move – particularly among those unable to be fully active and independent. The term 'sheltered housing' is currently used to mean 'accommodation . . . with a resident warden, alarm system and some communal facilities such as a residents' lounge, guest room and laundry' (Age Concern, 1990); but in some usages it may mean only that there is an alarm system. The MHLG Circular 82/69 (1969b) introduced

Category One and Category Two sheltered housing. In Category One, communal facilities and a warden were optional, and the accommodation could be bungalows or flats, usually without a lift. Category Two schemes are flats under one roof, with communal facilities, resident wardens and lifts, intended for more dependent people. They attract more subsidy. Occupants are exclusively elderly people, and all schemes have built-in alarm or communication systems. From April 1990 all builders registered with the National House Building Council have been expected to comply with a Code of Practice on the management of private sheltered housing (Oldman, 1990).

The standard of sheltered accommodation – particularly in the private sector – is usually high, and a warden and alarm system normally satisfy emergency needs. Because of heavy demand and government expenditure cuts in public sector housebuilding there was a boom in the construction of private sheltered housing development in the 1980s, although accommodation was often an option available only to those with substantial capital (Bookbinder, 1987). But as Oldman (1991) points out, 80 per cent of sheltered housing purchasers had equity surplus on the sale of their former homes, and 35 per cent had surpluses in excess of £5,000, while those trading up were often able to employ their savings – although this could reduce their income status. Recently, 'very sheltered housing' (Category 2½) schemes have been introduced – providing meals and home-help domestic assistance. For the less active and less independent elderly these schemes are clearly a compromise between conventional sheltered housing and residential homes. A further innovation has been the development of part-buy/part-rent schemes aimed particularly at those elderly lacking capital, while housing associations have introduced leasehold schemes whereby the elderly (since 1989) can buy sheltered accommodation at between 25 and 75 per cent of its market value with 25 per cent of the equity being met by an HAG. Households buying at 75 per cent of market value thus incur no rent at all.

During the 1960s and 1970s sheltered housing was the only realistic option for low-income elderly people who wished to improve the condition of their accommodation (Oldman, 1991). Sheltered housing (provided by local authorities and housing associations) not only was seen as a 'convenient and manageable solution for the health and welfare professions involved' (Cowan, 1984) but, by providing an answer to the housing problems of older people, was 'regarded as one of the most successful feats of social engineering over the . . . [previous] two decades' (Blytheway, 1984). Yet only 6 per cent of the elderly lived in sheltered housing by the early 1980s, and by 1989 there were nearly half a million sheltered dwellings in England, mainly local authority owned (Table 16.3).

Table 16.3 *Tenure of sheltered housing, England, 1989*

Local authority	303,061
Housing association	120,911
Other social sector	2,905
Private	38,798
Total	465,675

Source: DoE, Housing Investment Programme Statistics

But sheltered housing is not without its problems. In the private sector, development is seriously affected by property booms and slumps since the market for new accommodation is determined by buyers having to sell their existing houses in order to trade down. In general, the market for sheltered housing is

> the last to suffer in a housing recession but also the last to show signs of recovery. The decision-making process is also longer. This is an important move for . . . purchasers and they are keen to ensure that they choose the right home in the right location.
>
> (Herring, 1993)

During the recession of the late 1980s and early 1990s there was an over-supply of sheltered housing and many proposals for further schemes were abandoned. Some builders of sheltered housing went out of business while others began to diversify (Slaughter, 1993). In the local authority and housing association sectors, however, the supply of new sheltered housing was largely determined by government funding, and, like social housing in general, it was a victim of severe cuts in the 1980s and early 1990s. Overall, the number of starts peaked in 1987 at 8,513, but over the following five years output plummeted by 75 per cent to 2,028 (Figure 16.2). In 1988 the private sector alone accounted for 4,226 starts (a record), output dwindling to only 113 starts in 1997.

The demand for private sector sheltered housing is constrained by a number of inherent attributes associated with this form of accommodation. Compared to ordinary leasehold dwellings, annual service charges are high, and many older people, with the exception of some richer ones, cannot afford what the market has supplied. Another

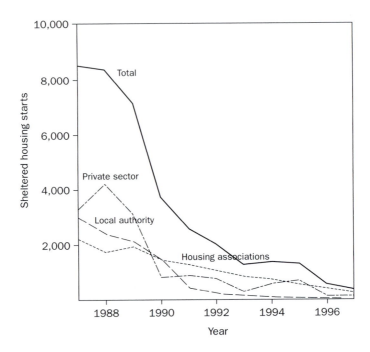

Figure 16.2 *Sheltered housing starts, England, 1987–97*
Source: DETR, *Housing and Construction Statistics*

problem was that in some cases the developers had constructed such very small rooms and dwellings that people did not want them. There were examples of very small flatlets in Frinton-on-Sea, Essex, that had been unoccupied for more than five years after completion. Developers tried to reduce costs and make their products affordable by reducing size, only to find that no one wanted the results. As with many other developments, demand has been miscalculated, and, as a consequence, there was an over-supply of sheltered housing.

A final problem was that many leases contained a clause compelling the resident to vacate the property at the onset of a serious disability – highlighting the difference between sheltered housing and all other forms of private and social sector housing.

Many elderly people, and particularly those who are relatively frail, have a preference for moving in with relatives, but although this could work well, it could equally be unsatisfactory for all parties involved. The provision of 'granny flats' is also fraught with difficulties – granny flats being 'an uneasy compromise between living alone and being independent' (Bookbinder, 1987). Elderly people opting for these solutions to their housing problems invariably suffer income loss since they may no longer be eligible for help towards rent or council tax and their income support and housing benefits may consequently be forfeited.

In 1999 the government announced plans – introduced under the title Supporting People – to reform the system of funding sheltered housing. It was intended that a number of existing funding streams – used to defray the costs of support services such as counselling, case recording and the co-ordination of care services by wardens – would be replaced by a single cash-limited pot, administered by local authorities. However, there was much uncertainty among RSLs and housing benefit officers over which services would be eligible under the new funding regime when it starts in April 2003 (Belcher and Williams, 1999), and concern among existing and potential residents about the general appeal and future affordability of housing in the sheltered sector. Clearly, it will be necessary to overcome cultural differences between housing providers and local authority social service departments if around half a million people in sheltered housing are not to lose services essential for their independence.

These concerns were expressed by the Local Government Association (LGA) in May 2000, which argued that sheltered housing should not be included in Supporting People when it comes into effect in 2003 and that it should remain part of normal housing provision (*Housing Today*, 2000) – a view not accepted by the DETR. The government clearly believed that sheltered housing not only offered the elderly more choice about where to live, but could also provide them with quality services that were flexible, responsive and user focused.

For the most frail and least independent, the only realistic option available, short of moving into the geriatric ward of a hospital, is a move into a residential home or a nursing home. It was estimated that '4 per cent of people over 65 years of age live in communal establishments offering personal or nursing care' (Laing, 1988). But the percentage rises very rapidly with increasing age so that from the age of 85 the proportion reaches about 20 per cent (Laing). Victor (1991) showed that about 3 per cent of those aged 65 and over lived in some form of institution, either a residential home, nursing home or long-stay hospital bed, but of those aged 85 and over, about 19 per cent lived in institutions. If projections by Laing (1988) were correct, by 2000 the number of elderly residents in communal establishments in England would have increased to 358,000, compared to 291,000 in 1986.

Funding of residential care in the 1990s

With increased demand, the private care home sector expanded rapidly in the last two decades of the twentieth century, increasing its share of long-term care beds from 10 per cent of the total in the early 1980s to around 70 per cent in 2000 (Brindle, 1999). Growth was greatly compounded by the withdrawal of the national health service from non-acute geriatric care, and by the privatisation of a large number of local authority residential homes. Over this period, private investors injected around £15 billion into residential and nursing homes, but by the late 1990s the sector was in decline (Brindle, 1999). Despite an ageing population, over-capacity had emerged, and unit costs rose as needs were increasingly satisfied by alternative community care services. With 70 per cent of care home residents funded by local authorities, moreover, many homes were no longer adequately profitable, particularly when social service departments were obliged to impose a squeeze on the fees they paid for state-funded residents. With property prices soaring in the late 1990s, many care homes were closed and redeveloped for housing (Brindle, 1999).

Because of market pressure, private residential and nursing care is very expensive. Charges for residential care ranged from £8,000 to £15,000 per annum even in the late 1980s (Hamnett *et al.*, 1991) and increased substantially throughout the 1990s, while nursing homes were even more expensive. In the depleted public sector, the cost of providing residential and nursing care would, of course, have been less. Local authorities and housing associations, moreover, are more likely to satisfy the need for housing for the disabled elderly; for example, 'mobility housing' dwellings incorporating wide doorways, level or ramped access and specially designed bathrooms and kitchens.

The staying put option

Among the elderly, moving home is by no means a panacea. Most elderly people prefer to stay put – with as many as 91 per cent of those aged 50–74 and 92 per cent of those aged 75 years and over very or quite satisfied with their homes, according to Social and Community Planning Research (1990). They have an attachment to their present home, surroundings and neighbours, and wherever possible they would prefer to 'live and die in their own homes, within their own familiar networks and where the need for independence is respected' (Wheeler, 1986). They dislike the upheaval of a move and they may be aware that there are financial penalties in moving as well as some advantages. Elderly low-income owner-occupiers, for example, might be obliged to forgo income support on the release of equity on moving, but could be faced with higher costs (ground rent and service charges) if they moved into a leasehold flat or sheltered accommodation (Oldman, 1991).

In the long term, rising house prices and rents suggest that the demand for new housing among elderly households, like the demand from households in general, is not matched by supply. What is clear is that for demographic and economic reasons it is very unlikely that the housebuilding boom of the 1960s, with about 200,000 new private dwellings per annum and 168,000 social sector dwellings constructed per annum in the United Kingdom, 1961–69, will ever be repeated. In the 1970s in the UK the annual construction of dwellings was about 315,000 per annum and in the 1980s about 218,000 per annum (CSO, 1994); and the proportion of those amounts that were accounted for by social housing in the 1980s had fallen to about 25.7 per cent of the total for an average year from about 45.7 per cent of the total in an average year in the 1960s. In the years 1990–92 the total new construction each year was about 189,000 dwellings, of which only about 17.5 per cent was social housing. In those three years of the 1990s the proportion of housing association dwellings had increased, but the proportion of local authority dwellings constructed each year was less than a third, approximately, of the number constructed per annum in the 1980s, and less than a tenth of the number constructed each year in the 1960s. In the case of one-bedroom houses and flats and sheltered housing, the decrease in supply was particularly apparent.

Because of deficiencies in the supply of housing for the elderly in relation to demand, successive governments in the 1980s began to promote a variety of 'staying put' schemes, the conversion of *zero/minimal dependency housing* into *low dependency* accommodation. But although it could be argued that this was influenced by the notion that wherever possible elderly people would prefer to stay put, cuts in public expenditure offered another, more financially stringent motive for the emphasis the government was now beginning to place on helping the elderly to stay put. In the 1980s there were substantial cuts in local authority capital expenditure on sheltered housing, local authorities finding it cheaper instead to adapt, modernise and repair an old person's existing home than to provide an alternative (Larkin, 1982). Tinker (1984), moreover, revealed that whereas the public sector cost of keeping an old person in sheltered housing amounted to £5,000 per annum, the comparative cost of a staying put scheme could be as little as £3,000 per annum (at 1984 prices).

In the 1980s, therefore, a number of schemes were developed by both the social and the private sectors to encourage elderly people to stay put. These ranged from the use of mobile local authority wardens and the distribution of dispersed alarm systems, to the provision of home equity release schemes by insurance companies to enable elderly owner-occupiers to unlock the capital on their homes to facilitate improvement, to the 'Care and Repair' and 'Staying Put' schemes of local authorities and housing associations respectively, with agency schemes in, for example, Bradford, Bristol and Newcastle, and the Anchor Housing Trust's staying put scheme which pioneered developments in this area in the early 1980s (Mackintosh *et al.*, 1990). Under both the Care and Repair and Staying Put schemes, the whole repair, improvement or adaptation process was undertaken by a local authority agency or housing association on behalf of the elderly owner-occupier or private tenant, and although the process was basically concerned with housing needs, it progressed beyond the needs of the house to the total situation of the occupant (Shelter, 1993b). It was a cause of concern, however, that the government failed to translate the many diverse local schemes into a national system (Leather and Mackintosh, 1989).

By the end of the 1980s, however, the government had become firmly committed to a policy of promoting 'staying put'. Both the Griffiths Report (1988) and the White Paper *Caring for People: Community Care in the Next Decade and Beyond* (Secretaries of State, 1989) were concerned with enabling the elderly to remain in their homes for as long as possible, supported, where necessary, by services designed to meet their needs and choices. The subsequent National Health Service and Community Care Act of 1990 (which came into effect on 1 April 1993) reorganised the system of local community care, broadly as recommended by the Griffiths Report, and gave local authorities responsibility for care of the elderly. The legislation was welcomed, particularly 'if adequate funds [were forthcoming] to create alternatives to residential, nursing home and hospital care' (McKechnie, 1993). But there was justifiable fear that the government would severely reduce the proportion of funding for housing for older people, leading to funding shortfalls in the foreseeable future (Shelter, 1993b).

The 1990 Act, moreover, had introduced substantial changes to the system of funding support to the residents of long-stay institutions. Not only were these changes designed to tighten the control of expenditure on long-term residential care, but they were also intended to tilt the balance of advantage away from residential care to care solutions in the community and involve a variety of providers in addition to the social service departments of local authorities (Gibb *et al.*, 1999). The Act consequently reduced the ability of local authorities to place people in residential care, required more of those in residential care to contribute to the costs of their care, and took into account not only a person's income but also their assets when means-testing for help with residential care costs – a practice that often necessitated the recipient to sell their home and to use the proceeds to pay for care.

Housing rehabilitation and the elderly

The condition of housing occupied by the elderly put a considerable strain on Care and Repair and Staying Put schemes. The *English House Condition Survey 1986* (DoE, 1988) showed that households aged 75 and over were more likely than other groups to have homes which were unfit or which lacked amenities. Although constituting only 10 per cent of all households, this age group occupied as much as 33 per cent of dwellings lacking amenities and 16 per cent of those that were unfit. As many as 19 per cent of single pensioners had houses in poor condition compared to 14 per cent of pensioner households of more than one person (DoE, 1988). The survey also showed that in 1986 those aged 60 and over, and especially those aged 75 and over, spent less on repairs and improvements to their home than younger age groups. The oldest group spent about a third of what was spent by those aged 40–59. Such figures not only provided a challenge to local authorities and housing associations operating Care and Repair and Staying Put schemes, but also

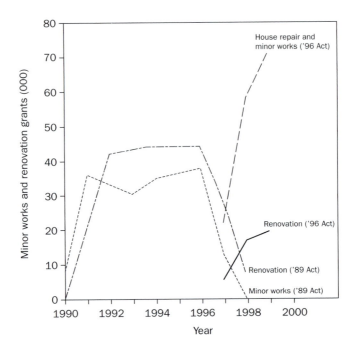

Figure 16.3 *Minor works and renovation grants awarded, England and Wales, 1990–99*
Source: DETR, *Housing and Construction Statistics*

did not encourage builders to target elderly people as a group despite the condition of their houses and the need for rehabilitation.

The Local Government and Housing Act of 1989 was, in part, intended to target means-tested grant aid to low-income elderly households living in housing in need of repair, modernisation or adaptation. But although the Act (in contrast to previous legislation) introduced greater flexibility in terms of financing small repairs, doubts were initially expressed about whether local authorities would be able to fund the new discretionary minor works grants and whether the elderly would apply for the larger (mainly mandatory) renovation grants (Leather and Mackintosh, 1989). As Figure 16.3 indicates, the number of minor works grants awarded in England and Wales, 1990–99, began to diminish after 1991, while the number of renovation grants continued to rise. Although this might suggest that local authorities were indeed beginning to exercise increasing discretion in awarding minor works grants, it does not necessarily follow that elderly applicants constituted a growing proportion of applicants for renovation grants. It seemed likely, however, that 'keeping frail and solitary older people in their own homes . . . [would] give issues of repair and condition increased prominence in the 1990s' (Warnes, 1992).

Future housing need

As Table 16.4 shows, the percentage of people in the age bracket 65–74 will decrease over the period 1987 to the year 2007, but will thenceforth increase over the next 20 years and possibly beyond. The percentage of the population aged 75 and over, however, will increase markedly throughout the whole period 1987–2027. It can be estimated that with a projected population of 55,353,000 in Great Britain in 2027, there will be 6,144,000 people (or 11.1 per cent of the population in 2027) aged 65–74, compared to 4.8 million people in the same age group in 1997, and 5,314,000 people aged 75 and over in the year 2027 compared to 4.2 million people of that age in 1997. There will thus have been an increase of about 1.3 million people aged 65–74 by the year 2027, and an increase of about 1.1 million people aged 75 and over at the end of the 30-year period.

Demographically, there will thus be an increase in the need for housing among the elderly well into the twenty-first century. But will elderly people have the financial means to create enough effective demand to satisfy their housing needs? Will there continue to be a demand for social housing? The answers are dependent on economic conditions and government policies. Nevertheless, there is homelessness, and a need for a 14 per cent increase in housing by the year 2012 (Gummer, 1994). This is due to the reduction in the size of households,

Table 16.4 *Projected size and age structure of the total population, Great Britain, 1987–2027 (%)*

Age	1987	1997	2007	2017	2027
0–4	6.4	6.0	5.5	5.5	5.5
5–15	13.7	14.0	13.0	12.2	12.3
16–64	64.3	63.9	65.3	63.9	61.7
65–74	8.9	8.6	8.7	10.5	11.1
75 and over	6.7	7.4	7.8	7.9	9.6
Total population (000)	55,355	56,198	56,251	56,093	55,353

Source: Ermisch (1990)

and it should be remembered that housing that is suitable for older people may also be suitable for younger people. A particular fact for housebuilders to note is the continuing trend for more and more people to live alone.

In the late 1980s and mid-1990s it was a cause for concern that, in the context of a growing number of people aged 75 and over, the supply of both new one-bedroom dwellings and sheltered accommodation decreased substantially, while assistance to staying put appeared to have been cut. But within a period of twenty years there is time for a very considerable change in social and political values, economic effects, land policies and governments. What may happen is that housing 'need' will once again become the determinant of housing policies (Clapham *et al.*, 1990) and there may be yet another attempt to introduce policies to make land available at less than free market prices in selected areas, perhaps in new settlements, some of which could be focused on housing for elderly people. Such policies would be likely to stress the need for affordable housing, not least for elderly people, and thus a demand for such housing may grow.

Conclusion

At the beginning of the new millennium there was not only much concern about the need to house an ageing population but also increasing concern about the amount of money that the elderly should contribute towards the cost of long-term care in nursing and residential homes. In determining eligibility for assistance, social workers sometimes, inappropriately, assessed the joint assets of a couple rather than their individual assets, even if only one of the partners was seeking care; assessed the value of the family dwelling, even though one partner continued to live in it; and assessed the whole of an assessee's occupational or personal pension, even though the other partner was dependent on a 50 per cent share of the pension. In addition, many elderly people in need of care had to face long delays before their financial position was assessed.

A further cause for concern was the possibility that local authorities would increasingly compel those in residential and nursing homes to pay all the costs of care until their assets had diminished to a negligible level, even though – under social security rules – councils had a responsibility to start providing help when assets were £16,000 or less and assume all responsibility at £10,000. In the mid-1990s this practice could no longer be generally assumed when Sefton Council granted assistance only when a person's capital fell below £1,500 (a course of action upheld by a High Court ruling in March 1997, but subsequently overturned in the Court of Appeal later in the year).

Inevitably, the total costs of residential and nursing care will rise in the future, at least in line with the ageing population. However, although half of all people over 80 have assets in excess of £16,000, arguably they should not necessarily bear the whole of the costs of their residential and nursing care. Already fees are a very contentious issue, with 40,000 houses having to be sold each year to pay for long-term residential and nursing care, and – other things being equal – this number will escalate since the proportion of the over-80s who are owner-occupiers is likely to rise from 46 to 60 per cent by 2015. The government therefore considered it necessary to set up a Royal Commission in 1998 to sort out the problem of care for the elderly, and its report a year later recommended that: all personal and nursing care should be free, and a National Care Commission should be set up to oversee standards – at a combined cost of around £1.2 billion per annum.

In response to the Royal Commission's report, the government announced in July 2000 that – under a £1.4 billion plan – all elderly people living in residential homes would receive free medical care (from October 2001), but that they would continue to incur 'hotel charges' for personal care. Social security assistance would continue to be available for these latter services, with the eligibility ceiling raised to £30,000, but local authorities would be empowered to provide interest-free loans for care fees (repayable from the recipient's estate on death), as an alternative to the elderly having to sell their homes.

These provisions, although welcomed to an extent, provoked criticism. The distinction between medical care and personal care is, for many elderly people, meaningless (particularly for sufferers of Alzheimer's disease, dementia and arthritis). Both forms of care are essential for the amelioration of their poor health. It is uncertain, moreover, how local authorities will be able to continue to foot the bill if their budgets become depleted – a situation that would compel many elderly people to sell their homes to meet the cost of personal care.

It is of note that in Scotland, as an outcome of the recommendations of the Sutherland Royal Commission, the Scottish Assembly in 2001 approved the provision of free nursing *and* personal long-term care – in contrast to practice south of the Border.

Further reading

Askham, J., Nelson, H.,Tinker, A. and Hancock, R. (1999) *To Have and to Hold: The Bond between Older People and the Homes They Own*, York: York Publishing Services. Provides evidence that although older home-owners prefer to own rather than rent their homes, they are aware of the many burdens that accompany this status.

Clapham, D., Kemp, P. and Smith, S.M. (1990) *Housing and Social Policy*, Basingstoke: Macmillan. The best general introduction to the relationship between housing and social policy, although possibly a little dated now.

Davey, J.A. (1996) *Equity Release: An Option for Older Home-Owners*, York: Centre for Housing Policy, University of York. A useful study that suggests that although equity release enhances people's living standards it does not resource expenditure on housing disrepair or care services.

Griffiths, R. (1988) *Community Care: An Agenda for Action*, London: HMSO. A detailed report recommending important changes to the system of community care.

Grundy, E. (1993) 'Moves into supported private households among elderly people in England and Wales', *Environment and Planning A*, 25.

Hancock, R., Askham, J., Nelson, H. and Tinker, A. (1999) *Home-Ownership in Later Life: Financial Benefit or Burden?*, York: York Publishing Services. A useful examination of home-ownership and housing wealth among the elderly, the effect of equity release, and a consideration of whether home-ownership is cheaper than renting in later life.

Harrison, L. and Heywood, S. (2000) *Health Begins at Home: Housing Interface for Older People*, Bristol: Policy Press. Provides an analysis of community care plans and Director of Public Health reports in three health regions that showed that little provision was made for housing services.

Leather, P. (2000) 'Grants to home-owners: a policy in search of objectives', *Housing Studies*, Vol. 15, No. 2. Argues the case for phasing out renovation grants and encouraging the elderly to meet the cost of renovating their homes through equity release.

Mackintosh, S., Means, R. and Leather, P. (1990) *Housing in Later Life: The Housing Finance Implications of an Ageing Society*, Bristol: School of Advanced Urban Studies, University of Bristol. Contains a useful account of the financial aspects of 'staying put' in both the private and the social sectors.

Oldman, C. (1986) 'Housing policies for older people', in Malpass, P. (ed.) *The Housing Crisis*, London: Croom Helm. A comprehensive survey and critique of housing policies in relation to the need of elderly people.

Oldman, C. (1990) *Moving in Old Age: New Directions in Housing Policy*, London: HMSO. Contains a useful account of the financial aspects of elderly people moving into sheltered housing.

Oldman, C. (2000) *Blurring the Boundaries: A Fresh Look at Housing and Care Provision for Older People*, Brighton: Pavilion Publishing. Provides an overview of new forms of enhanced sheltered housing and an examination of the claims that they could reduce reliance on or even replace residential care.

Wagner, G. (1988) *Residential Care: A Positive Choice*, London: HMSO. A useful report examining the many issues involved in the elderly selecting residential care rather than 'staying put'.

17 Gender and housing

The term 'gender' is not only about biological differences between men and women but also about what – in social terms – it means to be male or female (Dowling, 1998). Gender applies to men as well as women, and home-ownership is based on the reproduction of traditional views of femininity and masculinity (Connell, 1995). The term is therefore a component of identity as well as synonymous with sexuality.

Housing policies in Britain have traditionally been based around the family. Women without a male partner who are seeking housing may encounter discrimination, abuse and/or harassment within any form of tenure. Although the Sex Discrimination Act 1975 legislates against direct and indirect discrimination, there exists evidence that discrimination on the grounds of gender continues in all housing tenures (Sexty, 1990). Women are faced with having to deal disproportionately with issues such as poor-quality housing, discrimination from private and social landlords, and racial and sexual harassment. Clearly, they will not as individuals encounter all these issues on the same scale. However, it is important to recognise the effects of sexism and racism on women and to confront the problems in an appropriate and constructive manner.

This chapter therefore:

- examines the way in which the direction of housing policy has affected female-headed households with regard to the opportunity to gain access to affordable and good-quality housing;
- explores the impact of housing policies in each tenure; and
- investigates and discusses how housing policies impact on different groups of women with regard to the provision of housing.

The social rented sector

Lone parents

According to Haskey (1998), there were approximately 1.6 million lone parents in Britain in 1996, the majority of whom were female. Lone parenting affects about a quarter of all children. The proportion of lone parents caring for their children has increased from 8 per cent in 1971 to 21 per cent in 1996. Approximately 92 per cent of lone parents are female and around three-fifths of female lone parents were divorced or separated, while a third had never been married. Only one in 20 were widows (Webster, 2000).

Half of lone female parents are aged over 33 while half of lone male parents are aged over 42 (Pullinger and Summerfield, 1997). Teenage lone mothers make up less than one in 25 of all lone parents. The ONS (1998) found that three-quarters of homeless lone parents had been given accommodation by their local authority, while 14 per cent of lone parents were renting privately compared to 9 per cent of all households. The great majority of lone parents are in this situation because of relationship breakdown or widowhood; lone female parents who have never had a stable relationship with their child's father are relatively rare. As many as 49 per cent of lone parents were renting from local authorities or housing associations compared to 22 per cent of all households. About one in eight lone parents were in rent arrears, which is less than for couple families. However, this could be due to the fact that the majority had their full rent paid through benefit (Webster, 2000).

Access to local authority housing for women is based on factors which include the quality of accommodation, location and affordability (Hallett, 1996). Lone parents have the lowest level of income of all applicants for local authority housing. In one survey, as many as 80 per cent of single women considered themselves to be in urgent need compared to 70 per cent of two-parent households with children. Single parents were found to spend the shortest time on the housing waiting lists of any applicants, which is a likely reflection of their high-priority need (Prescott-Clarke et al., 1994).

According to Webster (2000), two-thirds of lone parents living in poor housing and neighbourhoods wanted to move compared to half of all households in similar conditions. Lone parents tend to live in the worst parts of the rental sectors. In England in 1996, 18 per cent were living in 'poor' housing conditions, such as in unfit properties and properties needing essential modernisation compared to 14 per cent for all households. One-quarter of unemployed lone parents with children were found to be living in poor conditions. In addition, lone parents are more likely to live in poor neighbourhoods as well as poor housing. In England in 1996, around 13 per cent of lone parents with dependent children were living in poor-quality neighbourhoods, which is twice the proportion for households taken as a whole. In addition, they were twice as likely to be dissatisfied with their housing compared with any other type of family (DETR, 1998b; Pullinger and Summerfield, 1997).

Box 17.1

Scotland

In Scotland 40 per cent of lone parents were found to have some form of dampness or condensation within their home. This is well above the 25 per cent for households in general.

Source: Scottish Homes (1997)

Divorced or separated women and men

Table 17.1 shows that in 1990, 44 per cent of divorced or separated women occupied a council house, with a further 4 per cent renting from a housing association. In 1998 the figure had reduced to 32 per cent of divorced and separated women renting from the council, while the percentage renting from a housing association had increased to 10 per cent. The corresponding figures for divorced or separated men are much lower, with 29 per cent renting from the council and 4 per cent from housing associations in 1990. In 1998, 25 per cent of divorced or separated men rented from their local authority and 6 per cent from a housing association. The figures clearly indicate the relative importance of renting in the social rented sector for male and female households. Women are over-represented in social rented housing.

Table 17.1 *Profile of heads of households by sex and marital status: social housing tenants, Great Britain, 1990 and 1998 (%)*

Marital status	Local authority		Housing association	
	1990	1998	1990	1998
Men				
Married	16	9	1	3
Cohabiting	n.a.	16	n.a.	5
Single	20	17	3	6
Widowed	39	26	3	7
Divorced/separated	29	25	4	6
All men	18	12	2	4
Women				
Married	22	n.a.	9	n.a.
Married/cohabiting	n.a.	n.a.	n.a.	n.a.
Single	35	27	7	12
Widowed	39	27	4	7
Divorced/separated	44	32	4	10
All women	39	29	5	9
All men and women	24	16	3	5

Sources: ONS (1991, 1999)

The owner-occupied sector

The heavy reliance of women on gaining access to housing in the social rented sector means that any changes imposed by government on this sector have particular effects on women (Hallett, 1996). The RTB legislation introduced in the Housing Act of 1980 is one example of a policy which enabled tenants to purchase their council homes. With lower average earnings than men, women are less likely than men to be able to purchase a property. Furthermore, RTB sales have reduced the number of good-quality homes and the overall number of council houses that are available to those who rely on the social rented sector to meet their housing needs.

Accounting for 68 per cent of the housing stock of Great Britain, owner-occupation is still, to some extent, seen as a privileged sector providing a degree of security and investment for later life, despite the large number of repossessions and poor housing conditions highlighted during the late 1980s and early 1990s. However, women have been disadvantaged in this sector, because formerly, building societies preferred to grant mortgages to males, as they were seen as more likely to have an uninterrupted pattern of employment. The mortgages obtained were usually at three to four times the average male salary, which disadvantages women, who on average earn approximately two-thirds of the incomes of males and thus have to commit more of their earnings to repaying their mortgages.

It is evident that women still earn significantly less than men. In 1997, men working full-time were earning an average of £408.70 per week, whereas women working full-time were earning £297.20 per week. By 1999, men working full-time had increased their average earnings to £440.70 per week and women's earnings had increased to £325.70 (ONS, 2001). However the gap between women's and men's earnings have remained relatively constant. In 1997, women earned 72.7 per cent of men's earnings; by 1999 this average had increased marginally to 73.9 per cent. However, earnings of women are often much lower, since a large number work part-time.

In 1997 the new Labour government introduced its Social Exclusion Unit to examine factors such as the lack of employment, low pay, low self-esteem, low status and discrimination. Gender issues can be integral to each of these expressions of exclusion. Given that women in general are still seen as the primary carers within the home, access to good-quality childcare has a major effect on women's ability to enter the workforce.

The emphasis of the Conservative government between 1979 and 1997 on owner-occupation discriminated against women, who were not able to participate within the sector on an equal basis to men. Women tend to be found in low-paid employment and frequently occupy junior positions within organisations, while men are disproportionally to be found in the more senior and well-paid posts within hierarchies. Hallett (1996) indicates that data from the General Household Survey show that 42 per cent of married women with dependent children and 24 per cent of lone mothers work part-time. In addition, 21 per cent of married women and 15 per cent of lone parents work full-time. Women are restricted by domestic and caring responsibilities in terms of the amount of time that they can spend at work. Thus, women are less likely than men to earn sufficient money to enable them to obtain a mortgage and become part of the owner-occupied sector (Beechey, 1986). Only 37 per cent of lone parents were owner-occupiers compared to 69 per cent of all households. About one in six lone-parent owner-occupiers are in arrears with their mortgage – four times the proportion for couples with dependent children (DETR, 1998b).

Table 17.2 indicates the extent to which owner-occupation is related to marital status. In 1990 as many as 54 per cent of married men were mortgagors, whereas only 30 per cent of married women were buying a property with a mortgage. However, in the same year, a higher proportion of married women than married men owned properties outright. In 1998 these disparities probably remained, although they cannot be verified because of lack of data. Table 17.2 further indicates that there are large differences in tenure patterns with regard to divorced and separated men and women. In 1990, 40 per cent of divorced or separated men were buying a property with a mortgage, while in 1998 the corresponding figure was 37 per cent. For women, the percentage figure was 14 per cent in 1990 and 34 per cent in 1998.

Table 17.2 *Profile of heads of households by sex and marital status: owner-occupiers, Great Britain, 1990 and 1998 (%)*

Marital status	Owned outright (as % of all houses)		With a mortgage (as % of all houses)	
	1990	*1998*	*1990*	*1998*
Men				
Married	24	31	54	52
Cohabiting	n.a.	5	n.a.	61
Single	14	15	39	37
Widowed	42	53	9	9
Divorced/separated	12	16	40	37
All men	23	27	49	49
Women				
Married	35	n.a.	30	n.a.
Married/cohabiting	n.a.	n.a.	n.a.	n.a.
Single	19	22	16	26
Widowed	44	6	54	6
Divorced/separated	12	30	14	34
All women	30	16	32	20
All men and women	25	28	41	41

Source: ONS (1991, 1999)

Widowed men and women show a similar tenure pattern in both 1990 and 1998. Table 17.2 indicates that in 1990, 42 per cent of men owned their property outright compared to 44 per cent of widowed women. In 1998, 53 per cent of men owned their property outright compared to 6 per cent of women. However, women tend to outlive men, and many in this category would have inherited their homes from male partners. Research has shown that widowed women tend to encounter problems in maintaining their property as a result of their previous low incomes during their working lives. When women do own their own home, they tend to buy flats or terraced houses and are likely to consign a higher proportion of their income to mortgage repayments compared to men (Sexty, 1990).

Thus, at a time when the supply of local authority housing is being reduced, owing to government policies, an increasing number of women are demanding accommodation within the social rented sector. It is important for women to be able to acquire housing which is based on need in the public sector, rather than on the ability to pay, as in the private sector. The government sees housing associations as the preferred landlord within the social rented sector, rather than local authorities. Women increasingly will be accommodated within housing association properties, and normally at rents higher than those charged by local authority landlords for similar housing.

The private rented sector

Hallett (1996: 73) states that 'Most of the criticisms of the private rented sector concentrate on five areas: insecurity of lettings, availability, cost, possible harassment from landlords and lack of support for those with particular care/social needs.' The stock that is available for renting in the private rented sector is often too expensive for tenants on low incomes, and such low-rent properties as there are, are often of poor quality.

The government considers the private rented sector an additional housing choice for those households who do not want, or are not ready, to buy their own property. For England, DTLR figures indicate that in 2001, 2 million households (one in ten) reside in this sector. London has the highest proportion of households renting in the private rented sector with 14 per cent (420,000).

Table 17.3 shows that single women accounted for 15 per cent in 1990 and 18 per cent in 1998 of tenancies in this sector. Lone parents accounted for 2 per cent of women within this sector and 8 per cent were single women (below the age of 60). Many women were found to be outside the protection of the Rent Act (GLC, 1986). Given that almost half the number of households within this sector are headed by predominantly unemployed women, access to this sector is based solely on the ability to pay, and so local housing benefit ceilings on rent levels will affect the type of accommodation that can be acquired.

The Housing Act of 1988 (which introduced assured and assured shorthold tenancies for all new lettings in the private sector with the use of market rents) was intended to reverse the decline within the private rented sector. Only 7 per cent of total households were renting privately in 1989, compared to 15 per cent in 1971 (Sexty, 1990). During the early and mid-1990s the number of households within this sector did indeed begin to increase, owing to the decline of house sales within the owner-occupied sector. Many owners chose to rent their properties while waiting for house sales and prices to increase again. However, the tenancies that were available within the sector were mainly shorthold. The rent levels of these properties were invariably linked to the level of the monthly mortgage repayments that were due on the property. As a result, they offered little scope for the unemployed or for households with low incomes (Douglas and Gilroy, 1994).

Table 17.3 *Profile of heads of households by sex and marital status: private sector tenants, Great Britain, 1990 and 1998 (%)*

Marital status	Unfurnished		Furnished	
	1990	*1998*	*1990*	*1998*
Men				
Married	2	5	1	1
Cohabiting	n.a.	9	n.a.	4
Single	8	11	14	14
Widowed	5	4	0	0
Divorced/separated	7	13	6	3
All men	3	6	2	3
Women				
Married	0	n.a.	0	n.a.
Married/cohabiting	n.a.	n.a.	n.a.	n.a.
Single	7	12	8	6
Widowed	6	7	1	2
Divorced/separated	6	8	3	2
All women	6	8	3	2
All men and women	4	7	2	2

Source: ONS (1991, 1999)

Gender and homelessness

There is an increase in the official numbers of homeless women, many of them young. There is an estimate that between 10 and 25 per cent of all homeless people in Britain are women, and half of this number will have children (Shelter, 1993c). Thirty per cent of lone parents with dependent children had experienced homelessness within the previous 10 years, compared to 6 per cent of all households. Three-quarters of these homeless lone parents had been given accommodation by their local council (ONS, 1998).

Somers (1992) found that the increase in the numbers of women who are homeless could be a direct result of the increase in reported violence against women in the home generally. Violence against women is more likely to occur in the home and be perpetrated by close friends or family members. Research suggests that many women experience violence in the home but keep it hidden in order to ensure that they remain housed (Austerberry and Watson, 1986).

Homeless women were found to be more likely to report higher rates of major mental illness and more difficulty in current life circumstances. They were also likely to have a greater prevalence of chronic health problems and a higher rate of attempted suicide. Women were found to be more likely than men to have had histories of sexual or physical abuse. More women than men stated that they had become homeless for social reasons such as family break-up and marital disputes (Darke, 1982).

Local authorities are obliged to provide housing only to those people who fall within particular categories, such as households with young children, pregnant women and people vulnerable through old age, mental health problems or illness and physical and/or learning disabilities (Sexty, 1990). Local authorities are permitted to impose their own interpretation of the degree of vulnerability of these groups, and this has resulted in a wide difference of interpretation between local authorities, with some landlords being much more generous in their interpretation of whom they are willing to assist (Niner, 1989).

The RTB policy of the Conservative government arguably achieved its aims but, coupled with a marked reduction in council housebuilding, it led to a reduction in the supply of council stock. The Conservative government assumed that there was no housing shortage and instead attempted to blame 'victims' such as 'single parents'. In the mid-1990s lone parents were singled out and, in effect, accused by the Conservative government of jumping waiting lists ahead of other applicants.

The causes of homelessness among women are closely related to those affecting the population as a whole. A large number of women may therefore become homeless because parents and friends are no longer willing or able to house them. Relationship breakdown is the second most common reason given by the population as a whole for becoming homeless. Consequently, it is probable that many more women than men have to leave the marital home owing to domestic violence, abuse or harassment (Morris and Winn, 1990). These two main reasons given for homelessness are closely linked in the case of women, as often after a relationship breakdown women will initially turn to friends and relatives for accommodation (given the small number of refuges that exist in Britain). Once a relative or friend is unable to provide further support, they will either try to move to accommodation offered by others or try to obtain housing in the social rented sector. Again, women are differentially affected in this: first, because of a lack of choice in other tenures due to their poor economic status; and second, by the contraction in the amount of local authority housing. Also, women will be in a much more vulnerable position than other households and so are more likely to be offered and to accept less desirable accommodation than are other households.

Box 17.2

The Homelessness Act 2002

Relationship breakdown is one of the main reasons that women give for applying as homeless to local authorities for accommodation. The housing options that women have available to them rely heavily on the policies of their local authority. There are a large number of variations in policies across local authorities in the treatment of women in these circumstances. The policies of some local authorities are sometimes related to the consideration of family law regarding who will get custody of the children and the family home. The policies of other local authorities will vary according to how women with or without children are treated if they are fleeing domestic violence, how women with a tenancy are treated, and the quality of advice they receive regarding their housing options.

The 2002 Act strengthens the protection provided by local authorities for people who become homeless through no fault of their own and have a priority need for

continued

accommodation, such as families with dependent children. The government in 2001 said it will make an order under existing legislation to extend the groups of people who are considered to have a priority need for housing. The current priority groups are:

- families with dependent children or someone who is pregnant;
- people who are vulnerable in some way (for example because of old age or their mental or physical disability); and
- people who are homeless as a result of a disaster such as flood or fire.

Local authorities must arrange suitable short-term accommodation for people in priority need groups if they have become homeless through no fault of their own. The new groups are as follows:

- people the council consider to be vulnerable as a result of fleeing violence or threats of violence;
- homeless 16- and 17-year-olds except those for whom a council has responsibility under the Children Leaving Care Act 2000;
- care-leavers aged 18 to 21; and
- people the council consider to be vulnerable as a result of an institutionalised background such as local authority care, the armed forces or prison.

Older women

Hallett (1996) points out that it is important when looking at older women and their housing needs, to realise that their housing experiences will not all be alike, as their housing situation will be a reflection of their past housing history. Some older women will find that they are able to remain within the family home while others may find that their circumstances require them to trade down or sideways, and others may find that they have to move into sheltered or supported housing.

The problems which women face generally in terms of discrimination and inequality continue into old age. The DoE (1992b) found that 46 per cent of women aged over 65 owned their home outright. A further 36 per cent were in council accommodation, 6 per cent were in housing association properties, and 12 per cent were in the private rented sector. The 1991 Census survey indicated that two-thirds of the elderly aged 65 or above were women, and 47 per cent of women in this age group were living alone compared to 24 per cent of men. The lower mortality rate among women in middle and later life would account for this pattern. However, this explanation could be misleading, as the majority of these women may have inherited their home following the death of a partner. Sykes (1994) found that older female owner-occupiers were more likely to be concentrated in older housing stock with a poor record of repair. Nineteen per cent of older people living alone are likely to be living in poor accommodation compared to 14 per cent of older couples. Many women in this group may not be able to afford repairs, may not have had to manage financial matters by themselves before and may be wary of allowing workmen into their home in order to carry out repairs (Coleman and Watson, 1987).

Local authorities and housing associations have a disproportionate number of households consisting of older people. In 1991, 44.6 per cent of households consisting of at least one older person were housed by local authorities, while 48.9 per cent of similar households were accommodated by housing associations, compared to 33.4 per cent of households with older people in all tenures. This particularly affects older women, who on average live longer than men. Given the high proportion of older women within the social rented sector, the supply of sheltered accommodation in this sector has not met demand and so many have been forced into sheltered accommodation within the private sector, even though women are disadvantaged in terms of their earning capacity during their working lives (Morris and Winn, 1990). It is therefore evident that many older women within this sector are unlikely to be able easily to afford private sheltered accommodation without recourse to income support.

A survey in 1983–84 showed that 21 per cent of older women lived in the private rented sector. These women frequently had long-established lettings but their accommodation often lacked basic amenities (GLC, 1986). The problem for older women renting within the private sector is that 'the real choice of moves is likely to be limited' (Sykes, 1994). Any move would be linked to the availability of adequate affordable housing within the area, which may not be an option for the women within this category.

Community support should be available for people (including older people, of whom the majority are women) who require support and help from professionals, relatives, friends and neighbours either within their own home or in the home of their carer. Data from the *General Household Survey* (CSO, 1992) has confirmed that 17 per cent of women are carers compared to 13 per cent of men. It has also been found that 83 per cent of carers do not receive any support from outside the home, which has meant that a high proportion of women are having to cope on their own and so are restricted in their ability to work outside the home

(Local Government Information Unit, 1990). Thus 'central government's policy of community support is only a thin disguise for the reality of care by female family members' (Gilroy, 1994). In its present form, it is also an ingenuous justification for cutting back on the funding of sheltered housing staffed by professional carers.

Black and Asian women

According to Hallett (1996), black and Asian women suffer harassment in all tenures. Rao (1990) found in her study that 32 per cent of black and Asian women said they had endured racial abuse, the same number had encountered racial assaults and 23 per cent had had some form of damage to their property.

Surveys carried out between 1985 and 1987 indicate that two-thirds of West Indian and Indian women were employed on a full-time basis, compared to approximately half of white women (Department of Employment, 1988). However, the social rented sector is the principal form of housing for many black and Asian women, many of them with low incomes. Moreover, a high proportion of black and Asian households are headed by women. Many black and Asian women have experienced racism and sexism in relation to their tenure. However, access to each sector is also restricted by the well-researched and documented discriminatory practices that affect black and Asian women (see CRE, 1984, 1987, 1988, 1989a, b).

Black and Asian women were twice as likely as white women to suffer long periods of homelessness. However, their homelessness often remains hidden, as they tend to reside with friends and relatives rather than become 'properly' homeless. It was found that 60 per cent of homeless black and Asian women were living with relatives and friends, compared to 42 per cent of white women. The most common reasons for black and Asian women becoming homeless include relationship breakdown, evictions or disputes within the private sector, and racial and sexual harassment (Rao, 1990).

The use of temporary accommodation by many local authorities pending the completion of investigations into eligibility for housing, coupled with the wait for an offer, results in many women being forced to spend long periods of time in accommodation that is often overcrowded and lacks adequate facilities. Often they have to endure sexual and racial

harassment in such accommodation (Rao, 1990). Research has shown
that there exist different perceptions of black and Asian women. Asian
women tend to be seen as passive and vulnerable while black women are
perceived as aggressive and sexually accessible. This perception further
exposes these women to racially motivated sexual harassment (Dhillon-
Kashyap, 1994).

Black and Asian women tend to be allocated the least desirable properties
in the less popular locations, and they often have to wait longer than
other groups of people to acquire a property. Research has confirmed that
75 per cent of white women reside in semi-detached or terraced
dwellings compared to 62 per cent of black and Asian women, and 24 per
cent of black and Asian women live in flats compared to 11 per cent of
white women. Furthermore, 90 per cent of black and Asian women with
three or more children live in one-bedroom flats or maisonettes compared
to 65 per cent of white women (Rao, 1990).

During the 1980s many local authorities had equal opportunity
policies which dealt with either racial or sexual harassment, but not
both. Thus, many black and Asian women who encountered racial
and sexual harassment found that race units would deal only with the
racial harassment cases while women's units would deal only with
the sexual harassment cases rather than combining and dealing with
both problems together (Rao, 1990). The situation has now improved
and there is often an equal opportunities section/department that will
deal with any type of harassment. In addition, it is now common practice
for all staff to receive training in how to deal with any complaint of
harassment regardless of its type, and front-line staff in housing
departments often have a specific role in this respect.

Asian women may leave the parental or marital home as a result of
harassment, or family and cultural conflicts. A Policy Studies Institute
study in 1982 showed that 46 per cent of Asian women were in unskilled
or semi-skilled jobs compared to 33 per cent of white women (Brown,
1984). Thus, Asian women may still be more likely to have a low income
and therefore will have more difficulty in acquiring housing in the private
sector and may not be aware of the options available within the social
rented sector, especially if English is their additional language. Rao
(1990) also found that many Asian women who left their parental or
marital home were afraid of being identified if they approached an
agency for assistance with housing.

A study carried out in 1983–84 indicated that black women head 3 per
cent of households within the private rented sector and, of these, over half

were single women. The report suggested that black women encountered poor housing conditions, the sharing of amenities and lack of security of tenure (GLC, 1986). As access to this sector is based solely on the ability to pay, black women will be differentially affected compared to white women.

The majority of racial violence encountered by black families is suffered by black women, who are often in the home for longer periods than men are (Amin, 1987). There is therefore a need to supply refuges specifically for black and Asian women fleeing domestic violence, as research has found that they often encounter racism from white female residents and refuge workers within predominantly white refuges (Mama, 1989). Since many black and Asian women have consequently felt isolated and have encountered harassment, a total of 14 refuges for black and Asian women have been set up in Britain in recent years. There also exist approximately 30 housing associations which deal specifically with the needs of black women (Dhillon-Kashyap, 1994).

Conclusion

Clearly, there is evidence that housing policies such as the promotion of owner-occupation by the Conservative government were at the expense of other tenures and so did not help the majority of women. The current (Labour) government has attempted to revive the social rented sector to a limited extent, but this still leaves many women without the ability to acquire good-quality and affordable housing for their households.

Further reading

Brion, M. (1995) *Women in the Housing Service*, London: Routledge. A good introduction to the study of women as users and managers within the housing profession.

Gilroy, R. and Woods, R. (eds) (1994) *Housing Women*, London: Routledge. Provides an appraisal of the impact of housing policy on different groups of women.

Hallett, C. (ed.) (1996) *Women and Social Policy: An Introduction*, Hemel Hempstead: Prentice Hall/Harvester Wheatsheaf. An introduction to the main social policies and their effect on women.

Roberts, M. (1991) *Living in a Man Made World*, London: Routledge. An excellent book that examines the relationship between gender and housing design.

18 Black and Asian minorities and housing

Economic disadvantage, social exclusion and institutional racism have all helped to perpetuate relatively poor housing for the majority of black and Asian households in Britain, and have inhibited both market forces and public policy from satisfying housing needs across much of this segment of the population. This chapter:

- begins by exploring some of the consequences of discrimination that black and Asian households have experienced when attempting to gain access to appropriate housing, and the effect of housing policies;
- analyses institutional and structural racism;
- discusses issues concerning black and Asian households and home-lessness; and
- examines aspects of racial harassment within the context of previous and more recent legislation and research.

Access to housing: discrimination, institutional and structural racism and harassment

The problems and obstacles faced by black and ethnic minority people seeking accommodation in all tenures have been well researched. There is clear evidence that many providers of housing lack the ability to confront and deal with the needs of black and ethnic minority people. However, some housing organisations do have equal opportunity policies that are followed closely and monitored regularly for their effectiveness.

Access to social housing

According to Harrison with Davis (2001), black and other minority ethnic people in Great Britain are disproportionately likely to be poorly housed despite the fact that many of them were born in the United Kingdom. Figure 18.1 shows that black and Asian people have a higher dependence on social rented accommodation than the majority white population, with 44 per cent of black households and 39 per cent of Pakistani and Bangladeshi households renting from local authorities or housing associations. In contrast, only 9 per cent of Indians and 22 per cent of white tenants were within this tenure category.

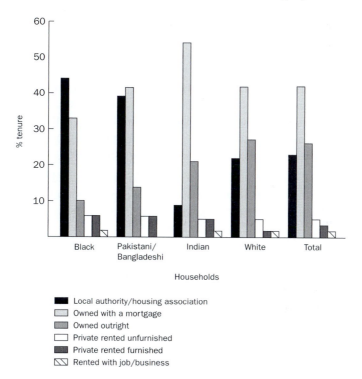

Figure 18.1 *Housing tenure and race, Great Britain (aggregate of 1995, 1996 and 1998)*
Source: ONS (1999)

Racial discrimination

A large number of studies have identified the racial discrimination which exists in the social rented sector (see CRE, 1984, 1987, 1988a, b, 1989a, b,

1990a,b) in terms of the quality of properties that are generally offered to black and Asian households. These studies also found that many black and other ethnic minority households tended to receive older properties than white applicants. They were also more likely to be allocated a flat than were white households, and were more likely to be offered properties that were not within their areas of preference.

Data from the 1991 Census show that within the local authority sector 28 per cent of white tenants occupied semi-detached houses, compared to only 6 per cent of black Caribbean, 4 per cent of Bangladeshi and 3 per cent of black African tenants (Sim, 2000). In addition, 73 per cent of black African, 67 per cent of Bangladeshi and 62 per cent of black Caribbean tenants lived in purpose-built flats compared to 35 per cent of white tenants (Howes and Mullins, 1997).

The 1991 Census data show that the average size of white households was 2.4, compared to 2.6 for black households, 3.0 for Chinese households, 3.8 for Indian households, 4.8 for Pakistani households and 5.2 for Bangladeshi households. The situation for some black and other ethnic minority households is further complicated by the need to include an extended family within the household, or by religious practices that in some way restrict the use of the property (Sim, 2000).

Table 18.1 shows that 15 per cent of Pakistani/Bangladeshi households are one bedroom below their required occupation standards compared to 8 per cent for Indian households, 7 per cent for black households and 2 per cent for white households. In addition, 38 per cent of white households and 36 per cent of Indian households had one bedroom above required standard, compared to 29 per cent for black households and 19 per cent for Pakistani and Bangladeshi households.

Table 18.1 *Occupation standards and race, Great Britain (aggregate of 1995, 1996 and 1998)*

	Bedroom standard				
	2 or more below standard	1 below standard	Equals standard	1 above standard	2 or more above standard
White	0	2	27	38	33
Black	1	7	45	29	19
Indian	1	8	32	36	24
Pakistani/Bangladeshi	7	15	48	19	12
Other	1	5	45	29	21
All ethnic minorities	2	8	42	29	20
Total	0	2	28	38	32

Source: ONS (1996, 1997, 1999)

Within the social rented sector few council properties are larger than three bedrooms (Sim, 2000). The RTB policy has reduced the number and quality of properties that are available. The Conservative government's housing policy wanted housing associations to become the main providers of properties within the social rented sector. However, research by Karn and Sheridan (1994) has found that the space standards within newly built housing association properties have fallen over the years, which disadvantages some black and ethnic minority groups, who tend to require larger properties, as indicated above.

Within the social rented sector, black and ethnic minority people are over-represented in the worst or most unpopular estates, in high-rise buildings, in older properties and in those of poorer physical quality (MacEwan, 1991). A substantial number of studies have shown that unless local authorities and housing associations monitor, analyse and review ethnic minority statistics concerning areas such as allocations and transfers, they will continue to be unaware of discrimination which may be taking place. In addition, there is also a need to ensure that staff training and consultation with black and other ethnic minority groups takes place, and that the workforce is representative of the community which it serves in order for workers to gain a better awareness of the issues that black and ethnic minorities have to face.

Institutional and structural racism

The Policy Studies Institute survey in 1982 found that 43 per cent of West Indian households and 23 per cent of Asian households resided in the inner city areas of London, Birmingham or Manchester compared to 6 per cent of white households in those cities (Brown, 1984). Many researchers have indicated the discrimination which black households faced when first trying to acquire council housing and which they are still facing (see CRE, 1984, 1987, 1988a, b, 1989a, b). One study found that only 16 per cent of black households had obtained a transfer compared to 26 per cent of white households, while 49 per cent Asian households had not even heard of transfers and so were not able to use the system at all (Henderson and Karn, 1987).

The substantial amount of research that has been undertaken indicates that institutional racism within housing departments is part of the reason for the poor treatment which black people have received from social landlords. Ginsburg (1992) has identified three types of institutional racism:

1 relative inadequacy in the physical standards of dwellings in relation to black applicants' needs, particularly the numbers of bedrooms, due to a failure to take into specific account the needs of black applicants in the housing construction and acquisition programmes of the past;

2 formal local policies creating differential access for black applicants such as dispersal policies, residency requirements and exclusion of owner-occupiers, cohabitees, joint families and single people from housing waiting lists; and

3 managerial landlordism involving racialised assessment of the respectability and deserving status of applicants, and assumptions about preferred areas of residence for different racial groups and about the threat of racial harassment by whites, which may cause avoidable trouble for managers.

Thus, while it can be argued that institutional racism is based on everyday policies, practices and procedures which are part of the normal functioning of housing departments and other institutions, it can also be seen as a 'process' involving many organisations whose practices coalesce to disadvantage black households (Phillips, 1987).

Structural racism within the social rented sector is evident through the cuts in public expenditure that this sector has had to endure. These have resulted in the reduction of the supply of housing at a time when large numbers of black and other ethnic minority households are having to rely on this sector in view of the many types of discrimination which they encounter in other tenures. The Housing Act of 1980, which introduced the right to buy council properties, has differentially affected black households. Those properties that have been sold are more likely to be found on the more desirable estates in the more popular areas. They have been sold mostly to white households, which owing to their normally longer period of residence than black people, have qualified for the largest discounts. Given that black people were prevented in the past from acquiring the more desirable council properties and are still facing discrimination in gaining access to the social rented sector, many black households have seen the supply of the more desirable council properties further reduced.

MacEwan (1991) considers that to tackle the expression of racism effectively, there must be an attack on the underlying economic and social structures which facilitate such expression. It is argued that institutional and structural racism have worked against black households: first, in reducing the supply of housing because local authorities are unable to build; second, in the reduced maintenance of council properties as many local authorities lack the finances to carry out programmed maintenance to their properties; and finally, RTB has reduced the supply of council housing while compounding the racism which prevented black households from acquiring desirable dwellings in popular areas (Ginsburg, 1992).

Racial harassment

Within the context of housing, research indicates that black people have endured racial attacks since approximately 1919. However, the issue of racial harassment of black and other ethnic minority residents on council estates was not recognised as a problem by local authority landlords until the 1980s (CRE, 1987). There are different types of harassment: 'The major problems appear to be verbal abuse, personal attacks and attacks on the home, but other identified problems may include the throwing of objects, rudeness, graffiti and attacks on household possessions such as cars' (Sim, 2000: 101). It was the Select Committees and the Home Office that brought the problem to the notice of the police, local authorities, and independent campaigning and advocacy groups (Lemos, 2000). Access to good housing is seen as the gateway to other important areas such as education and health. Thus if black people are consistently allocated the poorer-quality accommodation on the least desirable estates, other provisions that are received such as education will invariably tend also to be of poor quality.

With regard to housing associations, an investigation into Liverpool City Council's nominations found that white households were twice as likely to be offered a house, four times as likely to be offered a newly built property and a garden, and twice as likely to have central heating compared to black and other ethnic minority groups (CRE, 1989b). Further research into 40 housing associations found that while some operated good non-discriminatory practices, many had room for improvement. It was found that although a large number of housing associations monitored areas of service delivery such as lettings, several housing associations did not take any further action to change or improve their service in response to the information that they collected. The possibility for discrimination was found with regard to allocation policies. Officer discretion often resulted in new tenants being offered only flats or maisonettes. The study confirmed that only 26 of the 40 housing associations investigated maintained records of nominations and referrals, and thus a number of associations were unable to ascertain whether their housing was being provided on an objective basis. No black and Asian groups were represented on two-thirds of housing association management committees, and where there was ethnic minority representation it was mostly to be found on subcommittees and regional committees instead of on the main decision-making board (CRE, 1993).

Access to owner-occupation

Many ethnic minority households have encountered restrictions in obtaining mortgage finance. The CRE (1988a, b, 1990a) found discrimination by estate agents, and Sarre *et al.* (1989) found that often estate agents had a negatively stereotyped image of Asian people involved in property transactions. Often ethnic minority people have no other choice than to purchase older properties in inner city areas where other large groups of ethnic minority populations are based. These properties are often costly to maintain and slow to gain value (Ward, 1982). Many people belonging to ethnic minority groups have to rely on an informal network of friends and relatives in order to raise the finance required to buy a property (Sim, 2000). The statistics indicate that many Indian households may prefer to seek the avenue of owner-occupation at any cost rather than allow their households to become homeless or have to reside in poor-quality housing (Smith, 1989).

Various agencies within the owner-occupied sector have been seen to discriminate in the past against potential ethnic minority owner-occupiers. Estate agents, for example, who are keen to gain the highest price for a property, and thereby the highest commission, will often seek to match potential purchasers to suitable properties in order to speed up the process. As a result, some estate agents may assume that ethnic minority buyers want certain types of properties, such as terraced properties, in particular areas (Ward, 1982).

With regard to owner-occupation, the extent to which different ethnic minority groups own their home outright is partly determined by the age structure of the population, since older people are more likely to have paid off their mortgage or to have inherited the property. Thus, Figure 18.1 shows that while 27 per cent of white households were outright owners, the corresponding figure for Indian households was 21 per cent and that for Pakistani or Bangladeshi households 14 per cent. Black households were the least likely to be outright owners, with only 10 per cent falling within this category. Fifty-seven per cent of Indian households were owners with a mortgage, compared to 42 per cent for white and Pakistani or Bangladeshi households and 33 per cent of black households. Within the private rented sector the percentages were relatively similar, with 12 per cent of black and Pakistani or Bangladeshi households residing within this sector compared to 10 per cent of Indian and 7 per cent of white households.

Figure 18.2 indicates that 22 per cent of white residents reside in detached houses, compared to 14 per cent of Indian households and only 7 per cent of Pakistani or Bangladeshi and 3 per cent of black households. With regard to semi-detached houses, Indian households accounted for 36 per cent compared to 33 per cent for white, 21 per cent for Pakistani or Bangladeshi and 17 per cent of black households. However, 50 per cent of Pakistani or Bangladeshi households, 37 per cent of Indian and 32 per cent of black households lived in terraced

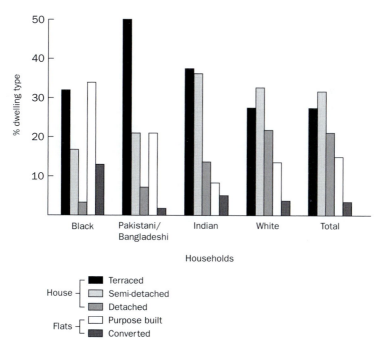

Figure 18.2 *Dwelling type and race, Great Britain (aggregate of 1995, 1996 and 1998)*
Source: ONS (1999)

properties, while white households were the lowest in this category at 27 per cent.

Thirty-four per cent of black households were found to reside within purpose-built flats, the highest proportion for any ethnic group, compared to 21 per cent of Pakistani or Bangladeshi and 14 per cent of white households. Indian households accounted for the lowest percentage within this category at only 8 per cent.

Those belonging to ethnic minority groups residing in the owner-occupied sector are often to be found at the bottom of the market, and their properties are often in a state of serious disrepair (Sim, 2000). A 1982 Policy Studies Institute study showed that standards of basic amenities rose between 1974 and 1982 for some households, both ethnic minority and white. However, 18 per cent of Bangladeshi households were found to lack adequate access to facilities such as inside toilets (Brown, 1984). It was found in the *English House Condition Survey 1986* (DoE, 1988) that 9.3 per cent of homes occupied by Caribbean and Asian heads of households were unfit and 21.9 per cent were in need of substantial repairs, compared to 3.8 per cent and 12.5 per cent respectively for United Kingdom-born heads of households.

Thirty-five per cent of West Indians, 28 per cent of Asians and 22 per cent of white people applied as owner-occupiers for improvement grants. However, only 45 per cent of West Indians who applied received grants, compared to 64 per cent of Asian and white owner-occupiers (Brown, 1984). Poor home-owners have experienced a decline in renovation grants, which are now worth less than 25 per cent of their mid-1980s value.

Access to private rented housing

Generally, black households pay more rent for poorer-quality accommodation in the private rented sector compared to white households. A survey found that 5 per cent of guesthouses and 20 per cent of accommodation agencies discriminated against black people when choosing a resident (CRE, 1990). The recommendations on good practice from the CRE included the introduction and monitoring of an equal opportunity policy, that all staff should receive guidance and training, that the policy should be reviewed at regular intervals, and that the DoE should require agencies to be licensed by the local authority.

Homelessness among the black and Asian minorities

At the same time as the production of public sector housing has fallen steeply, there has been a growth in the number of homeless people. In the 12 months to December 1998, local authorities accepted responsibility for housing 105,800 homeless households (26,500 in London). This compares with 53,000 accepted in the 12 months to December 1979. Inadequate records are kept of homelessness among the black and Asian population, but increased homelessness is likely to have a serious impact on black and Asian households, who are more likely to be discriminated against when trying to obtain accommodation. A survey in the London Borough of Brent in 1986 found that while black and Asian people made up 50 per cent of the borough's population, they represented 70 per cent of the borough's homeless figures. It was also found that people of African or Caribbean origins were twice as likely as Asians and three times as likely as whites to be homeless in the borough (Bonnerjea and Lawton, 1987).

Research has also found that the type of accommodation offered to homeless ethnic minority households tends to be inferior to that offered to white households, owing to the preferential treatment which white households receive (Bonnerjea and Lawton, 1987). A survey in the London Borough of Tower Hamlets indicated that Bangladeshi households were more likely to be offered low-grade temporary accommodation and spent longer periods in temporary accommodation than did white households (CRE, 1988a).

The CRE (1993) found that many housing associations had a number of discriminatory practices, including 'sons and daughters' policies, residence qualifications, the giving of preference to transfer applicants and the allocation of new tenants only to flats or maisonettes. Although significant numbers of housing associations no longer have many of these practices, some policies, such as a 'sons and daughters' policy, still exist within the sector and are part of the wider debate about maintaining sustainable communities.

The Housing Corporation began to develop strategies for black-led housing associations from 1986 (Sim, 2000) as a result of the large number of studies mentioned above regarding the discrimination experienced by black and Asian communities in housing organisations. Black-led housing organisations have often developed out of the need to attempt to redress the discriminations of the past and to ensure that the needs of black and ethnic minority households are met. There are currently approximately 63 black-led housing associations (managing about 12,000 properties) registered with the Housing Corporation out of a total of about 140 black-led housing associations. Their main aim has been the housing of black residents, although there are some that house all types of residents regardless of ethnic background, including families, older people and the young homeless. Black-led housing associations are generally small, with almost 80 per cent owning or managing fewer than 250 properties and half managing fewer than 100 properties. Just over half are located in the London region.

There are two main groups of black-led housing associations: those that provide general needs services, and properties that are similar to those offered by other housing organisations; and those associations which are culture-specific, providing property and services to a particular ethnic group. However, given the small numbers of black housing associations which exist in Britain, the majority of black households are more likely to be rehoused by housing associations which cater for all residents regardless of ethnic background. It is therefore important that the service delivered within all housing associations is monitored closely.

Race relations policy and housing

Local authorities have traditionally had responsibilities under the Race Relations Act of 1976 to tackle racial discrimination. The new Race Relations (Amendment) Act finalises the legal obligations on local authorities and registered social landlords that arise from the Macpherson Report (see Box 18.1). The Act, which became applicable in April 2001, places new duties on local authorities, including the enforceable duty to make arrangements to promote racial equality. The activities of all public authorities, such as the police, the Housing Corporation and contractors, are now subject to the prohibition of racial discrimination. The new Act places a clear duty on local authorities to deliver race relations in an integrated way. This is likely to be a challenge for many organisations, as race-related issues will need to move higher up the list of priorities. Many organisations will be required to examine the services they provide to their customers and to investigate their own employment practices.

Since the publication of the Macpherson Inquiry recommendations, feedback from Housing Corporation returns and Housing Inspectorate

---

Box 18.1

Macpherson recommendations

- The definition of a racist incident should be one that is perceived to be racist by the victim.
- Racist incidents should be reportable to agencies other than the police 24 hours a day.
- Information on racist incidents and crime should be made available to all agencies.
- Systems should be in place to access local contacts with minority communities.
- Joint racism awareness training should be considered with agencies including the police.
- Local initiatives aimed at addressing racism should be supported.
- Performance indicators should be developed around the recording, investigation and prosecution of racist incidents and these should also apply to multi-agency working, information exchange and training.

---

reports indicate that a large number of social landlords have not acted on the recommendations. This is evident in terms of:

- failing to attempt to anticipate the new duties in the Race Relations (Amendment) Act;
- the assumption that existing equal opportunity policies will be sufficient;
- a failure to benchmark and learn good practice from others; and
- the assumption that the Macpherson recommendations apply only to communities with large numbers of ethnic minority people.

The Macpherson Report, which was published in 1999, defined institutional racism and how this hindered the investigations into the murder of Stephen Lawrence. Within the report, institutional racism was defined as 'the collective failure of an organisation to provide services to people because of their colour, culture or ethnic origin'. In the wake of this report, many housing organisations stated that they would be reviewing their policies and practices in order to eliminate institutional racism. However, research by Blackaby and Chahal (2000) found that only 20 per cent of local authorities had in fact reviewed their policies and procedures. In addition, while 62 per cent of local authorities had written racial harassment policies and procedures, 29 per cent were unable to provide information regarding the number of incidents that had taken place in the past three years.

Box 18.2

Seeking racial equality: key events

- 1976 Race Relations Act;
- 1982 Lord Scarman's report on British inner city disturbances and policing;
- 1998 Housing Corporation launches black and minority ethnic policy;
- 1999 Macpherson Report into the handling of the investigation of the murder of Stephen Lawrence;
- 2000 Race Relations (Amendment) Act obliges public bodies to monitor the ethnic composition of their workforce and the effect of new policies and services on racial equality;
- 2001 Race and Housing Inquiry established by the Commission for Racial Equality, the National Housing Federation, the Housing Corporation and the Federation of Black Housing Organisations, chaired by Leroy Phillips. The aim was to produce a draft report by July 2001 and to have a code of practice in place by April 2002.

Studies have clearly shown that many housing staff have allocated black and Asian people to certain areas (see CRE, 1984, 1988a). Many of these officers may have done so in order to protect prospective black and ethnic minority tenants from possible racial harassment on mostly white estates. In addition, many black and ethnic minority households may be wary of accepting a property in a predominantly white area owing to the lack of support they feel they would receive from the housing organisation if they did encounter racial harassment from their neighbours. Racial attacks, racial harassment and the threat of them are thus very significant in perpetuating racism and racial inequalities in housing in contemporary Britain (Ginsburg, 1992).

Some housing organisations have a culture that sees racial harassment as something that is outside their main area of work and is just an exercise in box-ticking. Even with the introduction of racial harassment policies within housing organisations, it is the staff who can make a policy ineffective by treating racial harassment as a marginal issue. Inter-agency working is crucial for the effective tackling of harassment cases. However, there are often problems in this area with inter-departmental rivalries and poor joint working.

There are studies which indicate the level and extent of racial harassment in the social rented sector (see Bonnerjea and Lawton, 1988; CRE, 1987). Research has shown that once local authorities investigate racial harassment, they are more likely to transfer the black and ethnic minority household that has survived the harassment than to take legal action against the perpetrator. Thus, the intended aim of the perpetrator is achieved at a cost to the black and ethnic minority household (CRE, 1987). Many local authorities have amended their tenancy agreements to ensure that legal action can be taken against perpetrators on the grounds of racial harassment, and the different types of legal action available have been clarified for the use of housing officers and legal sections within housing organisations. However, the use of legal powers remains rare (Forbes, 1988).

Lemos (2000) found that where legal proceedings had taken place they were based on anti-social behaviour that had a racial element. However, while the procedure was seen as slow, housing professionals found that suspended possession orders or threats of evictions had the desired deterrent effect on racist residents.

In order to combat racial harassment, housing organisations need to monitor carefully the process of dealing with racial harassment cases and ensure that staff are adequately trained to deal sensitively and constructively with such cases. Better support systems both within housing organisations and through the use of voluntary agencies need to be established and publicised to tenants. The housing organisation also needs to gain the support of other white residents by explaining the policies to them and dealing with any negative or racist attitudes which may arise.

Box 18.3

Tackling racial harassment: code of practice for social landlords

The Junior Housing Minister Sally Keeble (representing Lord Falconer, the Housing Minister, in the House of Commons) launched a code of practice for social landlords in June 2001. The code is intended to be a 'toolbox' for registered social landlords and local authorities to make black and other ethnic minority people feel safe in their homes from racial harassment. It covers five key areas:

1 working with other agencies on harassment;
2 prevention and publicity;
3 encouraging reporting of harassment;
4 supporting victims and witnesses; and
5 taking action against perpetrators.

Box 18.4

The Macpherson Report and racial harassment policies

- Training should be provided on a rolling basis for all staff within the housing organisation. The training should attempt to address issues such as valuing cultural diversity and the raising of racism awareness. Board members of registerd social landlords, councillors and residents should be included within the training and all participants should be encouraged to develop a victim (or survivor) centred approach.
- It should be ensured that staff are aware that good and complete records should be kept of all harassment cases. It is important that there are full records in the event of a case being taken to court.
- People who have been harassed and possible witnesses should be protected through the use of equipment such as emergency alarms, 24-hour telephone assistance and inter-agency working, which would include the police.
- Consideration should be given as to whether it is appropriate to transfer or rehouse on a temporary basis households who have been harassed. It is important that the decision is based on the wishes of the victim, after ensuring that the household is aware of the support that would be available to protect them if they desired to remain within their home.
- There should be close monitoring of cases by staff with resident groups and the police as appropriate.
- Legislation should be used where required. Although racial harassment is not recognised as a specific offence in law, legal remedies which are available include the following:
  1 An injunction can be obtained under the Housing Act 1996.
  2 It is also possible to obtain an order that would require the perpetrator to stop insulting and/or harassing, under the Protection from Harassment Act 1997.
  3 An antisocial behaviour order can be obtained under the Crime and Disorder Act 1988. Only local authorities and the police can apply.
  4 Possession of the perpetrator's property can be sought.
- Work should be undertaken with groups such as present and previous victims, staff, residents and agencies to review the harassment policy on a regular basis.
- It should be ensured that tenancy and licence agreements contain specific clauses that prohibit harassment.
- It should be ensured that policies are clear and easy to follow, with the definitions of terms such as 'harassment' and 'neighbour dispute' precisely explained.
- It is important that harassment policies are advertised and promoted by the organisation.
- Black and other ethnic minority agencies and groups able to assist with aspects such as victim support and the review of harassment policies should be identified.

Source: Blackaby and Chahal (2000)

Conclusion

It is evident that housing policies have always been based on the assumptions that women and black and Asian minorities should fit into the existing structures and organisations, which have been developed by and for white males. Thus, studies have shown that housing policies were developed which have directly and indirectly discriminated and are still continuing to discriminate against these groups of people.

It is clear that not all women and ethnic minority people encounter discrimination to the same extent or have the same housing needs. It is important, however, that housing organisations and agencies have staff who have a clear understanding and awareness of the issues faced by these groups of people and are trained to provide a sensitive but effective service.

The adoption of equal opportunity policies by housing organisations to address the issues faced by women and black people is worthless unless it is supported by a timetabled plan for implementation of such policies.

Further reading

Bagihole, B. (1997) *Equal Opportunities and Social Policy*, New York: Addison Wesley Longman. Examines the key areas of social policy with regard to equal opportunities issues.

Blackaby, B. and Kusminder, C. (2000) *Black and Minority Ethnic Housing Strategies: A Good Practice Guide*. Coventry: CIH, and London: Federation of Black Housing Organisations. Provides guidance for local authorities and registered social landlords on developing strategies that are linked to the local housing strategy.

Forbes, D. (1988) *Action on Racial Harassment: Legal Remedies and Local Authorities*, London: Legal Action Group and London Housing Unit. A useful book that provides information about the wide range of powers available to local authorities to tackle racial harassment.

Harrison, M. with Davis, C. (2001) *Housing, Social Policy and Difference: Disability, Ethnicity, Gender and Housing*, Bristol: Policy Press. Provides an overview of gender, disability and ethnicity within the context of housing policies and practices.

Law, I. (1996) *Racism, Ethnicity and Social Policy*, Hemel Hempstead: Prentice Hall/Harvester Wheatsheaf. Provides an analysis of the issues concerning race across the main social policy sectors.

19 Homelessness

The most serious indictment of housing policy in recent years has been the appallingly high level of homelessness. With the failure of market forces to match housing need with the supply of affordable housing, government policy in recent years has significantly failed to reduce an unacceptably high level of homelessness and avert an increase in the number of homeless households in temporary accommodation. In reviewing this scenario, this chapter:

- quantifies the extent of the homelessness in the 1990s;
- identifies the causes of homelessness;
- examines the economic efficiency of policy;
- considers the need for action;
- discusses the policy options available to government in the early 1990s; and
- concludes by reviewing the effects of Labour's policy post-1997.

Extent of homelessness in the 1990s

Whereas there were 'only' 62,920 statutorily homeless households in England in 1980, the number increased to a peak of 137,250 in 1991 (equivalent to around half a million people). Thereafter it fell to around 105,000 in 1999 (Table 19.1) before rising to nearly 111,000 by 2000, to an extent because of inadequate investment in affordable housing in the late 1990s. Similarly – since only a proportion of the homeless were subsequently accommodated in council dwellings – the number of homeless households in temporary accommodation in England increased from 4,710 in 1980 to 62,170 in 1999 (Table 19.1) and continued to rise in 2000. There were also marked regional disparities, with concentrations of homeless households and homeless households in temporary accommodation in London (Table 19.1). As in England, homelessness in Wales similarly peaked in the early 1990s and then diminished, but in Scotland it continued to rise – albeit gradually – throughout the decade as an outcome of a lingering recession north of the Border and the slow growth of affordable rented housing. Whereas there were 17,300 homeless households in Scotland in 1991, the number increased to 18,200 by 1998 (Table 19.1).

Table 19.1 *Homeless households, Great Britain, 1991 and 1999*

	Households accepted as homeless		Number of homeless households in temporary accommodation	
	1991	1999	1991	1999
London	36,310	27,840	37,130	36,330
West Midlands	17,280	13,360	2,120	1,570
South East (excluding London)	13,750	12,380	7,890	8,220
North West	18,890	10,860	2,100	1,830
South West	9,050	9,490	2,630	4,870
Eastern	8,560	8,550	3,940	4,250
Yorkshire & Humberside	12,480	8,210	1,620	1,740
East Midlands	9,730	7,300	1,810	1,920
North East	7,870	4,830	430	1,110
Merseyside	3,330	1,950	260	370
England	137,250	104,770	59,930	62,170
Wales	8,843	4,171	n.a.	n.a.
Scotland	17,300	18,200[a]	n.a.	n.a.

Sources: DETR; Scottish Executive; Welsh Executive

Note: [a] 1998

Causes of homelessness

Aware that the rise in homelessness was of increasing public concern, the government was very quick to suggest that homelessness was largely due to a change in human behaviour – Mr Nicholas Ridley (Environment Secretary), for example, citing young people leaving home, marital break-ups and an increase in illegitimate births (Travis, 1988), and Mr Michael Spicer (Housing Minister) similarly attributing increased homelessness to social changes, particularly the break-up of the family and too many people leaving home prematurely (Brindle, 1990). In addition, the government attempted to lay the blame on resource-constrained local authorities for having too many empty dwellings, and on immigration policies where localised housing crises occurred such as among the Bangladeshi community in Tower Hamlets.

Undoubtedly, 'blame-the-victim' explanations can be borne out statistically. As Figure 19.1 shows, most homeless people leave their home either because parents or friends are no longer willing to accommodate them, or because of breakdowns with a partner. Other 'official' reasons for homelessness include the loss of private rented accommodation, and repossession because of mortgage and rent arrears. But while it is possible to identify the reasons for people losing their home, it is quite another matter to explain why they were unable to obtain alternative housing. Homelessness is clearly not just about needs, it is equally about supply.

But before we turn to supply, it is important to recognise that the increase in the number of households during the last two decades of the twentieth century is quite substantial – the number of households in England and Wales, for example, being expected to increase by 2.2 million between

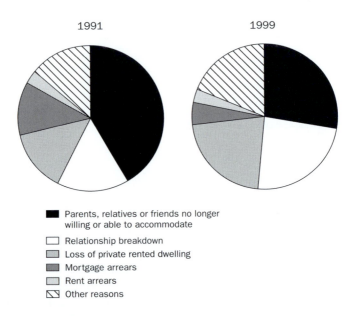

Figure 19.1 *Causes of homelessness, England, 1991 and 1999*
Source: DETR, *Homelessness Statistics*

1983 and 2001 (CSO, 1987). The 'baby boom' of the 1960s led to a high level of formation of households in the 1980s and 1990s; there is an increase in household formation among the under-30 and 30–44 age cohorts; divorce and separation is on the increase; and there is a growing percentage of the population over the age of 75 (Malpass, 1986).

In contrast, the supply of relatively low-cost rented housing markedly decreased in the 1980s and early 1990s. The decline in the rate of housebuilding in the social sector and the selling off of council houses have been principal contributory factors in the increase in homelessness (Malpass and Murie, 1990). The number of housing starts in the sector decreased from 56,300 in 1980 to 43,300 in 1993, with local authority starts plummeting from 41,500 to only 2,000 in the same period. By 1989 some local authorities were no longer building houses at all, for example Bradford, Liverpool, Nottingham, Plymouth, Portsmouth, Southampton and many London boroughs. Simultaneously, a total of 1,270,914 council houses were sold under RTB provisions in the period 1980–91. Thus, the local authority stock fell dramatically from 6,485,500 to 4,874,000 dwellings, or from 31 to 20.9 per cent of the total stock of housing in Great Britain, 1980–93. Since local authorities had been forced to pay back £20 billion of debts to the Treasury in the 1980s (out of receipts raised by the privatisation of much of their housing) and had up to £6 billion of receipts 'frozen' in their bank accounts in the early 1990s, there was no question of local authorities replacing any more than a very small proportion of their lost stock to meet the needs of the homeless. In the small private rented sector, there was also a decrease in supply, from about 2,400,000 dwellings in 1980 to 2,206,000 in 1993, despite repeated attempts by the government to revive the sector (see Chapter 6). In total, there was thus a loss of over 1.8 million rented dwellings in these sectors in the 1980s and early 1990s.

Although the *English House Condition Survey 1991* (DoE, 1993) reported some slight improvement in the condition of housing, 1986–91, it must be acknowledged that about 30 per cent of England's housing was built before 1919, and that the cost of putting right all the defects in the total housing stock would amount to at least £50 billion (Walentowicz, 1992), indicating that supply is inadequate in qualitative as well as quantitative terms.

Clearly, 'demographic changes, in the context of declining housebuilding [and qualitative deficiencies in the housing stock] are the first factor undermining the notion that homelessness is a residual problem'

(Malpass and Murie, 1990), but a number of other factors also contribute to the large deficiency of supply in relation to need. In a report produced for the DoE, Niner (1988) indicated that whereas Labour-controlled authorities were more responsive to homelessness, some Conservative authorities took a deliberately harsh line. While, for example, Birmingham, Newcastle and Nottingham aimed to be generous (with Birmingham trying to house accepted households within a day), Westminster adopted a 'tough but legal' approach and tried to find lawful reasons for not housing applicants rather than look positively at what the 1985 Act permitted. There is little evidence that these differences in attitude changed in the 1990s.

The large number of empty houses is also an important supply constraint. Although the total number of dwellings standing empty in England had decreased by 12 per cent from 867,300 in 1993 to 767,000 in 1997 (largely owing to a reduction in the number of empty private dwellings), there was a marked increase in the number of empty houses in the social sector (see Table 19.2). This to an extent could have been attributable to the Conservative government giving priority to the revival of the private rented sector rather than to public investment in the social housing sector. However, after nearly three years of Labour government, the number of empty dwellings at the beginning of 2000 still exceeded 750,000, of which 637,000 were in the private sector and 112,000 in the social sector, with the remainder owned by government departments (Hetherington, 2000g). In London the problem of empty houses was particularly acute. As many as one-third of the capital's 100,000 empty private houses at the beginning of the new millennium had been empty for more than one year, while at the same time there were 40,000 homeless families in temporary

Table 19.2 *Empty dwellings, England, 1993 and 1997*

	1993		1997	
	Number	% tenure stock	Number	% tenure stock
Private dwellings	764,000	4.3	640,000	4.0
Local authority dwellings	70,000	1.4	81,200	2.4
Housing association dwellings	17,500	2.2	26,800	2.7
Government dwellings	15,800	8.5	14,605	20.0[a]
Total	867,300	4.3	767,000	3.7

Source: DoE/Empty Homes Agency

Note: [a] The high percentage of government dwellings empty in 1997 compared to 1993 is attributable to the large reduction in the stock of government dwellings resulting from privatisation in the intervening years.

accommodation. It was clear that owners preferred to incur a reduced rate of council tax on their empty houses rather than to make them habitable.

Although limited resources are the principal reason for local authority dwellings being left empty (many are in a severe state of disrepair), a high proportion of empty private properties are kept off the market in anticipation of a significant upturn in house prices, while other private dwellings (particularly those formerly in the rented sector) are in a particularly poor condition and require renovation, but their owners either have to wait a long time in the queue for grants or are ineligible for assistance.

The economic inefficiency of policy

Through relying on bed and breakfast hotels, hostels and other temporary accommodation to house an increasing number of homeless households, the government generated substantial diseconomies in public expenditure. By 1991, the number of households in bed and breakfast accommodation (in England) peaked at 12,150, a sixfold increase in seven years, whereas the total number of households in temporary accommodation reached a record 71,510 by 1999 (Table 19.3). The NAO (1989) estimated that it cost an average of £15,500 a year to keep a family in bed and breakfast accommodation in contrast to only £8,200 to supply a new council dwelling, and Shelter (1992) reported that whereas the current total cost of bed and breakfast accommodation amounted to £146 million per annum, the first-year cost of providing new council housing was 'only' £78 million.

The cost of temporary housing is also borne by the households accommodated. All too often the condition of short-stay accommodation is very poor and results in respiratory problems, overcrowding, noise,

Table 19.3 *Temporary accommodation for homeless households, England, 1991–99*

	Bed and breakfast	Hostels	Private sector leasing	Other	Homeless at home[a]	Total
1991	12,150	9,990	23,740	14,050	8,700	68,630
1995	4,500	9,660	11,530	18,450	8,890	53,030
1999	8,120	8,920	22,390	22,750	9,330	71,510

Source: DETR, *Homelessness Statistics*

Note: [a] Households accepted as homeless but remaining in their existing accommodation pending rehousing.

lack of play space, lack of cooking facilities and comparatively poor access to health care (FPHM, 1990), but ultimately these intangible costs are borne by society as a whole. Clearly, bed and breakfast accommodation is 'often very expensive and unsuitable for families, and at a cost of £15,000 per family a year is bad value for money' (House of Commons, 1991).

A further diseconomy resulting from the failure to build more houses was the increasing practice of local authorities leasing private rented houses to accommodate homeless households – a practice particularly prevalent in Conservative-controlled authorities in Greater London. In the aftermath of aggressive privatisation programmes in the 1980s, several London boroughs had little council housing left – Wandsworth, for example, paid £2.95 million to private landlords in 1989/90 to house homeless families (Bowcott, 1990), a sum greater than the cost of supplying council housing.

The supply of affordable housing for owner-occupation was also a major cause of homelessness, particularly in London. Since the average price of a house in the capital was around £155,000 in March 2000, a household without savings would have needed an income of as much as £50,000 to have obtained a mortgage to purchase such a property. For this and other reasons, at the beginning of the new millennium 187,000 households were on waiting lists for social housing in London and 40,000 families were in temporary accommodation (Perry, 2000b). High house prices in London also tended to reduce the availability of private rented housing (the first rung on the housing ladder for many households) since landlords were encouraged to sell off their properties rather than let.

Demand constraint: policies of the Department of Social Security

The demand for housing among poorer households is determined quite substantially by the provision of various welfare payments: mainly housing benefits in the case of tenants, and income support and other payments in the case of the homeless housed in temporary accommodation. It is, however, with regard to homeless single people that demand has been severely constrained in recent years, contributing quite substantially to the increase in the numbers of people sleeping rough and the increase in begging in the streets of London and other cities in the late 1980s and early 1990s.

Before April 1988, homeless single people were normally eligible for an *urgent needs payment* which could enable them to book into cheap bed and breakfast accommodation. With a booking receipt, they could then claim *supplementary benefit* which was paid in advance. But after April 1988, urgent needs payments were replaced by *crisis loans* (for which young single people were generally ineligible), and it was the intention that *income support* (which replaced supplementary benefit) would be paid two weeks in arrears. Clearly, this prevented many single homeless people from obtaining accommodation, since most bed and breakfast hotels required payment in advance. However, Mr Nicholas Scott, Social Security Minister, subsequently announced that DSS officers would be advised of the potential risks to health and safety when dealing with claims for crisis loans from young people, and that full income support entitlement would be paid as soon as possible. But the biggest obstacle to young people obtaining accommodation was the withdrawal of income support to 16- and 17-year-olds not on youth training schemes (thereby saving the Exchequer initially £88 million per annum) and the reduction in income support paid to 18- to 25-year-olds.

Box 19.1

The rough sleepers strategy: an assessment

By 2000 the Rough Sleepers Strategy was well on the way to meeting its target of decreasing the number of people sleeping on the streets by two-thirds over the period 1998–2002. Whereas in 1998 there were an estimated 1,851 people sleeping rough each night in England, in August 2000 the estimated number had fallen to 1,180, a 36 per cent reduction (*Housing Today*, 2000), and by August 2001 the number had plummeted to 703 (Perkins), 62 per cent less than in 1988.

Unlike earlier attempts to tackle the problems of rough sleeping, the strategy recognises that homelessness is about more than not having a roof over one's head; it's about not being part of the community, not maintaining relationships and not having a job. Whereas it was possible that most of the street homeless were not getting the help they needed under the Rough Sleepers Initiative (RSI), the new strategy, in contrast, attempted to provide more of a holistic solution than hitherto to the needs of the most vulnerable.

However, although the numbers sleeping rough have fallen under the strategy, they had been decreasing continuously since 1990 as a result of the £250 million injected into the system by the RSI over the decade, and because of unemployment plummeting from its high level in the 1980s (Brooks, 2000). The principal weakness of the strategy is that it

continued

ignores the reluctance of many rough sleepers to accept a place in a hostel, where the new services are provided. They fear, rightly or wrongly, that in hostels they will become victims of bullying, violence, drug abuse and excessively restrictive regimes. The strategy is essentially a 'top-down' approach, and a more 'user-friendly' service could emerge if the homeless could be encouraged to set up self-help groups to provide a link between individual rough sleepers and the various charities that supply specialised services.

To help end rough sleeping, the government must recognise that young people need access to accommodation so that they can then concentrate on finding work. The Housing Green Paper (DETR/DSS, 2000) proposed that one way of realising this aim would be to introduce a range of eligible rents for shared accommodation in the private sector (i.e. shared bedsitters, flats and houses) and abolish the current means of assessment based on a single room in non-self-contained accommodation. In this way, the rent that housing benefit would pay 'might better reflect the type of shared accommodation which is available . . . [and] help young people obtain and maintain accommodation which can give them a secure base for job search and job security' (DETR/DSS, 2000: 114).

Supply constraint: policies of the Department of the Environment

The substantial decrease in the size of the local authority housing stock in recent years has imposed a very considerable constraint on the ability of the government to house the homeless. In an attempt to tackle this problem (albeit on a limited scale), a number of initiatives were introduced in response to increasing public concern.

The government introduced a Homeless Initiative in November 1989 – allocating an extra £250 million to reduce homelessness in England over a two-year period, and targeted particularly at London and the South East, the main stress areas. Of this sum, £177 million went to local authorities to renovate their empty properties and lease houses in the private sector for short-term accommodation, and £73 million was allocated to housing associations for broadly the same purpose. In addition, an increased number of voluntary organisations in England received grants to help homeless people. Whereas there were only 31 such organisations receiving grants (totalling £2 million) in 1990/91, the number increased to 147 (receiving £6.1 million) in 1992/93.

A Rough Sleepers Initiative was launched in June 1990 with an allocation of £96 million over three years, with a further £86 million becoming available in November 1992 to extend the initiative until 1996. Funds were thus made available to provide hostel spaces and create more permanent move-on accommodation for people leaving hostel accommodation. Extra money was also made available to local authorities to increase the number of places for former rough sleepers in housing leased from private owners. By mid-1992 the Rough Sleepers Initiative had provided about 850 places in hostels and over 1,200 in move-on accommodation, and by December 1993 a further 1,300 permanent places. It is probable that the initiative was largely responsible for reducing the number of people sleeping rough in central London from 1,046 in January 1991 to 440 in March 1992 (Central Office of Information, 1993).

Notwithstanding this welcome improvement, the initiative had only a very minor effect on the total problem of homelessness. At the time of its introduction, Shelter (1990b) dubbed it a 'sticking plaster over the wound of Britain's growing housing shortage', while the Institute of Housing (1990) said the initiative was 'calamine lotion to cover the spots'.

In Wales extra measures to deal with homelessness were announced in December 1989. Utilising the Homelessness Reserve of over £4 million, local authorities were encouraged to submit bids involving partnership schemes with housing associations, and support for voluntary organisations was increased to £580,000 in 1992/93. In addition, a three-year £800,000 programme was launched to help young single homeless people in Cardiff to find and retain permanent accommodation.

In Scotland additional resources amounting to £15 million were made available in the early 1990s to fund projects to combat homelessness. In 1992/93, for example, £7.5 billion was spent funding 44 projects, including the provision of emergency hostels, follow-on accommodation, and furnished tenancies in Edinburgh and elsewhere – providing, in total, accommodation for some 700 homeless people.

Other measures to reduce homelessness included the Flats over Shops Initiative and a scheme to use homes repossessed by mortgage lenders for accommodating homeless families. The first of these was launched by the government in October 1991, whereby a sum of £25 million was made available to housing associations over three years both to nominate homeless families in need of accommodation and to pay their rent to landlords of the property. The second proposal, announced by the DoE and Council of Mortgage Lenders (CML) in November 1991, involved lenders putting repossessed properties under the management of housing associations, which would then let them to families accepted as homeless by local authorities. The families would claim housing benefits to pay market rents, which would be passed on to the mortgage lender (less management costs). Within a year, such a scheme was under way. The National & Provincial and the Bristol & West building societies both planned to spend £5 million buying up properties whose owners had defaulted on their mortgages, and to use the accommodation to house homeless families nominated by local authorities or housing associations.

But all these policy measures involved comparatively little public expenditure. The largest of them, the Homelessness Initiative, involved expenditure of only £250 million – 'a paltry sum when measured against the scale of the problem' (Gould, 1989). It was clear that the government was 'skimming the surface of the problem, instead of tackling the root cause by investing more money in housing' (Shelter, 1989).

The need for action

Although it seemed essential for the government to remove obstacles blocking the payment of social security benefits to young single people in need of housing, and to make payments to the under-25s equal to those received by every other recipient, the problem of homelessness is more the result of the deficiency of supply than a shortage of 'effective' demand.

It is therefore important to increase the supply of low-cost housing by putting empty houses back into use and to embark upon new housebuilding programmes. Outside of central government there have been several initiatives to reduce the number of empty dwellings. Funded by voluntary organisations, the Empty Homes Agency was set up in 1991 in an attempt to accommodate homeless families in some of the country's (then) 760,000 empty houses and flats. The agency aimed to put the owners of empty houses in touch with housing associations willing to lease them, with the intention of housing associations subsequently letting them to homeless people nominated by local authorities. Also in 1991, the Association of London Authorities (ALA) launched its Home Front programme to encourage London boroughs to buy up empty houses in the private sector to provide permanent homes for homeless families. The ALA pointed out that 97,000 houses were empty in the private sector in London, while more than 3,000 people slept rough, 8,000 families were in bed and breakfast accommodation and 36,000 people were in other temporary homes.

Calls for an increased rate of housebuilding were both numerous and vociferous. The Faculty of Public Health Medicine (FPHM, 1990) argued that the plight of the homeless and those living in substandard accommodation was a national disgrace and stated that 2 million homes were needed immediately and a further 4 million by 2000. Greve (1990) similarly estimated that up to 4 million extra homes would be needed in Britain by the end of the century to counter the rapid increase in homelessness, while both Lady Anson (1990), Vice Chairman of the Conservative-controlled ADC, and Shelter (1993a) suggested that 100,000 new socially rented houses needed to be supplied annually, in contrast to only 7,000 social sector starts for 1990/91 and 1991/92. The Bishops' Conference of England and Wales, defending the Church's right to speak out on the question of homelessness, not only reported that the government had failed to supply enough decent housing in the social sector, but recommended that if mortgage interest relief (currently worth £7 billion) was targeted towards first-time buyers, this would release substantial sums for investment in low-cost accommodation.

Policy options: the early 1990s

At the General Election of 1992 the manifestos of the main political parties contained a plethora of measures to reduce homelessness. The Labour Party proposed investing more in housebuilding through the phased release of receipts from council house sales, allowing local authorities to lease or buy empty houses to provide accommodation for homeless people, and similarly allowing housing associations to lease or buy empty houses to accommodate the homeless. The Social Democrats favoured paying income support to claimants in advance, assisting with the payment of initial deposits on accommodation, extending the duty of local authorities to provide accommodation for 16- to 18-year-olds, funding the provision of short-term rented accommodation to reduce the need for bed and breakfast accommodation, and legislating to bring into use dwellings left empty without reasonable cause for more than a year.

The Conservatives, in contrast, pledged that they would introduce a new pilot scheme to promote homesteading as part of Estate Action. Those in housing need would be offered the chance to restore and improve council dwellings in return for a lower rent. In addition, a task force would be set up (headed by an independent chairman) to bring the relatively small number of empty government-owned residential properties back into use to be let on short leases, and there would be a continuation of the Rough Sleepers Initiative. But none of these proposals involved any significant increase in public expenditure. There were no proposals to increase effective demand among young single homeless people, nor to fund either large-scale acquisitions of empty housing or housebuilding on the required scale.

On returning to office after the 1992 election, the Conservative government continued to preside over an almost record level of homelessness, but subsequent to a review of homelessness legislation, *Access to Social Housing* (DoE, 1994), it proposed ending the automatic right of the 'statutory' homeless to claim permanent social housing. It was intended that henceforth the homeless would join a single 'streamlined' waiting list and would not gain priority over other households with possibly comparable housing requirements. As a temporary solution to the needs of the homeless, local authorities would have to provide accommodation for at least a year, in many cases in private rented accommodation or in hostels, but critics feared that homeless families would never get permanent homes and that temporary accommodation would only exacerbate the problems of certain priority groups such as one-parent families. The cost of the scheme, moreover, was estimated at over £20 billion over 20 years – the extra amount of housing benefits that would have to be paid if local authorities were to rely on private rented housing to accommodate many of the homeless. There were also fears that housebuilding in the social sector would cease altogether, with local authorities having to depend upon private rented housing to provide a permanent rather than temporary solution to the needs of the homeless and others on council waiting lists (Shelter, 1994). This would require even greater subsidisation, and as with housing policy in general, the government would thus be distorting demand rather than facilitating an adequate supply of new social rented housing.

Conclusion

Soon after the 1997 General Election, the new Labour administration adopted a high profile in tackling homelessness by giving unintentionally homeless families the same priority in the allocation of local authority housing as the disabled and other vulnerable groups – an eligibility that had been withdrawn by the previous Conservative government. This improved the prospects of up to 100,000 people securing permanent housing rather than having to rely on bed and breakfast accommodation. Based on the Housing Green Paper (DETR/DSS, 2000), the Homes Bill of 2001 proposed that unmarried ex-service personnel and others who had been institutionalised – such as single women who had fled from domestic violence, adolescents living in council care and prisoners released from jail – would become eligible for priority treatment rather than continuing to languish at the bottom of housing lists.

Unintentionally homeless households would also benefit if many of the houses left empty by their owners were brought back into use. In London the Empty Homes Agency (believing that speculators were taking advantage of the property boom in 2000 by keeping their vacant houses off the market in anticipation of realising substantial capital gains) proposed that local authorities be empowered to impose the full council tax on owners of long-term empty property to encourage more housing to be offered for sale or rent in areas of need (Hetherington, 2000g). However, it is to be hoped that if private landlords have left their property empty mainly because of its poor condition, the reduction (in 2001) in value added tax from 17.5 per cent to 5 per cent on conversion of houses into flats or its abolition on property empty for 10 years or more will do much to encourage the rehabilitation of such property and increase its supply.

Most single homeless people, however, have to resort to sleeping on the streets. The Labour government therefore spent, initially, £8 million in 1997/98 funding a range of schemes designed to reduce the need to sleep rough. In London alone, grants totalling more than £1.2 million were awarded to the Big Issue Foundation and to voluntary groups and charities running emergency hostels, mediation services, and rent and deposit guarantee schemes to help the under-25s obtain and hold on to affordable flats (Waugh, 1998).

Following a report by the Social Exclusion Unit in 1998, a £200 million Rough Sleepers Strategy was announced in an attempt to reduce the number of people sleeping on the streets. A 'street tsar' was appointed to head the newly established Rough Sleepers Unit, which was charged with the task of increasing the amount of hostel accommodation available for rough sleepers. Its aim was to reduce the number of people living on the street by two-thirds by 2002 and to zero after that. More than 2,500 new beds were to be provided to meet these targets, including 550 in London, where more than 600 people slept rough each night. However, hostel places would be offered to rough sleepers only if they signed up to a New Deal job-finding package, a contentious condition since finding a job is normally much easier after, rather than before, more permanent housing and a more settled way of life have been secured (see Box 19.1).

By 1999 the problem of homelessness had taken on a new dimension. In addition to the need to apply solutions to the problem of homelessness among UK families and single people, there was a need to ensure that the growing number of asylum seekers were adequately housed, particularly as access to housing through conventional means was strictly limited (see Box 19.2).

Box 19.2

Asylum seekers and housing

The Immigration and Asylum Act of 1999 excluded asylum seekers from the mainstream welfare and housing system in the United Kingdom. Unlike other homeless people in Britain, asylum seekers are not permitted to use the local authority homeless persons unit for assistance, cannot apply for social housing through the local housing register, and cannot afford accommodation in the open market since they are not entitled to housing benefit. They nevertheless can apply to the Home Office through the National Asylum Support Service (NASS) for a 'voucher-only' service, which places the onus on seekers to make their own arrangements to stay with friends or relatives already housed.

Alternatively, through NASS, asylum seekers might wish to apply for rented housing, even though, when NASS considers applicants, no detailed assessment of the family's housing need is undertaken and any preference for type of housing or its location is ignored. Thus many families have been dispersed across the country – often from London and Kent to northern or Scottish cities such as Glasgow, Newcastle and Liverpool, where the lack of employment has reduced the demand for housing. Between April 2000 (when NASS was set up) and July 2001 a total of 32,000 asylum seekers were found accommodation across the United Kingdom. In August 2001, however, the Home Office was intent on reinforcing its dispersal policy by appealing for 5,000 homes to rent in the South East to ease the high concentration of asylum seekers (and associated pressures on services) in Inner London. Private landlords – often through the medium of letting agents – were to be given four-year contracts, with rents being paid by NASS even if the accommodation was unoccupied some of the time. NASS would also assume the responsibility of managing and maintaining the properties at the Exchequer's expense.

In addition to the problems of dispersal (unfamiliar surroundings and neighbours, racial harassment, violence, etc.), asylum seekers and their families often have to live in housing that is not of a decent standard. This is due to arrangements that NASS has with a number of agencies to obtain and manage accommodation on its behalf. Some of these agencies are social housing providers but the majority are private accommodation agencies without any housing stock of their own. These agencies subcontract accommodation from local lettings agents and private landlords in the NASS dispersal areas, which means that most of the accommodation occupied by asylum seekers is privately owned.

Research, however, has shown that the private rented sector includes some of the worst housing conditions in Britain, with half being more than 75 years old and approximately 19 per cent classified as unfit for human habitation. The estimated cost of remedying unfitness in the sector is around £2.3 billion. For this reason, NASS was insistent that homes rented through housing providers or letting services acting on behalf of private landlords would need to be of a reasonable standard, and that contracts would be terminated if standards were not met.

The harassment of occupants by landlords is common and can range from entering the property without permission to intimidation and/or physical assault. The use of assured shorthold tenancies has increased from 140,000 (8 per cent of the private rented sector) to 1.22 million (54 per cent) and has resulted in threats to tenants – not least if they are asylum seekers – if they solicit advice or begin legal action to improve their living conditions.

Taking into account problems of dispersal, the condition of housing and landlord–occupant relationships, Shelter recommends that:

- realistic and effective mechanisms are introduced to ensure that the accommodation of asylum seekers is planned, implemented and monitored by local agencies; and
- asylum seekers are provided with information about their legal and contractual rights to decent accommodation and support, and how to enforce these rights.

Further reading

Carter, M. (1997) *The Last Resort: Living in Bed and Breakfast in the 1990s*, London: Shelter. A study of the extent and nature of self-placement in bed and breakfast accommodation in England and Wales.

Fitzpatrick, S., Kemp, P. and Klinker, S. (2000) *Single Homelessness: An Overview of Research in Britain*. An essential review of contemporary research on homelessness in Britain that emphasises that there has been a shift of emphasis away from explaining homelessness as a 'housing problem' to a more complex explanation weaving together social, economic and individual factors.

McCluskey, J. (1997) *Where There's a Will . . . Developing Single Homelessness Strategies*, London: CHAR (Housing Campaign for Single People). Examines the extent of strategic working for single homeless people, and sets out the benefits and other outcomes of adopting such an initiative.

Pleace, N. (1998) *The Open House Programme for People Sleeping Rough: An Evaluation*, York: Centre for Housing Policy, University of York. An examination of low-cost emergency accommodation for people sleeping rough outside of the major conurbations.

Smith, J., Gilford, S. and O'Sullivan, A. (1997) *Young Homeless People and Their Families*, London: Family Policies Studies Centre. A study comparing the background and attitude of young homeless people and their parents with those of other families living on council estates.

20 Conclusions

Although good-quality, secure housing is essential for the well-being of every household, housing professionals witnessed a substantial increase in housing problems faced by individuals and families during the last two decades of the twentieth century. With the return of a Conservative government in 1979 under the premiership of Mrs Margaret Thatcher, the uneasy consensus of the previous half-century or more had broken down. Thatcherism, involving a lurch back to the free market, dominated housing policy for a generation – in latter years under the nostrums of the Major administrations, 1991–97. A neo-liberal welfare regime emerged in which public expenditure on housing was reduced, housing investment in the public sector plummeted, local authorities were superseded by housing associations as the main providers of new social housing, the local authority stock of dwellings was increasingly privatised, rents were permitted to rise to market levels within both the social and the private rented sectors, while mortgage repayments imposed an increased burden on owner-occupiers, and mortgage foreclosures and homelessness increased. Against this backdrop, this concluding chapter:

- analyses Labour's housing policy, 1997–2001, considering some of its achievements and failures;
- notes the aims of housing policy in Scotland, Wales and Northern Ireland;
- examines the contents of Labour's manifesto of 2001 in respect of housing;
- reviews Labour's housing policy in the aftermath of the 2001 election; and
- refers critically to post-election policy omissions.

Labour's housing policy: the record, 1997–2001

During its first two years in office, the Blair government – for macroeconomic reasons – generally failed to grapple with the worst excesses of the housing policy that it inherited. Total housing expenditure, capital spending on local authority housebuilding and acquisitions, housing starts in the social sector and the size of the local authority housing stock all continued to plummet, while rents (except in the local authority sector) and mortgage payments continued to rise ahead of average earnings. Only in respect of mortgage repossessions and homelessness were previous trends reversed (Table 20.1).

Table 20.1 *Key housing policy indicators, United Kingdom, 1996–99*

	1996	*1999*	*% change*
Public expenditure: housing (£bn at 1998/99 prices)[a]	4.9	3.5	−28.6
Capital investment: local authority new-build and acquisitions (£m at 1997/98 prices)[b]	67	47	−29.9
Housing starts:[b] Local authority	1,656	482	−70.9
Housing association	30,304	23,592	−22.1
Local authority housing (% total)[c]	18.0	17.0	−5.6
Average rents (£ per week):[b]			
Local authority	40.10	42.24	+5.3
Housing association	48.25	54.51	+13.0
Private	50.65	59.24	+17.0
Mortgage payments (£ per week):[c]	54.87	70.63	+28.7
Average male earnings (£ per week):[d]	302.8	336.0	+11.0
Mortgage repossessions	42,560	30,000	−29.5
Local authority homeless acceptances:[e]	148,339	115,921	−21.9

Source: Wilcox (2000)

Notes:
[a] 1996/97–1999/2000
[b] 1996/97–1998/99, England
[c] 1996–98
[d] 1996–99, England
[e] 1996–99, Great Britain

The achievements

By the year 2000 – in its attempt to ameliorate the impact of Thatcherite policy – the Blair government met with some success in achieving its policy aims as set out in Labour's 1997 manifesto. First, it halted and then reversed the reduction in social housing investment. Although there had been a £1.6 billion cut in investment in this sector (including a £547 million reduction in Housing Corporation funding) as a result of the commitment by the Chancellor (Mr Gordon Brown) to adhere to the previous adminstration's spending plans for the first two years of the new government, Labour subsequently released £900 million of capital receipts in its first budget in 1997 – albeit in the form of credit approvals – in compliance with its long-standing pledge to release £5 billion of capital receipts. This was followed in 1998 by a further £3.6 billion in back receipts under the *Comprehensive Spending Review* (Treasury, 1998). It was clear that the bulk of this money would be invested in renovations and improvements, rather than in new housebuilding, in an attempt to reduce the £19 billion backlog of repairs to the social housing stock, particularly in areas of greatest need. The second spending review in July 2000 (Treasury, 2000) increased DETR expenditure limits from £3,096 billion in 2001/02 to £4,647 billion in 2003/04, including a 50 per cent increase in local authority borrowing approvals and a major boost for RSLs through the Housing Corporation's development programme, albeit with an emphasis again on renovation and improvement rather than on new homes.

Second, the government ensured that – in order to attract a substantial volume of private sector funds into the renovation and improvement of social housing – the stock transfer programme was accelerated, and by May 2001 as many as 117 local authorities (with a total of 583,000 homes) had completed sales to new landlords.

Third, to the benefit of people still living and working in local authority housing, Labour abolished the Conservatives' discredited system of compulsory competitive tendering (CCT) and replaced it with Best Value (a hybrid between competitive tendering and public service provision). Since April 2000, 'best value has offered councils the opportunity to compete on an even footing with the private sector to provide services – as long as services are up to scratch' (Blake, 2001: 21),

Fourth, Labour attempted to address the whole 'sustainable communities' debate. It not only insisted that until at least 2015 around 60 per cent of all new housing development should be on brownfield land, but also lowered value added tax on repairs and improvements from 17.5 to 7.5 per cent, and to 5 per cent on homes empty for more than three years and on flat conversions.

Fifth, Labour undoubtedly promoted the expansion of home-ownership, the preferred tenure of seven out of ten households by the year 2000. Although mortgage interest tax relief was abolished in April 2000 after years of piecemeal cuts, mortgage interest rates fell to their lowest level since the mid-1960s, an indirect consequence of the Bank of England's response to government inflation targets.

Finally, Labour recognised the need to resolve the many problems of social exclusion. A new social exclusion unit was established in 1997 and was responsible for promoting cross-departmental initiatives to tackle deprivation in the poorest neighbourhoods. A total of 18 policy action teams were set up, and £800 million was allocated to the 'new deal for the communities' fund for combating social exclusion. At the start of 2001 the unit launched its 'strategy for neighbourhood renewal' for reversing the decline in Britain's poorest areas. Among its many detailed targets, the strategy aimed to eliminate a third of substandard council housing by 2004 and narrow the gap between the most deprived areas and the rest. Only time will tell whether the work of the unit has had a positive impact on neighbourhood renewal.

However, because of the lack of parliamentary time, the government was unable to get a number of important pieces of proposed legislation on to the statute book. The first – the Homes Bill – would have imposed a duty on local authorities to protect homeless people in priority need, including (for the first time) 16- and 17-year-olds fleeing domestic violence; would have offered applicants greater choice in the allocation of social housing; and would have introduced a 'seller's pack' for the home ownership market. The second – the Commonhold and Leasehold Reform Bill – would have offered long-anticipated changes for leaseholders. The third – concerned with the licensing of houses in multiple occupation (HMOs) – did not even reach the Bill stage.

The failures

Overall – in the view of *Roof* (2001b) – the Blair government's first-term housing policy was 'promising', but it could have been better. In terms of its 1997 manifesto pledges, Labour's principal weakness was its inability to reform housing benefit, a complex task made more difficult 'by piecemeal regulation and a verification framework designed to tackle fraud in the system, but which in practice has just added to the delays faced by claimants' (Blake, 2001: 21). Similarly, there has also been a failure to reform the system of state help to poorer owner-occupiers. Although a public–private scheme (involving the lending institutions) was intended to replace limited income support for mortgage interest payments, by 2001 the government seemed 'content with the efforts being made by the industry to boost individual take up of private insurance schemes' (Blake, 2001: 21). There was also a failure to reform private renting. Possibly more fundamentally, Labour adopted policies not explicitly included in its manifesto; for example, it halved social housing construction and – through stock transfers – doubled the rate of privatisation since ousting the Conservatives in 1997, to the detriment of households most in need of affordable housing, many of whom were condemned to temporary bed and breakfast accommodation.

In its Housing Green Paper (DETR/DSS, 2000), the government was nevertheless concerned that too many English people (nearly 3 million) continue to live in poor-quality housing; that many people live on run-down estates where ill health, crime and poverty are endemic; that most social sector tenants have been denied choice and have been charged rents that are not comparable for comparable homes; that some households who have bought have not been able to afford to keep up with their mortgage repayments after becoming unemployed; that home-owners – including many retired people – have not been able to afford to maintain their own homes; that many people – including the elderly and vulnerable – live in housing that is difficult to keep warm in winter; and that some people are homeless or even sleeping rough.

However, although the government, in its Green Paper, acknowledged that a start had been made on raising the quality of housing and management in the social sector, tackling problems in the home-ownership market, and improving the services and protection available to vulnerable people, it recognised that further initiatives were required to:

- improve the delivery of affordable housing;
- improve the quality of social housing and housing management;
- promote a stronger role for local authorities to reflect local conditions;
- provide tenants of social housing with a real choice concerning the homes they live in;
- review tenure arrangements for social housing to retain security for long-term tenants;
- maintain or restructure rents in the social housing sector to ensure that they are affordable, fair and at a sub-market level;

- raise the standards of reputable private landlords and the quality of housing in the private rented sector;
- develop housing benefit measures to improve claimant service, tackle fraud and error, and improve incentives to work; and
- strengthen the protection available to homeless families.

It was clearly the government's declared aim 'to offer everyone the opportunity of a decent home and so promote social cohesion, well-being and self-dependence' (DETR/DSS, 2000: 16).

Housing policy in Scotland, Wales and Northern Ireland, 1997–2001

The Green Paper was, of course, not directly relevant to Scotland, Wales and Northern Ireland. In 1999 these countries secured devolved powers to formulate their own housing policies (although in the case of Wales this applied only to secondary legislation). Within two years – as an example of primary legislation – the Housing Act of 2001 emerged from the Scottish Parliament and created a single social housing tenancy incorporating the Right to Buy (RTB); introduced a single framework for the regulation of social landlords to be adminsistered by Scottish Homes; enhanced the role of Scottish Homes by converting it to an executive agency of the Scottish Executive; promoted the development of the strategic role of local authorities, for example by assuming the responsibility for producing single housing plans embracing all social landlords in an area; updated and redefined repair and improvement grants; and changed homelessness legislation to regulate and monitor the role of local authorities in tackling homelessness. Without having powers to frame primary legislation, Wales was at a disadvantage in developing policy since it could only alter regulations. Thus, despite enthusiasm in the Welsh Assembly for the transfer of stock from local authorities to registered social landlords (RSLs), transfer was only in its embryonic stage by 2001, way behind England and Scotland. Despite the Northern Ireland Assembly having powers to produce primary legislation, the political situation in the province slowed down moves to review the role of the Northern Ireland Housing Executive to address such issues as transfers, and the needs of the different communities.

Labour's manifesto, 2001

Within the context of the Housing Green Paper on the one hand and the devolution of policy-making on the other hand, Labour's manifesto at the 2001 General Election set out the party's intended housing policy for the period of the following parliament (Labour Party, 2001). Based on additional investment of £1.8 billion over three years, the party committed itself to reducing by one-third the backlog of substandard housing by 2004, with all social housing brought up to a decent standard by 2010.

Although many local authorities would continue to provide high-quality council housing, Labour would encourage local government – subject to tenant agreement – to transfer 200,000 dwellings per year to housing associations and arm's-length council housing companies to bring all social housing up to a decent standard by 2010, and would also examine ways in which 4 million council and housing association tenants could be helped to gain – at a discount – an equity stake in the value of their home. It was expected that equity stakes would be proportionate to the length of time the tenant had been in the property and paying rent, that tenants would be able to pay a higher than normal rent with the excess rent being matched by funding from the state or the landlord, and that tenants would be able to take the equity with them when they moved. Although it might be argued that the introduction of equity stakes would represent the most radical threat to the sustainability of social housing since the Thatcherite RTB scheme was launched in the 1980s, Labour saw the proposal as a way of tackling the increasing polarisation of owner-occupation and social sector renting, and would revive the spirit of RTB without reducing the total supply of social rented property.

With regard to the private rented sector, the manifesto confirmed that Labour would honour its long-standing commitment to introduce a licensing scheme for HMOs.

As well as emphasising that low interest rates enable more people to own their own homes, the manifesto committed Labour to make it easier for people buying and selling homes through the introduction of a new seller's pack (as proposed in the previous parliament), through the development of a modern basis of land registration to make conveyancing faster and cheaper, through grants for low-income home-owners, through reforms to leasehold and commonhold law, and through help for 10,000 key workers in high-cost areas to tackle recruitment problems.

Concerned that housing benefit had not been reformed in the previous parliament, Labour promised in its manifesto that it would, as a priority, work with local authorities to spread best practice in administration and tackle fraud and error, simplify housing benefit and its administration, distinguish between claimants of working age and pensioners, reform the provision of benefit to private tenants, and examine the case for longer awards. The manifesto also declared that in the long term, rents would be restructured to ensure that for people of working age, housing benefit would strengthen work incentives.

The manifesto also made it clear that Labour – in tackling social exclusion – would reduce homelessness and the use of costly bed and breakfast accommodation, and was committed to cutting rough sleeping to two-thirds of its 1998 level by 2002 and to keeping the number as low as possible thereafter.

Labour's housing policy in the aftermath of the 2001 election

Soon after Labour's election win on 7 June 2001, Tony Blair's second-term government began to deal with the unfinished business of the 1997–2001 administration – as set out in the manifesto. However, only time will tell whether the declared aims of housing policy – under a newly established Department of Transport, Local Government and the Regions (DTLR) – will or will not be largely achieved. It was probable that progress would be slow. For example, in the Queen's Speech of 20 June 2001 there was reference to only two housing bills that would be published in the following session of parliament, both concerned with home-ownership. The first was the Commonhold and Leasehold Reform Bill (resurrected from the 2000/01 session) and the second was the new Land Registration Bill, intended to speed up registration through the process of electronic land searches and conveyancing. However, in the aftermath of the Queen's Speech, two further issues appeared on the housing agenda: the problem of substandard housing and the high cost of housing for key workers in London and the South East.

It was clear to the new Secretary of State at the DTLR, Mr Stephen Byers, that the abolition of substandard housing was central to the government's intention to reduce child poverty. In total – according to the DTLR – as many as 2.4 million children lived in substandard housing in 2001, and two-thirds of this number were in the private rented sector (Wintour, 2001). The Secretary of State therefore pledged to improve the living conditions of about 250,000 children living in council dwellings by 2004, and an additional 50,000 in the private sector through a combination of renovation and better heating. To help a further 1.5 million children living in substandard private rented dwellings after 2004, Mr Byers suggested that it would be 'appropriate to attach conditions to the payment of housing benefit. One such condition could be a requirement to improve the state of the property so that it could meet [the DTLR's] definition of decency' (Byers, 2001). However, although this approach was mooted in the Housing Green Paper (DoE/DSS, 2000), it provoked criticism not only from landlords (inevitably), but also from some housing specialists who argued that tenants could be faced with eviction if the landlord had to forgo benefit, and that the stock of private rented accommodation would decrease as properties would be taken off the market (Wintour, 2001).

Other specialists claimed that Mr Byers's approach did not go far enough. Mr Frank Field MP, for example, argued that whereas the Secretary of State had direct control over the budgets of local authorities and could thus use his powers against poorly performing councils, he had far less control over the spending of housing associations. While housing associations might be good at building new properties, many appear to be not very good at using their resources to maintain their stock in reasonable condition. The Secretary of State thus needed to 'act with equal rigour against rogue housing associations [and to] transfer part of the stock and budget of the large, poorly performing housing associations to smaller, more efficient and humane housing providers' (Field, 2001). A further cause for concern was that in many run-down areas, owner-occupied housing was being sold at auction to private landlords at knock-down prices, and then let to housing benefit tenants 'whose rent payments from the public purse . . . all too quickly [covered] the cost of buying and [presented] the landlord with one of the best returns on capital available anywhere in the country' (Field, 2001). Since an increasing number of private landlords apparently cared 'little for their property and less for their neighbourhood' (Field, 2001), and were concerned only with exploiting the housing benefit system at taxpayers' expense, local authorities clearly needed powers to refuse housing benefit payments to landlords unwilling to keep their properties in reasonable condition.

Regarding the issue of affordable housing in high-cost areas, the DTLR announced in August 2001 that as from the following September around two thousand teachers in London and the South East would be the first beneficiaries of the £250 million Starter Homes Initiative. Assistance of up to £25,000 per applicant was available in the form of interest-free loans, grants towards deposits on homes or help to take part in shared ownership schemes. Eligible households had to contain at least one person who was employed as a teacher before an award could be made, and the subsidy would only be given out once mortgage agreements had been signed. Grants were repayable if the household subsequently sold the property. Bidders – mainly RSLs – were responsible for running the initiative, and provided an interface between the owner-occupied and social housing sectors. It was intended that in the second phase of the scheme, other key workers such as nurses, police officers and other public sector workers would also be eligible for assistance (see p. 268).

Policy at the crossroads

Because of Treasury constraints, the DETR increasingly promoted public–private financial arrangements as a supplementary means of providing resources for the maintenance and improvement of social housing. However, since local authorities were barred from borrowing on the value of their stock, they found it increasingly necessary to increase the pace and scale of LSVTs and set up more arm's-length housing management companies – since RSLs were permitted to borrow in the capital market. In the expanding RSL sector, rents were inevitably raised to market levels to provide an adequate return on investment (although the rate of increase might decelerate since social housing grants (SHGs) were raised from 60 per cent in 2001/02 to 68 per cent in 2002/03). Public–private financial initiatives were increasingly condemned by Labour MPs and the public sector union UNISON. The latter, at its annual conference in June 2001, also opposed the use of arm's-length housing companies on the grounds that these could form the basis of full-scale privatisation at a later stage when opposition had died down (Chatterjee, 2001). Possibly as a result of these criticisms and as an indication of new thinking, the Secretary of State, Mr Stephen Byers, declared at the Labour Party Conference in October 2001 that the government would be giving local authorities the opportunity of raising funds in the private market for investment in their housing stock. If the amount of permitted borrowing was substantial, this could delay or reverse the decline of local authority housing during the first decade of the twenty-first century.

Policy omissions

Labour's 2001 manifesto failed to include any policy proposals concerning the 'North–South divide' (other than the Starter Homes Initiative); housing problems relating to people in care, the elderly, women, and black and Asian households; and the continuing issue of homelessness. As in other key areas of public policy, the neo-liberal nostrums of the free market (involving the use of increasing amounts of private capital, and market pricing) coupled with private self-interest seem destined – in the early twenty-first century – to supersede a social democratic approach to housing welfare that emerged during the first half of the twentieth century and remained largely intact until the late 1970s.

In housing, as in many other fields, Blair governments after 1997 seemed determined to consolidate policies introduced under successive Thatcher administrations in the 1980s rather than adopt a left-of-centre radical agenda more in keeping with Labour's social democratic past.

Further reading

Since at the time of writing – there are few, if any, publications on Labour's housing policy following its election victory in June 2001, the following publications are recommended to enable housing researchers to analyse policy as it is implemented during the early years of the new millennium:

Council of Mortgage Lenders (quarterly) *Housing Finance.* Contains a mass of useful tables on house prices and mortgages, as well as useful quantitatively flavoured articles on private housing.

DTLR (quarterly/annually) *Housing and Construction Statistics.* An invaluable source of data on new housebuilding and renovation.

To gain a more qualitative understanding of the future development of policy and its application, reference should be made to the following academic journals: *Area, Housing Studies, International Journal of Urban and Regional Research, Journal of Social Policy, Policy and Politics* and *Urban Studies.* Professional journals often provide a more immediate response to policy, and the following could usefully be consulted: *Agenda, Housing Review, Housing Today, Inside Housing, Roof* and *Social Housing.*

Organisations such as the Chartered Institute of Housing and Shelter also produce very useful publications on aspects of housing policy, while occasional papers and working papers on housing issues are published by academic institutions such as the Centre for Housing Policy, University of York; CURS (the Centre for Urban and Regional Studies, University of Birmingham); the Department of Urban Studies at the University of Glasgow; and SPS (the School for Policy Studies, University of Bristol).

● **Postscript**

Notwithstanding the policy omissions alluded to at the end of Chapter 20, a number of new initiatives were announced by the government in the autumn of 2001 which signposted the future direction of policy, and there were also changes in the home-ownership market unforeseen at the time of the 2001 election that had an impact on policy.

Emanating from a recommendation in the Housing Green Paper (DETR/DSS, 2000), around 3 million council tenants were beginning to face rent increases of up to 16 per cent above the rate of inflation because the DTLR was insisting that from April 2002 rents were to be more closely aligned with private property values. However, alignment would be phased over 10 years, with annual increases of no more than 1.5 per cent in real terms, and in some areas where property values were low (such as in parts of the north) rents would actually decrease in relation to inflation. In contrast in London many key workers could be disadvantaged by rising rents simply because house prices had doubled in the capital in recent years, and/or they were living in areas which had become fashionable. Although the government was willing to cap rent increases in the capital (so that no rent would exceed £100 per week in 2002–03), there was no provision for capping to continue.

It is very probable that the extra revenue derived from rent increases (possibly amounting to around £1.2 billion per annum) will only in part be invested in improvement, and therefore more than 1.5 million council dwellings (half the public sector stock) will need to be transferred to RSLs by 2010 if the government is to meet its 10-year target to tackle the repairs backlog without increasing public spending.

Clearly the government was consolidating the role of housing associations as the principal providers of social housing. It not only expanded the size of this sector by promoting transfers, but in November

2001 announced that the SHG would be increased from 60 to 68 per cent for new housing association homes, the highest it had been for 10 years.

Although only indirectly the result of public policy, interest rates fell substantially in the latter part of 2001 to the benefit of mortgagors. In October, following a reduction in bank base rates, mortgage lenders cut their variable interest rate from 5.75 per cent to 5.5 per cent or less, the lowest since the late 1950s; and in November, following a further cut in base rate, most mortgage lenders cut their rates by a further 0.5 per cent. However, in contrast to the late 1980s, it was unlikely that lower interest rates would fuel inflation since the housing market showed signs of a slowdown – possibly as a response to a lower rate of economic growth, worsening employment prospects and a reduced level of consumer confidence after the terror attacks in the United States on the 11 September.

Owner-occupation, however, is of concern to the government in respect to market disparities across Britain. In London (where average house prices had tripled in 10 years), owner-occupation became unaffordable for many households where the price of even the cheapest one-bedroom flat had risen to about £80,000 by 2001 (out of reach of most key workers), while as many as 210,000 families were on waiting lists in the capital with little chance of securing accommodation in the foreseeable future. It would thus seem that the £250 million available to fund the Starter Homes Initiative – designed to help 10,000 key workers get housed by 2003–04 – is inadequate, and might fuel inflation at the lower end of the market since it will not create any new homes. The government thus indicated in the autumn of 2001 that it would review the planning guidance rules that stated that developers – in order to receive planning permission – only had to supply a certain percentage of affordable housing in any housing development scheme. In London, for example, the review could mean that the proportion has to rise from the current 25 per cent to say 50 per cent on the same site, or that developers could use different sites to provide the same number of cheaper dwellings. The government was also intent on increasing the supply of affordable housing for low-income households in parts of the country that had attracted many second home-owners. In November 2001 it thus signalled the end of half-price council tax for second home-owners, so that a doubling in tax would both deter the future acquisition of second homes and help fund the building of new houses in the countryside for the rural poor.

In parts of the north, however, the supply of housing is in excess of demand, rather than vice versa, and prices have plummeted in recent years. In late 2001 the government was attracted to the idea that local authorities (from a targeted 'market renewal fund' and within specific zones) should guarantee to purchase a house from a vendor if the value of the property fell below, say, £15,000, and then renovate it and either sell it or offer it for rent. With regard to the malfunction at the lower end of the market across Britain, the Chancellor of the Exchequer, Mr Gordon Brown, announced in his Autumn Statement of 2001 that stamp duty in 2,000 local government wards would be abolished on property sales of up to £150,000 from 30 November, but only time will tell whether such a concession will increase the number of transactions in the selected areas.

With fewer and fewer houses being built in the social sector, and the problem of affordability being largely unresolved in the home-ownership market (at least in London), homelessness remained a major issue, despite a recorded reduction in the number of rough sleepers. In London, for example, around 43,000 families were living in bed and breakfast accommodation in 2001, while at the same time over 104,000 London homes were empty. An initiative was therefore launched by the Empty Homes Initiative – a government quango – to convert around 66,000 properties (including empty shops and offices) into accommodation for the registered homeless and for other households living in substandard accommodation. More importantly, the Homelessness Act 2002 (based on the Homes Bill 2001) compelled local authorities to publish their homelessness strategies within 12 months of the introduction of the legislation. Strategies – updated every five years – would be based on a review that would not only involve the comprehensive mapping of a homelessness map, but would also include details about the resources available to the housing authority, other bodies (such as social services), landlords, voluntary organisations and other relevant services. Stemming from the review, the strategy – its format set out in Clause 3 of the Act – requires local authorities, in collaboration with other relevant bodies, to ensure sufficient accommodation for people who are or who may become homeless. Clearly, only time will tell whether this initiative will be successful. It is fairly certain, however, that few extra resources will be allocated to this initiative. The emphasis will be on the reorganisation of services rather than on the use of additional resources – not least in terms of providing more affordable housing.

With regard to the need to provide accommodation for asylum seekers, the Home Secretary, Mr David Blunkett, announced in October 2001 that his department would begin to dismantle the 'disastrous system' that had seen asylum seekers being sent to rundown housing in towns and cities across Britain. Instead, initially eighty reception centres would be opened where asylum seekers would live in hostel-type conditions (with legal and medical services provided) until their case had been decided. There was, however, concern among various housing bodies (such as the National Housing Federation) about how the housing needs of asylum seekers would be satisfied once they gained refugee status, and the extent to which their plight could add to the overall problems of social exclusion.

It was quite clear that housing policy – as it gradually evolved during the autumn of 2001 under the ministerial responsibility of Lord Falconer, was compatible with the policy-aims of Labour set out in its election manifesto of the same year and in the Queen's speech for the following session of Parliament – as discussed in Chapter 20. Only time will tell whether these aims will be realised.

References

Abrams, M. and Rose, R. (1959) *Must Labour Lose?*, Harmondsworth: Penguin Books.

Age Concern (1990) *Sheltered Housing for Sale*, Fact Sheet, London: Age Concern.

AMA (1982) *Building for Tomorrow: Housing Investment, Construction and Employment*, London: AMA.

AMA (1983) *Defects in Housing Part 1*, London: AMA.

AMA (1984) *Defects in Housing Part 2*, London: AMA.

AMA (1985) *Defects in Housing Part 3*, London: AMA.

AMA (1986) *The Case for Local Housing*, London: AMA.

AMA (1987) reported in Warman, C. (1987) 'Ridley releases his proposals for a new rented property era', *The Times*, 30 September.

AMC (1973) Memorandum M21, in *House of Commons Tenth Report of the Expenditure Committee, Environmental and Home Office Subcommittee, Session 1972–73*, London: HMSO.

Amin, K. (1987) 'Black women and racist attacks', *Foundation*.

Anson, Lady (1990) reported in Wolmar, C. (1990) 'Figures show sharp increase in council's priority lists', *Independent*, 23 June.

Archbishop of Canterbury's Commission on Urban Priority Areas (1985) *Faith in the City*, London: Church House Publishing.

Ardill, J. (1987) 'Ridley to allow more homes in the rural south', *Guardian*, 12 December.

Armstrong, H. (1998) Speech at the Annual Conference of the Chartered Institute of Housing, Harrogate, 16 June.

Audit Commission (1986) *Managing the Crisis in Council Housing*, London: HMSO.

Aughton, H. and Malpass, P. (1991) *Housing Finance: A Basic Guide*, London: Shelter Publications.

Austerberry, H. and Watson, S. (1986) *Housing and Homelessness*, London: Routledge & Kegan Paul.

Babbage, A.G. (1992) 'Housing associations: many a slip twixt cup and lip', *Housing and Planning Review*, February/March.

Balchin, P. (1981) *Housing Policy and Housing Needs*, Basingstoke: Macmillan.

Bank of England (1985) *Quarterly Bulletin*, 1, London: Bank of England.

Bank of England (1991) *Quarterly Bulletin*, 1, London: Bank of England.

Bank of England (1992) *Quarterly Bulletin*, 3, London: Bank of England.

Barlow, J. and Duncan, S. (1994) *Success and Failure in Housing Provision: European Systems Compared*, Oxford: Pergamon.

Barnes, C. (1997) *Care, Communities and Citizens*, Harlow: Longman.

Bassett, K. and Short, J. (1979) *Housing and Residential Structure: Alternative Approaches*, London: Routledge & Kegan Paul.

Bayley, R. (2000) 'Close to the edge', *Roof*, September/October.

Beechey, V. (1986) 'Studies of women's employment', in *Feminist Review*, London: Virago.

Belcher, J. and Williams, J. (1999) 'Early system warning', *Guardian*, 24 November.

Best, R. (1982a) reported in *Guardian*, 3 November.

Best, R. (1982b) reported in *Guardian*, 6 November.

Best, R. (1985) reported in Carvel, J. (1985) 'Duke and architect urge rethink on housing', *Guardian*, 25 January.

Best, R. (1992) 'The homeless generation', *Observer*, 5 July.

Bevan, A. (1945) Parliamentary Debates (Hansard), *House of Commons Official Report*, Vol. 414, London: HMSO.

Beveridge Report (1942) *Social Insurance and Allied Services*, Cmnd 6404, London: HMSO.

Blackaby, B. and Chahal, K. (2000) *Black and Minority Ethnic Housing Strategies: A Good Practice Guide*, Coventry: CIH.

Blair, T. (1997) Speech at Aylesbury Estate, Southwark, London, 2 June.

Blake, J. (2001) 'Hit, miss or maybe', *Roof*, May/June.

Blytheway, B. (1984) 'The lucky five per cent', *Roof*, March/April.

Boelhouwer, P. (1991) 'Convergence or divergence in the general housing policy in seven European countries?', Paper presented at the conference on Housing Policy as a Strategy for Change, European Network for Housing Research, Oslo, 24–27 June.

Boleat, M. (1992) 'Britons still dream of their own home', *Observer*, 13 October.

Boliver, D. (1992) 'The tower block tenants who are forced to live in misery', *Guardian*, 6 June.

Bonnerjea, L. and Lawton, J. (1987) *Homelessness in Brent*, London: Policy Studies Institute.

Bonnerjea, L. and Lawton, J. (1988) *No Racial Harassment This Week*, London: Policy Studies Institute.

Bookbinder, D. (1987) 'Housing options for the elderly', *Roof*, September/October.

Bowen, M. (1980) 'Slump in housebuilding heads for 30 year low', *Sunday Times*, 2 November.

Bradshaw, J. (1997) 'The concept of human need', in Fitzgerald, M., Halmos, P., Muncie, J. and Zeldin, D. (eds) *Welfare in Action*, London: Routledge & Kegan Paul.

Bramley, G. (1990) 'Bridging the affordability gap in 1990', London: ADC and House Builders Federation.

Bramley, G. (1991) 'Bridging the affordability gap', London: ADC and House Builders Federation.

Bramley, G. (1993) 'The enabling role for local authorities: a preliminary evaluation', in Malpass, P. and Means, R. (eds) *Implementing Housing Policy*, Buckingham: Open University Press.

Bramley, G. (1998) 'Why need does add up', *Roof*, November/December.

Bramley, G., Bartlett, W. and Lambert, C. (1995) *Planning, the Market and Private Housebuilding*, London: UCL Press.

Breheny, M. (2000) *The People: Where Will They Work?*, London: TCPA.

Brindle, D. (1990) 'Britain needs millions of homes', *Guardian*, 14 March.

Brindle, D. (1999) 'Lonely old age', *Guardian*, 20 March.

British Gas (1991) *The British Gas Report on Attitudes to Ageing 1991*, London: Burston Marsteller.

Brooks, E. (2000) 'More than just street wise', *Housing Today*, 31 August.

Brown, C. (1984) *Black and White Britain: The Third Policy Studies Institute Survey*, London: Heinemann.

Brown, P. (2001) 'Blueprint for rural renewal drawn up', *Guardian*, 23 January.

Bulmer, M. (1987) *The Social Bases of Community Care*, London: Allen & Unwin.

Burrows, R. and Wilcox, S. (2000) 'Half the poor? The growth of low income home-ownership', *Housing Finance*, No. 47.

Byers, S. (2000) Speech to the Social Market Foundation, London, 2 August.

Cabinet Office (1999) *Rural Economies*, London: The Stationery Office.

Cameron, S., Baker, M., Bevan, M., Hull, A. and Williams, R. (1998) *Regionalism, Devolution and Social Housing*, Coventry: National Housing Forum (Publications).

Cameron, A., Harrison, L., Burton, P. and Marsh, A. (2001) *Crossing the Housing and Care Divide*, Bristol: Policy Press.

Carvel, J. (1982a) 'The rents that went through the roof', *Guardian*, 12 October.

Carvel, J. (1982b) 'The minister's version of home insulation', *Guardian*, 26 November.

Carvel, J. (1984a) 'Homebuyers defaulting in record numbers', *Guardian*, 20 February.

Carvel, J. (1984b) '3m more houses will be needed by 2000', *Guardian*, 13 March .

Carvel, J. (1984c) 'Builders appeal for home grants', *Guardian*, 19 October.

Carvel, J. (1985) '£19bn needed to right council house defects', *Guardian*, 5 March.

Central Office of Information (1993) *Housing*, London: HMSO.

CHAR (1981) reported in *Guardian*, 15 June.

Chatterjee, M. (2001) 'Unison says no to arm's length option', *Housing Today*, 21 June.

Clapham, D., Kemp, P. and Smith, S.M. (1990) *Housing and Social Policy*, Basingstoke: Macmillan.

Clarke, S. and Ginsburg, N. (1975) 'The political economy of housing', in *Political Economy and the Housing Question*, London: Political Economy of Housing Workshop.

CML (1999) *Annual Finance Survey*, London: CML.

Coates, K. (1975) *The Labour Party and the Struggle for Socialism*, Cambridge: Cambridge University Press.

Cole, I., Kane, S. and Robinson, D. (1999) *Changing Demand, Changing Neighbourhoods: The Response of Social Landlords*, Sheffield: Centre for Regional, Economic and Social Research, Sheffield Hallam University.

Coleman, A. (1985) *Utopia on Trial*, London: Hilary Shipman.

Coleman, D. and Salt, J. (1992) *The British Population: Patterns, Trends and Processes*, Oxford: Oxford University Press.

Coleman, L. and Watson, S. (1987) *Women over Sixty*, Canberra: Australian Institute of Social Studies.

Connell, R.W. (1995) *Masculinities*, Sydney: Allen & Unwin.

Corbyn, J. (2000) *Council Housing*, House of Commons Early Day Motion 838, Session 1999–2000, London: House of Commons.

Cowan, H. (1999) *Community Care, Ideology and Social Policy*, Englewood Cliffs, NJ: Prentice Hall.

Cowan, R. (1984) 'Elderly ghettoes', *Roof*, November/December.

Cowling, M. and Smith, S. (1984) 'How to avert Labour's great housing disaster', *Guardian*, 19 November.

CRE (1984) *Race and Council Housing in Hackney: Report of a Formal Investigation*, London: CRE.

CRE (1987) *Living in Terror: A Report on Racial Violence and Harassment in Housing*, London: CRE.

CRE (1988a) *Homelessness and Discrimination: Tower Hamlets*, London: CRE.

CRE (1988b) *Racial Discrimination in a London Estate Agency: A Report of a Formal Investigation into Richard Barclay and Co.*, London: CRE.

CRE (1989a) *Race and Housing in Glasgow: The Role of Housing Associations*, London: CRE.

CRE (1989b) *Racial Discrimination in Liverpool City Council: Report of a Formal Investigation into the Housing Department*, London: CRE.

CRE (1990a) *Racial Discrimination in an Oldham Estate Agency: Report of a Formal Investigation into Norman Lester and Co.*, London: CRE.

CRE (1990b) *Sorry It's Gone*, London: CRE.

CRE (1993) 'Housing associations and race equality', *Black Housing*, Vol. 9, No. 8.

Crosland, A. (1971a) 'Twelve points for a Labour housing policy', *Guardian*, 15 December.

Crosland, A. (1971b) 'Housing and equality', *Guardian*, 15 June.

Crosland, A. (1974) Parliamentary Debates (Hansard) *House of Commons Official Report*, Vol. 873, London: HMSO.

Crosland, A. (1975) Speech to Housing Trust, June.

CSO (1987) *Social Trends, 17*, London: HMSO.

CSO (1992) *General Household Survey*, London: HMSO.

CSO (1993) *Social Trends, 23*, London: HMSO.

CSO (1994) *Social Trends, 24*, London: HMSO.

Curtice, J. (1991) 'House and home', *British Social Attitudes*, 8th Report, Aldershot: Dartmouth.

Daniel, T. (1987) Letter to *Guardian*, 3 October.

Darke, J. (ed.) (1982) *The Roof over Your Head*, Nottingham: Spokesman, for Labour Housing Group.

Davis, O.A. and Whinston, A.B. (1961) 'The economics of renewal', in Wilson, J.Q. (ed.) *Urban Renewal: The Record and the Controversy*, Cambridge, MA: Harvard University Press.

Dean, J. and Goodlad, R. (1998) *The Role and Impact of Befriending*, Brighton: Pavilion.

Denington Committee (1966) *Our Older Homes: A Call for Action*, Report of the Central Housing Advisory Committee, London: HMSO.

Department of Employment (1988) *Employment Gazette*, Vol. 93, No. 3.

Desai, M. (1983) 'Economic Alternatives for Labour, 1984–9', in Counterpoint, *Socialism in a Cold Climate*, London: Unwin.

DETR (1998a) *Rethinking Construction*, Report of the Construction Industries Task Force chaired by Sir John Egan, London: The Stationery Office.

DETR (1998b) *English House Condition Survey 1996*, London: The Stationery Office.

DETR (1998c) Circular 6/98, *Affordable Housing*, London: The Stationery Office.

DETR (1999) *Towards an Urban Renaissance*, Final Report of the Urban Task Force chaired by Lord Rogers of Riverside, London, E. & F.N. Spon.

DETR (2000a) Urban White Paper, *Our Towns and Cities: The Future. Delivering an Urban Renaissance*, Cm 4911, London: The Stationery Office.

DETR (2000b) *Low Demand Housing and Unpopular Neighbourhoods*, London: The Stationery Office.

DETR (2000c) *Public Expenditure Plans*, London: The Stationery Office.

DETR/DSS (2000) The Housing Green Paper, *Quality and Choice: A Decent Home for All*, London: The Stationery Office.

DETR/LPAC (1998) *Planning for the Communities of the Future*, London: DETR.

DETR/MAFF (2000) Rural White Paper, *Our Countryside: The Future. A Fair Deal for Rural England*, Cm 4909, London: The Stationery Office.

Dhillon-Kashyap, P. (1994) 'Black women and housing', in Gilroy, R. and Woods, R. (eds) *Housing Women*, London: Routledge.

Diamond Report (1976) *Report on the Royal Commission on the Distribution of Incomes and Wealth*, Report No. 4, Cmnd 6626, London: HMSO.

DoE (1971) *Fair Deal for Housing*, Cmnd 4728, London: HMSO.

DoE (1973a) *Widening the Choice: The Next Steps in Housing*, Cmnd 5280, London: HMSO.

DoE (1973b) *Better Homes: The Next Priorities*, Cmnd 339, London: HMSO.

DoE (1973c) *English House Condition Survey 1971*, London: HMSO.

DoE (1974) *Local Authority Housing Programmes*, Circular 70/74, London: HMSO.

DoE (1977a) *Housing Policy: A Consultative Document*, London: HMSO.

DoE (1977b) *National Dwelling and Housing Survey*, London: HMSO.

DoE (1978a) *English House Condition Survey 1976*, London: HMSO.

DoE (1978b) *National Dwelling and Housing Survey*, London: HMSO.

DoE (1979) 'The financial aspects of council house sales' (unpublished).

DoE (1980) *Appraisal of Financial Effects of Council House Sales*, London: HMSO.

DoE (1982) *Assistance with Housing Costs*, London: HMSO.

DoE (1983) *English House Condition Survey 1981*, London: HMSO.

DoE (1985a) *Physical and Social Survey of Houses in Multiple Occupation in England and Wales*, London: HMSO.

DoE (1985b) Green Paper, *Housing Improvement: A New Approach*, Cmnd 9513, London: HMSO.

DoE (1987) *Housing: The Government's Proposals*, Cmnd 214, London: HMSO.

DoE (1988) *English House Condition Survey 1986*, London: HMSO.

DoE (1989) *Local Authorities' Housing Role: 1989 HIP Round*, London: HMSO.

DoE (1991) *Planning and Affordable Housing*, Circular 7/91, London: HMSO.

DoE (1992a) PPG 3, *Housing*, London: HMSO.

DoE (1992b) *Housing and Construction Statistics 1981–1991 Great Britain*, London: HMSO.

DoE (1993) *English House Condition Survey 1991*, London: HMSO.

DoE (1994) *Access to Social Housing*, London: HMSO.

DoE (1995a) White Paper, *Our Future Homes*, Cm 2902, London: HMSO.

DoE (1995b) *Annual Report*, London: HMSO.

DoE (1996a) Green Paper, *Future Growth. Where Shall We Live?*, London: HMSO.

DoE (1996b) Circular 13/96, *Planning and Affordable Housing*, London: HMSO.

DoE (1997) Circular 1/97, *Planning and Compensation Act 1991: Planning Obligations*, London: HMSO.

Donnison, D.V. (1967) *The Government of Housing*, Harmondsworth: Penguin Books.

Donnison, D.V. (1987) 'New drama for a crisis', *Guardian*, 11 May.

Dorling, D. and Gentle, C. (1994) 'Mortgage change is step in right direction', *Guardian*, 28 March.

Douglas, A. and Gilroy, R. (1994) 'Young women and homelessness', in Gilroy, K. and Woods, R. (eds) *Housing Women*, London: Routledge.

Doyal, L. and Gough, I. (1991) *A Theory of Human Need*, Basingstoke: Macmillan.

Dowling, R. (1998) 'Gender, class and home ownership: placing the connections', *Housing Studies*, Vol. 13, No. 4.

DSS (1992) Circular 10/92, *Housing and Community Care*, London: HMSO.

DSS (1993) *Department of Social Security Statistics*, London: HMSO.

Durham, M. (1990) 'Families growing rich in inheritance bonanza', *Sunday Times*, 29 July.

Dwelly, T. (1999) 'Less to spend more to lend', *Roof*, January/February.

Dwelly, T. (2000a) 'New Labour, new homes?', *Roof*, July/August.

Dwelly, T. (2000b) 'Housing market healthcheck', *Roof*, May/June.

Earley, F. (2000) 'Scotland', *Housing Finance National Markets Review 2000*, London: CML.

Economist, The (1983a) 'How Britain voted', 18–24 June.

Economist, The (1983b) 'Subsidy for house buyers', 19–25 February.

Economist, The (1986) 'Undermining the council empires', 26 April.

Employment Gazette (1993), December.

Ermisch, J. (1990) *Fewer Babies, Longer Lives*, York: JRF.

Esping-Andersen, G. (1990) *The Three Worlds of Welfare Capitalism*, Cambridge: Polity Press.

Evans, A.W. (1988) *House Prices and Land Prices in the South East: A Review*, London: House Builders Federation.

Evening Standard (1985) 'How speculators threatened tenants', 14 January.

Ezard, J. (1987) 'Scarman's plea to use B and B cash for houses', *Guardian*, 6 October.

Field, F. (2001) 'How to beat the new slum landlords', *Guardian*, 7 August.

Flynn, N. and Walsh, K. (1982) *Managing Direct Labour Organisations*, London: ILGS.

Foot, M. (1973) *Aneurin Bevan*, London: Davis Poynton.

Forbes, D. (1988) *Action on Racial Harassment: Legal Remedies and Local Authorities*, London: Legal Action Group.

Forrest, R. (1982) 'The social implications of council house sales', in English, J. (ed.) *The Future of Council Housing*, London: Croom Helm.

Forrest, R. and Murie, A. (1982) 'The great divide', *Roof*, November/December.

Forrest, R. and Murie, A. (1984) 'Right to buy? Issues of need, equity and polarisation in the sale of council houses', *Working Paper 39*, Bristol: SAUS.

Forrest, R. and Murie, A. (1988) 'Fiscal reorientation, centralization and privatization of council housing', in van Vliet, W. (ed.) *Housing Markets and Policies under Fiscal Austerity*, New York: Greenwood Press.

Forrest, R. and Murie, A. (1990) *Selling the Welfare State*, London: Routledge.

Forrest, R., Kennett, P. and Leather, P. (1994) 'Mortgage equity withdrawal: causes and consequences', *Findings*, Housing, York: JRF.

Forshaw, R. (2000a) 'Revolutionary plan for biggest-ever HA', *Housing Today*, 13 April.

Forshaw, R. (2000b) 'Private finance boom for transfers goes on', *Housing Today*, 27 April .

FPHM (1990) *Housing and Homelessness: A Public Health Perspective*, London: FPHM.

Frankena, M. (1975) 'Alternative models of rent control', *Urban Studies*, 12.

Garratt, D. (2000) 'Affordability', *Housing Finance*, No. 47.

Garratt, D. (2001) 'London's housing market', *Housing Finance*, No. 49.

Gauldie, E. (1974) *Cruel Habitations*, London, Allen & Unwin.

Gibb, K., Munro, M. and Satsangi, M. (1999) *Housing Finance in the United Kingdom*, 2nd edn, Basingstoke: Macmillan.

Gilroy, R. (1994) 'Women and owner-occupation in Britain', in Gilroy, R. and Woods, R. (eds) *Housing Women*, London: Routledge.

Ginsburg, N. (1992) 'Black people and housing policies', in Birchall, J. (ed.) *Housing Policy in the 1990s*, London: Routledge.

GLA (2000) *Homes for a World City*, London: GLA.

GLC (1986) *Private Tenants in London: The Greater London Council Survey 1983–84*, Greater London Council Housing Research and Policy Report No. 5, London: GLC.

Goss, S. and Lansley, S. (1981) *What Price Housing? A Review of Housing Subsidies and Proposals for Reform*, London: SHAC.

Gould, B. (1984) 'Agenda', *Guardian*, 27 July.

Gould, B. (1989) reported in Travis, A. (1989) '£250 million to aid homeless', *Guardian*, 16 November.

Grant, C. (1992) 'Voluntary transfer: policy that never was', *Observer*, 5 July.

Greenwood, A. (1969) Parliamentary Debates (Hansard) *House of Commons Official Report*, Vol. 777, London: HMSO.

Greve, J. (1971) *Homelessness in London*, Edinburgh: Scottish Academic Press.

Greve, J. (1990) *Homelessness in Britain*, York: JRT.

Grey, A., Hepworth, N. and Odling Smee, J. (1980) *Housing Rents, Costs and Subsidies: A Discussion Document*, London: CIPFA.

Griffiths, D. (1982) 'Triple onslaught on council housing', *Labour Weekly*, 12 March.

Griffiths, D. and Holmes, C. (1984) 'To buy or not to buy . . . is the question?', *Marxism Today*, May.

Griffiths Report (1988) *Community Care: An Agenda for Action*, London: HMSO.

Grosskurth, A. (1986) 'A long wait is a short while', *Roof*, July/August.

Grundy, E. (1987a) 'Future patterns of morbidity in old age', in Caird, F.I and Grimley Evans, J. (eds) *Advanced Geriatric Medicine*, 6, Bristol: John Wright.

Grundy, E. (1987b) 'Household change and migration among the elderly in England and Wales', *Espace, Populations, Societies*, 1.

Grundy, E. (1992) 'The living arrangements of elderly people', *Reviews in Clinical Gerontology*, 2.

Grundy, E. (1993) 'Moves into supported private housing among elderly people in England and Wales', *Environment and Planning A*, 25.

Guardian (1982) 'Housing', 31 December.

Guardian (1988a) 'Mr Ridley's roof leaks', 25 May.

Guardian (1988b) 'Housing Bill built on shaky premises', 31 May.

Gummer, J. (1994) Speech at ADC Conference, 'Tackling empty houses', as reported in *Planning Week, The Journal of the Royal Town Planning Institute*, 28 April.

Hall, P. (2001) 'How much better than no bread', *Town and Country Planning*, January.

Hallett, C. (ed.) (1996) *Women and Social Policy: An Introduction*, Hemel Hempstead: Prentice Hall/Harvester Wheatsheaf.

Hamnett, C. (1983) 'From the foundations up', *New Statesman*, 14 October.

Hamnett, C. (1988a) 'Conservative government housing policy in Britain, 1979–85: economics or ideology', in van Vliet, W. (ed.) *Housing Markets and Policies under Fiscal Austerity*, New York: Greenwood Press.

Hamnett, C. (1988b) 'Housing the new rich', *New Society*, 22 April.

Hamnett, C. and Randolph, W. (1983) 'The great flat break up', *New Society*, 22 April.

Hamnett, C., Harmer, M. and Williams, P. (1991) *Safe as Houses*, London: Paul Chapman.

Harker, M. and King, N. (1999) *An Ordinary Home: Housing and Support for People with Learning Difficulties*, London: IDEA Publications.

Harloe, M. (1979) *Private Rented Housing in England and the USA*, London: CES.

Harloe, M. (1995) *The People's Home? Social Rented Housing in Europe and America*, Oxford: Blackwell.

Harrison, M. (1998) 'Theorising exclusion and difference: specificity, structure, and minority ethnic housing issues', *Housing Studies*, Vol. 13, No. 6.

Harrison, M. with Davis, C. (2001) *Housing, Social Policy and Difference: Disability, Ethnicity, Gender and Housing*, Bristol: Policy Press.

Harvey, D. (1974) 'Class monopoly rent, finance capital and urban revolution', *Regional Studies*, 8.

Haskey, J. (1998) 'One parent families and their dependent children in Great Britain', *Population Trends*, 91, Spring.

Hebden, P. (2000) 'Prudential plans worry City', *Housing Today*, 15 June.

Henderson, J. and Karn, V. (1987) *Race, Class and State Housing*, Aldershot: Gower Press.

Herring, L. (1993) reported in Slaughter, J. 'Home truths for retiring types', *Sunday Times*, 20 June.

Heseltine, M. (1979) reported in Young, J. 'Right to buy homes for five million', *The Times*, 21 December.

Hetherington, P. (1999) 'Time runs out for green and pleasant land', *Guardian*, 15 May.

Hetherington, P. (2000a) 'Town grouse and country grouse', *Guardian*, 1 March.

Hetherington, P. (2000b) 'Young lose out in housing boom', *Guardian*, 17 March.

Hetherington, P. (2000c) 'Prescott plans to abolish council housing', *Guardian*, 24 January.

Hetherington, P. (2000d) 'Councils can tax second homes at full rate', *Guardian*, 29 November.

Hetherington, P. (2000e) 'Urban horror story', *Guardian*, 5 April.

Hetherington, P. (2000f) 'North's sink estates are beyond saving', *Guardian*, 6 June.

Hetherington, P. (2000g) 'Speculators blamed as 750,000 homes left empty', *Guardian*, 14 February.

Hilditch, S. (1981) 'Labour's discussions', *Roof*, September/October.

Hilditch, S. (1985) 'Planning housing investment', in Labour Housing Group, *Right to a Home*, Nottingham: Spokesman.

Hillman, J. (1969) 'New houses for old – when?', *Observer*, 2 February.

Hills, J. (1999) *Reinventing Social Housing Finance*, London: Institute for Public Policy Research.

Holmans, A. (1995) 'Housing demand and need in England 1991–2001', *Housing Research*, 157, York: JRF.

Holmans, A. and Brownie, S. (2001) 'Demand for housing is set to exceed all expectations', *Housing Today*, 8 March.

House of Commons (1980) *The First Report of the Environment Committee*, Session 1980/81, London: HMSO.

House of Commons (1981a) *The Third Report of the Environment Committee*, Session 1980/81, London: HMSO.

House of Commons (1981b) *The Second Report of the Environment Committee*, Session 1980/81, London: HMSO.

House of Commons (1982) *The First Report of the Environment Committee*, Session 1981/82, London: HMSO.

House of Commons (1991) *The Third Report of the Public Accounts Committee*, Session 1990/91, London: HMSO.

Housing Today (2000) 'LGA insist sheltered housing is not supported housing', 25 May.

Howes, E. and Mullins, D. (1997) 'Finding a place: the impact of locality on the housing experience of tenants from minority ethnic groups', in Karn, V. (ed.) *Ethnicity in the 1991 Census*, Vol. 4, London: HMSO.

Hughes, G. (1994) 'Out of step and stumbling', *Guardian*, 12 August.

Hughes, J. (1999) 'Come North, poor are told', *Guardian*, 19 December.

Hutton, W. (1994) 'Bed times for the good life', *Guardian*, 2 August.

Insley, J. (2001) '. . . but house boom just carries on', *Observer*, 22 July.

Institute of Housing (1990) reported in Knewstub, N. and Pilkington, E. '£15m to get homeless off the streets', *Guardian*, 23 June.

Jackson, A. (2001) *Evaluation of the Homebuy scheme in England*, York: York Publishing Services.

Jacobs, J. (1962) *The Death and Life of Great American Cities*, London: Jonathan Cape.

Jenkins, J. (1988) 'Labour authorities make lousy landlords', *New Society*, 19 February.

Johnson, R. J. (1987) 'A note on housing tenure and voting in Britain, 1983', *Housing Studies*, Vol. 2, No. 2.

Johnston, L., MacDonald, R., Mason, P., Ridley, L. and Webster, C. (2000) *Snakes and Ladders: Young People, Transitions and Social Exclusion*, Bristol: Policy Press.

Jones, K. (1994) *The Making of Social Policy in Britain, 1830–1990*, 2nd edn, London: Athlone Press.

JRF (1990) 'Policy implications of current demographic trends', *Findings*, Social Policy 7, York: JRF.

JRF (1991) *Inquiry into British Housing, Second Report*, York: JRF.

JRF (1994) 'Filling England's empty homes', *Findings*, Housing 111, York: JRF.

Kaletsky, A. (1992) 'Governing means admitting that there is a choice of alternatives', *The Times*, 24 August.

Karn, V. (1993) 'Remodelling a HAT: the implementation of the Housing Action Trust legislation, 1987–92', in Malpass, P. and Means, R. (eds) *Implementing Housing Policy*, Buckingham: Open University Press.

Karn, V.A. and Sheridan, L. (1994) *New Homes in the 1990s: A Study of Design, Space and Amenity in Housing Association and Private Sector Housing*, York: JRF.

Kaufman, G. (1979) reported in *Guardian*, 21 December.

Kaufman, G. (1981) Speech in Blackpool, 13 February.

Kaufman, G. (1982) reported in *Guardian*, 24 September.

Kaufman, G. (1983) reported in *Labour Weekly*, 18 February.

Kemeny, J. (1995) *From Public Housing to the Social Market: Rental Policy Strategies in Comparative Perspective*, London: Routledge.

Kemp, P. (1987) 'The ghost of Rachman', *New Society*, 6 November.

Kempson, E. and Ford, J. (1997) 'Falling through the private net', *Roof*, May/June.

Kempson, E. and Whyley, C. (1999) *Kept Out or Opted Out? Understanding and Combating Financial Exclusion*, Bristol: Policy Press.

Kerr, M. (1988) *The Right to Buy*, London: HMSO.

Kilroy, B. (1978) *Housing Finance: Organic Reform?*, London: LEFTA,

Kilroy, B. (1979) 'Labour housing dilemma', *New Statesman*, 28 September.

Kilroy, B. (1980a) *The Financial Implications of Government Policies on Home Ownership*, London: SHAC.

Kilroy, B. (1980b) 'Housing finance gone cuckoo', *Roof*, January/February.

Kilroy, B. (1981a) 'The real competition for resources in housing', *Housing Review*, July/August.

Kilroy, B. (1981b) 'Why housing investment is being cut so much', *Housing Review*, March/April.

Kilroy, B. (1981c) 'Council house sales: was the government misled by DOE's evidence?', *Housing and Planning Review*, Summer.

Kilroy, B. (1982) 'The financial and economic implications of council house sales', in English, J. (ed.) *The Future of Council Housing*, London: Croom Helm.

King, M.A. and Atkinson, A.B. (1980) 'Housing policy, taxation and reform', *Midland Bank Review*, Spring.

Kirkwood, A. (2001) 'Feeble and complacent', *Roof*, January/February.

Knight, M. (1982) 'Mortgage rights advice', *Roof*, January /February.

Labour Party (1976) *Labour's Programme 1976*, London: Labour Party.

Labour Party (1981) *A Future for Public Housing*, London: Labour Party.

Labour Party (1982a) *Labour's Programme 1982*, London: Labour Party.

Labour Party (1982b) *12 Point Plan*, London: Labour Party.

Labour Party (1983) *The New Hope for Britain*, London: Labour Party.

Labour Party (1985) *Homes for the Future*, London: Labour Party.

Labour Party (1997) *Because Britain Deserves Better*, London: Labour Party.

Labour Party (2001) *Ambitions for Britain: Labour's Manifesto 2000*, London: Labour Party.

Laing, W. (1988) 'The mixed economy in long term care', in Wells, N. and Freer, C. (eds) *The Ageing Population: Burden or Challenge?*, Basingstoke: Macmillan.

Lansley, S. (1982) *Housing Finance: A Policy for Labour*, London: Labour Housing Group.

Larkin, A. (1982) 'Keeping the old folks at home', *Roof*, July/August.

Leather, P. (2000a) *Crumbling Cities? Helping Owners Repair and Maintain Their Homes*, York: JRF.

Leather, P. (2000b) 'Grants to home-owners: a policy in search of objectives', *Housing Studies*, Vol. 15, No. 2.

Leather, P. and Mackintosh, S. (1989) 'Means-testing improvement grants', *Housing Review*, Vol. 38, No. 3.

Leather, P. and Mackintosh, S. (1993) 'Housing renewal in an era of mass home-ownership', in Malpass, P. and Means, R. (eds) *Implementing Housing Policy*, Buckingham: Open University Press.

Leather, P., Mackintosh, S. and Rolfe, S. (1994a) 'Housing conditions and housing renewal in England', Paper presented to the European Network for Housing Research Conference, Glasgow, August.

Lemos, G. (2000) *Racial Harassment: Action on the Ground*, London: Lemos & Crane.

Levinson, S. (1988) 'Home prices still rising despite stock market fall', *Independent*, 9 January.

Levitas, R. (1996) 'The concept of social exclusion and the new Durkheimian hegemony', *Critical Social Policy*, Vol. 46, pp. 5–20.

Linton, M. (1987) 'How the Tories rule the roost', *Guardian*, 19 July.

Linton, M. (1994) '£36 million a year helps keep borough loyal', *Guardian*, 17 January.

Livingstone, K. (2000) 'The Government must be saved from the dead hand of the Treasury', *Independent*, 4 September.

Lloyd, G. (1972) 'Price of flats for sale rockets as new landlords move into big blocks', *The Times*, 5 April.

Local Government Information Unit (1990) *Caring for People: The Government's Plans for Care in the Community*, Special Briefing No. 32, Local Government Information Unit.

Long, N. (2000) Letter to *Guardian*, 23 January.

Lundquist, L.J. (1986) *Housing Policy and Equality*, London: Croom Helm.

McCarthy, R. (2000) 'Backdoor', *Roof*, July/August.

MacEwan, M. (1991) *Housing, Race and Law*, London: Routledge.

Macfarlane, R. (2000a) *Local Jobs from Local Development: The Use of Planning Agreements to Target Training and Employment Outcomes*, York: York Publishing Services.

Macfarlane, R. (2000b) *Using Local Labour in Construction: A Good Practice Resource Book*, Bristol: Policy Press.

McGurk, P. (1987) reported in *Daily Mirror*, 6 January.

McIntosh, A. and Utley, C. (1992) 'Transferring allegiance', *Roof*, July/August.

McIntosh, A., Aughton, H. and Kilroy, B. (1978) Letter to *Guardian*, 17 November.

McKechnie, S. (1987) Letter to *Guardian*, 17 August.

McKechnie, S. (1993) reported in Shelter (1993) *What Future for Older People*, London: Shelter Publications.

Mackintosh, S., Means, R. and Leather, P. (1990) *Housing and Later Life: The Housing Finance Implications of an Ageing Society*, SAUS Study No. 4, Bristol: University of Bristol.

McLaughlan, C. (1984) 'Death for public housing', *Tribune*, 21 December.

Maclennan, D. (1989) *The Nature and Effectiveness of Housing Management in England*, London: HMSO.

Maclennan, D., Gibb, K. and More, A. (1991) *Fairer Subsidies, Faster Growth: Housing, Government and the Economy*, York: JRF.

McLoughlin, J. (1979) 'High rate mortgage relief faces axe', *Guardian*, 26 July.

McMahon, 'Kit' (1981) Speech in London, 13 July.

Malpass, P. (ed.) (1986) *The Housing Crisis*, London: Croom Helm.

Malpass, P. and Murie, A. (1990) *Housing Policy and Practice*, 3rd edn, Basingstoke: Macmillan.

Malpass, P. and Murie, A. (1999) *Housing Policy and Practice*, 5th edn, Basingstoke: Macmillan.

Mama, A. (1980) 'Violence against black women: gender, race and state responses', *Feminist Review*, No. 32, Summer.

Marsh, A. and Mullins, D. (1998) 'The social exclusion perspective', *Housing Studies*, Vol. 13, No. 6.

Martin, J., Meltzer, H. and Elliott, D. (1988) 'The prevalence of disability amongst adults', *OPCS Surveys of Disability in Great Britain, Report 1*, London: HMSO.

Marx, K. (1981 edn) *Capital*, Vol. 3, Harmondsworth: Penguin Books.

Massey, D. and Catalano, A. (1978) *Capital and Land: Land Ownership by Capital in Great Britain*, London: Edward Arnold.

Matthews, R. and Leather, P. (1982) 'Housing in England: the view from the HIPS', *Roof*, May/June.

Maude, F. (1985) Parliamentary Debates (Hansard) *House of Commons Official Report*, Vol. 80, London: HMSO.

Meade Committee (1978) *Independent Committee's Report on Taxation*, London: Institute of Fiscal Studies.

Merrett, S. (1979) *State Housing in Britain*, London: Routledge & Kegan Paul.

Merrett, S. (1991) *Quality and Choice in Housing*, London: Institute of Public Policy Research.

Merrett, S. with Gray, F. (1982) *Owner Occupation in Britain*, London: Routledge & Kegan Paul.

MHLG (1953) *Housing: The Next Step*, Cmnd 2605, London: HMSO.

MHLG (1965) *Housing Programme 1965–70*, Cmnd 2836, London: HMSO.

MHLG (1968) *Old Houses into New Homes*, Cmnd 3602, London: HMSO.

MHLG (1969a) *House Condition Survey: England and Wales 1967*, London: HMSO.

MHLG (1969b) *Housing standards and costs: accommodation specially designed for old people*, Circular 82/69, London: HMSO.

Midwinter, E. (1994) *The Development of Social Welfare in Britain*, Oxford: Oxford University Press.

Miliband, R. (1972) *Parliamentary Socialism*, London: Merlin Press.

Milne, S. (1986) 'MPs to fuel council estates sales boom', *Guardian*, 24 April.

MoH (1948) *Building Licensing: Defence Regulations 56A*, Circular 40/48, London: HMSO.

Monck, E. and Lomas, G. (1981) *Housing Action Areas: Success and Failure*, CES Policy Series, 10, London: Centre for Environment Studies.

Moody, G. (2000) 'The £19 billion question', *Roof*, July/August.

Moorhouse, J.C. (1972) 'Optimal housing maintenance under rent control', *Southern Economic Journal*, 39.

Morris, J. and Winn, M. (1990) *Housing and Social Inequality*, London: Hilary Shipman.

Muellbauer, J. (1986) 'How house prices fuel wage rises', *Financial Times*, 23 October.

Muellbauer, J. (1988) 'Tax houses as part of income', *Sunday Times*, 12 September.

Muellbauer, J. (1990) *The Great British Housing Disaster and Economic Policy*, London: Institute of Public Policy Research.

Mullins, D., Niner, P. and Riseborough, M. (1993) 'Large scale voluntary transfers', in Malpass, P. and Means, R. (eds) *Implementing Housing Policy*, Buckingham: Open University Press.

Murphy, A. and Muellbauer, J. (1990) reported in Harris, R. (1990) 'Who dares storm the castle of the mortgaged monster?', *Sunday Times*, 28 January.

Murphy, E. (1991) *After the Asylum: Community Care for People with Mental Illness*, London: Faber & Faber.

Murphy, M.J. (1989) 'Housing the people: from shortage to surplus', in Joshi, H. (ed.) *The Changing Population of Britain*, Oxford: Basil Blackwell.

NAO (1989) *Homelessness*, Report by the Comptroller and Auditor General, London: NAO.

National Assembly (2000) *Reports of the National Housing Strategy Task Groups*, Cardiff: National Assembly.

Needleman, L. (1965) *The Economics of Housing*, London: Staples Press.

Nevin, B., Lee, P., Goodson, L., Murie, A. and Pillimore, J. (2000) *Changing Housing Market and Urban Regeneration in the M62 Corridor*, CURS, Birmingham: University of Birmingham.

NFHA (1985) *Inquiry into British Housing: Report*, London: NFHA.

NFHA (1994) *CORE Lettings Bulletin*, No. 17, London: NFHA.

NHIC (1979) *The Take-up of Private Sector Improvement Grants*, London: NHIC.

NHIC (1980) *The Market Value of Housing Improvement*, London: NHIC.

NHIC (1985) *Improving Our Homes*, London: NHIC.

Niner, P. (1988) DoE report on homelessness, unpublished.

Niner, P. (1989) *Housing Needs in the 1990s*, London: National Housing Forum.

Niner, P. and Forrest, R. (1982) 'Housing action area policy and progress: the residents' perspective', *CURS Research Memorandum*, 91, Birmingham: University of Birmingham.

North Islington Housing Rights Project (1982) *Room for Improvement: Resident Landlords and Islington Council*, London: NIHRP.

OECD (1998) *National Accounts*, Vol. 2, Paris: OECD.

Oldman, C. (1990) *Moving in Old Age: New Directions in Housing Policies*, SPRU, London: HMSO.

Oldman, C. (1991) 'Financial effects of moving in old age', *Housing Studies*, Vol. 6, No. 4.

Oldman, C. and Beresford, B. (1998) *Homes Unfit for Children: Housing Disabled Children and Their Families*, Bristol: Policy Press.

O'Leary, J. (1997) 'Town planning and housing development', in Balchin, P. and Rhoden, M. (eds) *Housing Finance. The Essential Foundations*, London: Routledge.

ONS (1980) *General Household Survey 1980*, London: HMSO.

ONS (1992a) *National Monitor 1991 Census Great Britain*, London: HMSO.

ONS (1992b) *General Household Survey 1992*, London: HMSO.

ONS (1994) *Family Expenditure Survey*, London: HMSO.

ONS (1996) *General Household Survey 1996*, London: The Stationery Office.

ONS (1997) *General Household Survey 1997*, London: The Stationery Office.

ONS (1998) *Social Trends 28*, London: The Stationery Office.

ONS (1999) *General Household Survey 1998*, London: The Stationery Office.

ONS (2000a) *Regional Trends 2000*, London: The Stationery Office.

ONS (2000b) *New Earnings Survey 1999*, London: The Stationery Office.

Page, D. (1987) 'What estates to get into', *Roof*, January /February.

Page, D. (2000) *Communities in the Balance: The Reality of Social Exclusion on Housing Estates*, York: York Publishing Services.

Parkinson, M. (1998) *Combating Social Exclusion: Lessons from Area-Based Programmes in Europe*, Bristol: Policy Press.

Patten, J. (1987) 'Housing: room for a new view', *Guardian*, 30 January.

Perry, J. (1998) 'Right turn', *Guardian*, 12 August.

Perry, J. (2000a) Letter to *Guardian*, 25 January.

Perry, J. (2000b) 'No man's land', *Guardian*, 22 March.

Phillips, D. (1987) 'The institutionalization of racism in housing: towards an explanation', in Smith, S.J. and Mercer, J. (eds) *New Perspectives on Race and Housing in Britain*, Glasgow: Centre of Housing Research, University of Glasgow.

Pitcher, G. (1987) 'Renaissance of the rented sector', *Sunday Times*, 28 November.

Planning Week, The Journal of the Royal Town Planning Institute (1994) 'Affordable housing shortage exacerbates homelessness', 28 April.

Power, A. (1987) *Property before People*, London: Allen & Unwin.

Power, A. (1997) *Estates on the Edge*, Basingstoke: Macmillan.

Power, A. (1999) 'Streets of shame', *Guardian*, 5 May.

Power, A. (2000) 'Giant step backwards', *Guardian*, 12 April.

Power, A. and Rogers, R. (2000) 'Inner cities: survival and revival', *Housing Today*, 19 October.

Power, R.M. (1982) 'Provision and choice in housing', in Counterpoint, *Socialism in a Cold Climate*, London: Unwin.

Prescott-Clarke, P., Clemens, S. and Park, A. (1994) *Routes into Local Authority Housing*, London: DoE.

Pullinger, J. and Summerfield, C. (1997) *Social Focus on Families*, London: ONS.

Quilgars, D. (1999) 'High and dry down Acacia Avenue', *Roof*, March/April.

Rahman, M., Palmer, G., Kenway, P. and Howarth, C. (2000) *Monitoring Poverty and Social Exclusion 2000*, York: JRF.

Rao, N. (1990) *Black Women in Public Sector Housing*, London: CRE.

Ratcliffe, P. (1998) '"Race", housing and social exclusion', *Housing Studies*, Vol. 13, No. 6.

Raynsford, N. (2000) 'Double your money', *Housing Today*, 21 July.

Regan, S. (2000) 'Houses in motion', *Guardian*, 31 March.

RICS (1985) *Better Housing for Britain*, London: RICS.

RICS (1992) *Housing the Nation*, London: RICS.

Ridley, N. (1986) reported in Lipsey, D. 'Ridley's home base', *New Society*, 7 November.

Roberts, A. (1982) 'Making housing an election issue', *Tribune*, 12 November.

Roberts, A. (1984) Letter to *Guardian*, 14 November.

Robine, J.M. and Blanchet, J.E.D. (1992) *Health Expectancy*, First workshop of the International Health and Life Expectancy Network (REVES).

Robson, B.T. (1999) 'Vision and reality: urban social policy', in Cullingworth, B. (ed.) *British Planning: 50 Years of Urban and Regional Policy*, London: Athlone Press.

Robson, B.T. (2001) *Are Northern Cities Losing Out?*, Manchester: University of Manchester Press.

Robson, B.T., Bradford, M.G. and Tomlinson, R. (1998) *Updating and Revising the Index of Local Deprivation*, London: DETR.

Rogers, R. and Power, A. (2000) *Cities for a Small Country*, London: Faber & Faber.

Rogers, R. (2001a) 'Save our cities', *Observer*, 13 February.

Rogers, R. (2001b) Letter to *Observer*, 20 February.

Roof (1997) 'Back to Basics, Number 5: CCT', March/April.

Roof (1998) 'Good news, bad news', September/October.

Roof (2001a) 'Benefit blues', January/February.

Roof (2001b) 'Plenty of promise', May/June.

Rooker, J. (1985) Speech to the annual conference of the Institute of Housing, Glasgow, June.

Room, G. (ed.) (1995) *Beyond the Threshold: The Measurement and Analysis of Social Exclusion*, Bristol: Policy Press.

RTPI (1973) Memorandum M25 in House of Commons, Tenth Report of the Expenditure Committee, Environmental and Home Office Subcommittee, Session 1972/73, London: HMSO.

Salman, S. and Bar-Hillel, M. (2000) 'Cutprice housing to stop the exodus', *Evening Standard*, 15 June.

Sarre, P., Phillips, D. and Skellington, R. (1989) *Ethnic Minority Housing: Explanations and Policies*, Aldershot: Avebury.

Saunders, P. (1990) *A Nation of Home Owners*, London: Unwin Hyman.

Scarman, Lord (1987) Speech at the launch of the United Nations International Year of Shelter for the Homeless.

Scottish Homes (1997) *Scottish House Condition Survey: Main Report*, Edinburgh: Scottish Homes.

Scottish Office (1996) Circular 12/96, *Planning Obligations*, London: HMSO.

Secretaries of State (1989) *Caring for People: Community Care in the Next Decade and Beyond* (the Wagner Report), London: HMSO.

Sexty, C. (1990) *Women Losing Out: Access to Housing in Britain Today*, London: Shelter Publications.

SFHA (1993) *Affordability*, Briefing Note No. 11, Edinburgh: SFHA.

SHAC (1981) reported in *Guardian*, 15 June.

Shelter (1979a) *Shelter's Election Manifesto*, London: Shelter Publications.

Shelter (1979b) *Roof*, September/October.

Shelter (1979c) reported in *Observer*, 23 December.

Shelter (1980) *And I'll Blow Your House Down: Housing Need in Britain – Present and Future*, London: Shelter Publications.

Shelter (1982a) *Housing and the Economy: A Priority for Reform*, London: Shelter Publications.

Shelter (1982b) *Homes Wasted*, London: Shelter Publications.

Shelter (1986) *Homes above All*, London: Shelter Publications.

Shelter (1987) *Shelter Campaign New*, London: Shelter Publications.

Shelter (1989) reported in Travis, A. (1989) '£250 million to aid homeless', *Guardian*, 16 November.

Shelter (1990a) *Forced Out*, London: Shelter Publications.

Shelter (1990b) reported in Knewstub, N. and Pilkington, E. (1990) '£15 million to get homeless off the streets', *Guardian*, 23 June.

Shelter (1991) *Building for the Future*, London: Shelter Publications.

Shelter (1992) *Homes Cost Less than Homelessness*, London: Shelter Publications.

Shelter (1993a) *Annual Report*, London: Shelter Publications.

Shelter (1993b) *What Future for Older People?*, London: Shelter Publications.

Shelter (1993c) *Facts and Figures on Homelessness*, London: Shelter Publications.

Shelter (1994) 'Housing's new hit-list', *Roof*, March/April.

Shore, P. (1977) Speech to the Housing Centre Trust, June.

Shucksmith, M. (2001) 'Exclusive countryside? Social exclusion and regeneration in rural Britain', Paper presented at the conference of the Institute of British Geographers, Plymouth, 5 January.

Silver, S. (1995) 'Three paradigms of social exclusion', in Rodgers, G., Gore, C. and Figueiredo, J.B. (eds) *Social Exclusion: Rhetoric, Reality, Responses*, Geneva: International Institute for Labour Studies, United Nations Development Programme.

Sim, D. (2000) 'Housing inequalities and minority ethnic groups', in Anderson, I. and Sim, D. (eds) *Social Exclusion and Housing: Context and Challenges*, Coventry: CIH.

Simpson, D. (1985) 'Report attacks housing trap', *Guardian*, 14 February.

Sims, P. (2001) *Releasing Brownfields*, York: JRF.

Slaughter, J. (1993) 'Home truths for retiring types', *Sunday Times*, 20 June.

Smith, D. (1990) *North and South*, Harmondsworth: Penguin Books.

Smith, M. (1998) 'Vale of cheers', *Guardian*, 20 May.

Smith, S.J. (1989) *The Politics of 'Race' and Residence*, Cambridge: Polity Press.

Social and Community Planning Research (1990) *British Social Attitude Surveys*, London: SCPR.

Social Security Advisory Committee (1995) *Memorandum to the Social Security Advisory Committee: Housing Benefit Changes for Private Sector Tenants*, London: HMSO.

Somers, A. (1992) 'Domestic violence survivors', in Robertson, M. and Greenblatt, M. (eds) *Homelessness: A National Perspective*, London: Plenum.

Somerville, P. (1998) 'Explanations of social exclusion: where does housing fit in?', *Housing Studies*, Vol. 13, No. 6.

Spittles, D. (1992) 'Homeowners under house arrest', *Observer*, 13 December.

Stanbrook, I. (1985) Parliamentary Debates (Hansard), *House of Commons Official Report*, Vol. 83, London: HMSO.

Stenhouse, C. and Henderson, P. (2001) *Caring Communities: A Challenge for Social Inclusion*, York: York Publishing Services.

Stevens, S. (2000) 'Northern Ireland', *Housing Finance National Markets Review 2000*, London: CML.

Stone, P.A. (1970) 'Housing quality: the seventies problem', *Building Societies Gazette*, June.

Sunday Times, The (1983) 'Developing a loophole in home improvement grants', 6 March.

Sykes, R. (1994) 'Elderly women's housing needs', in Gilroy, R. and Woods, R. (eds) *Housing Women*, London: Routledge.

Taylor, M. (2000) *Top-Down Meets Bottom-Up: Neighbourhood Management*, York: JRF.

TCPA (1996) *The People. Where Will They Go?*, London: TCPA.

Tinker, A. (1984) *Staying Home: Helping Elderly People*, London: HMSO.

Townsend, P. (1979) *Poverty in the United Kingdom*, Harmondsworth: Penguin Books.

Travis, A. (1988) 'Government plans legislation to tighten rules on homelessness', *Guardian*, 14 June.

Treasury (1947) *Capital Investment in 1948*, Cmnd 1268, London: HMSO.

Treasury (1980) *The Government's Expenditure Plans, 1980/81*, Cmnd 7746, London: HMSO.

Treasury (1988) *The Government's Expenditure Plans, 1988–89 to 1990–91*, Cm 288, London: HMSO.

Treasury (1998) *Comprehensive Spending Review*, Cm 4011, London: HMSO.

Treasury (2000) *Spending Review*, Cm 4807, London: HMSO.

Trustee Savings Bank (1993) *TSB Affordability Index*, 4, London: TSB.

Victor, C. (1991) *Health and Health Care in Later Life*, Buckingham: Open University Press.

Victor, C. with Munro, M., Tinker, A. and Warnes, T. (1992) 'Housing types, amenities, tenures and assets', in Warnes, T. (ed.) *Homes and Travel: Local Life in the Third Age*, Research Paper No. 5, London: Carnegie United Kingdom Trust.

Waldegrave, W. (1987) Speech to Institute of Housing Conference, Brighton, June.

Walentowicz, R. (1992) 'Developing trends', *Roof*, March/April.

Walker, P. (1978) 'The real tenant's charter', *Guardian*, 5 November.

Ward, R. (1982) 'Race, housing and wealth', *New Community*, Vol. 10, No. 1.

Warman, C. (1987) 'Prices of housing land rise by 27%', *The Times*, 16 November.

Warnes, A.M., Howes, D.R. and Took, L. (1985) 'Residential location and intergenerational visiting in retirement', *Quarterly Journal of Social Affairs*, Vol. 1, No. 3.

Warnes, T. (ed.) (1992) *Homes and Travel: Local Life in the Third Age*, Research Paper No. 5, London: Carnegie United Kingdom Trust.

Warrington, M. (1994) Paper presented at the Institute of British Geographers Annual Conference, London.

Waugh, P. (1998) '£1.2m handout to get young homeless off London streets', *Evening Standard*, 30 March.

Waugh, P. (2000) 'Billions to be spent on most deprived estates', *Independent*, 19 July.

Weaver, M. (2000) 'Going fast', *Guardian*, 2 February.

Webster, D. (2000) 'Lone parenthood: two views and their consequences', in Anderson, I. and Sim, D. (eds) *Social Exclusion and Housing: Context and Challenges*, Coventry: CIH.

Welsh Office (1997) Circular 13/97, *Planning Obligations*, London: HMSO.

Wheeler, R. (1986) 'Housing policy and elderly people', in Phillipson, C. and Walker, A. (eds) *Ageing and Social Policy: A Critical Assessment*, Aldershot: Gower.

Wiener, J.M. (1990) 'Measuring the activities of daily living: comparisons across national surveys', *Journal of Gerontology, Social Sciences*, Vol. 45, No. 6.

Wilcox, S. (1997) 'New depths', *Roof*, January/February.

Wilcox, S. (2000) *Housing Finance Review, 2000/2001*, York: JRF.

Wilcox, S. and Pearce, B. (1991) *Home Ownership, Taxation and the Economy*, York: JRF.

Williams, C.C. and Windebank, J. (1999) *A Helping Hand: Harnessing Self-help to Combat Social Exclusion*, York: York Publishing Services.

Williams, N.J., Sewel, J.B. and Twine, F.E. (1987) 'Council house sales and the electorate: voting behaviour and ideological implications', *Housing Studies*, Vol. 2, No. 4.

Williams, P. (1990) reported in Durham, M. (1990) 'Families growing rich in inheritance bonanza', *Sunday Times*, 29 July.

Williams, P. (2000) 'Wales', *Housing Finance National Markets 2000*, London: CML.

Williams, R. (1983) *Keywords: A Vocabulary of Culture and Society*, 2nd edn, London: Flamingo/Fontana.

Wilson, D. (1986) 'Where Cathy came in', *Observer*, 23 November.

Wilson, H. (1964) Speech at Town Hall, Leeds, 8 February.

Wintour, P. (1997) 'Ghetto busters to tackle poverty', *Observer*, 7 December.

Wintour, P. (2001) 'Landlords who neglect houses may lose benefit', *Guardian*, 2 August.

Wolmar, C. (1982) 'Shrinking the Green Belt', *New Society*, 21 June.

World Bank (1998) *World Development Report 1997*, Oxford: Oxford University Press.

Index